Advances in Sustainable Aviation

Tahir Hikmet Karakoç • C. Ozgur Colpan
Yasin Şöhret
Editors

Advances in Sustainable Aviation

Springer

Editors
Tahir Hikmet Karakoç
Department of Airframe
and Powerplant Maintenance
Anadolu University, Faculty of Aeronautics
and Astronautics
Eskisehir, Turkey

C. Ozgur Colpan
Department of Mechanical Engineering
Dokuz Eylul University, Faculty of
Engineering
Buca, Izmir, Turkey

Yasin Şöhret
Department of Airframe
and Powerplant Maintenance
Süleyman Demirel University
School of Civil Aviation
Isparta, Turkey

ISBN 978-3-319-67133-8 ISBN 978-3-319-67134-5 (eBook)
DOI 10.1007/978-3-319-67134-5

Library of Congress Control Number: 2017956935

Printed on acid-free paper

This Springer imprint is published by Springer Nature
The registered company is Springer International Publishing AG
The registered company address is: Gewerbestrasse 11, 6330 Cham, Switzerland

Preface

The rapid depletion of fossil fuels and increasing concerns regarding environmental problems are prompting us to develop more efficient energy systems for a sustainable future. The aviation sector, including air and cargo transportation, management of air traffic, management of airport terminals and associated buildings, numerous manned and unmanned aerial vehicle operations, and so on, is one of the fastest growing sectors in the last decade. The number of aircrafts in service, managed airports, and people using air transportation are increasing every day. Given this growth, the sustainable management of energy in the aviation sector is vital.

This book, *Sustainable Aviation: Advances in Sustainable Aviation*, presents the state-of-the-art sustainable aviation. The book compiles selected outstanding papers presented at the International Symposium on Sustainable Aviation 2016 (ISSA-2016). This event was held in Istanbul, Turkey, from May 29, 2016, to June 1, 2016, with the participation of researchers, scientists, and academics from all over the world. The contribution of leading aviation companies and distinguished scientists in the field of aviation turned the event into a platform where the future of the sector was discussed.

The book is composed of 4 parts and 20 chapters in total. Each chapter in the book aims to present an issue and solution on how to achieve more sustainable aviation. The first part of the book includes research associated with aviation fuels and combustion technologies. Recent studies on aerodynamics are presented in the second part of the book. The third part involves energy and environmental issues associated with the aviation. The last part of the book addresses the advances in employment of electronic systems and control technologies in aviation.

This book provides a unique source of sustainable aviation with a prime focus on advances in sustainable aviation. The contributions of the authors and reviewers and the assistance provided by the conference organizing committee members in the preparation of this book are sincerely appreciated.

Eskisehir, Turkey — T. Hikmet Karakoç
Izmir, Turkey — C. Ozgur Colpan
Isparta, Turkey — Yasin Şöhret

Contents

Part I
Advances in Fuel and Combustion

Chapter 1
Environmentally Clean Reformulated Aviation Gasoline

Olesia Kondakova and Sergii Boichenko

Production of fuel for aircraft engines is one of the priorities of world refining industry. Due to growing air traffic intensity, it is necessary to improve the fuel efficiency of aircraft and fuel quality. As soon as provision with high-quality fuel is a decisive issue of aircraft efficiency, saving and rational use of aviation fuels become of national importance.

Aviation gasoline blend components are obtained in various technological processes. The basic component in the production of gasoline is catalytic reforming at which the aromatization and isomerization of straight gasoline is the output. This increases its detonation resistance. Additional components with a high detonation resistance – alkylate, isooctane technical, pirobenzol, ethyl liquid, isopentane, and toluene – are entered into the composition of fuels (Table 1.1).

The main indicator of gasoline quality is its detonation stability. This is an ability of fuel to burn without detonation in piston engines with spark ignition; octane number has direct influence on operational and environmental performance of transport.

But today, all the known brands of aviation gasoline use tetraethyl lead as anti-knock additive in spite of its toxicity. Thus, for aviation piston engines, according to GOST 1012, aviation gasoline brands B-95/130 and B-91/115, whose oil fraction boils within 40–180 °C, are produced. Today known brands of aviation gasoline are B-92 (ТУ 38.401-58-47-92) and B-70 (ТУ 38.101913-82) with low concentration of TEL. DEF STAN 91-90 Issue 3 reflects requirements for AVGAS 100LL, AVGAS 100, and AVGAS 80. According to ASTM D 6227, new aviation gasoline 82UL, which already contains TEL, is introduced, but its testing has not yet been completed. Having analyzed these documents, we have compiled a comparative table (Table 1.2) presenting TEL content in aviation gasoline [3]. The table shows

O. Kondakova (✉) • S. Boichenko
National Aviation University, Kyiv, Ukraine
e-mail: izabellac@mail.ru

T.H. Karakoç et al. (eds.), *Advances in Sustainable Aviation*,
DOI 10.1007/978-3-319-67134-5_1

Table 1.1 Characteristics of high-octane components [19–21]

№	Indicator	TEL[a]	Isooctane	Toluene	Benzene	Pirobenzene	Isopentane	Alkylate
1	Density at 20 °C, kg/m^3	1650	691.8–692.1	865–867	878–880	850	626.2	698—715
2	Boiling temperature, °C	80	99.15–99.35	110,6	80.1	80–175	27.85	40–180
3	MON[b]	0.1% addition can increase TEL on gasoline by 10 units	100–99	102.1	111.6	88	90.3	90
4	Maximum allowable concentration of working area, mg/m^3	0.0001	300	150	5	No data	900/300	No data
5	Hazard class	1	4	3	2	No data	4	No data

[a]Tetra ethyl lead
[b]Motor octane number

Table 1.2 Content of TEL in aviation petrol [22, 23]

№	Aviation petrol brand	Regulation	Content of antiknock agents	Name of antiknock agents
1	B-95/130	GOST 1012-72	3.1 g/kg	TEL
2	B-91/115		2.5 g/kg	
3	B-92	ТУ 38.401-58-47-92	2.0 g/kg	
4	B-70	ТУ 38.101913-82	0	
5	Avgas 80	DEF STAN 91-90 Issue 3	0.14 g/l	
6	Avgas 100LL		0.56 g/l	
7	Avgas 100		0.85 g/l	
8	82UL	ASTM D6227	0	Similar to automobile gasoline but without automotive additives
9	85UL	–	0	Oxygenate-free
10	91/96	–	Almost negligible	TEL
11	91/96UL	ASTM D7547	0	Ethanol-free, antioxidant, and antistatic additives
12	100/130	–	1.12	TEL
13	G100UL	–	0	Aromatic compounds such as xylene or mesitylene
14	100SF	–	0	Mesitylene
15	115/145 ("Avgas 115")	–	1.29	TEL

the trend to reduction of TEL concentration in aviation gasoline, but almost all brands today contain toxic TEL. Therefore, the development of new environmentally friendly aviation gasoline is an urgent modern problem that needs solution.

One solution to this problem is the introduction of aliphatic alcohols, ethanol, methanol butanol, and other oxygenates into the aviation gasoline instead of tetraethyl lead.

Oxygenates include lower alcohols and ethers, which could be used as high-octane components of motor fuels. They are made of raw alternative materials, ethanol, methanol, and butanol, predominantly derived from coal, plant products, and heavy oil residues. The use of aliphatic alcohols expands resource potential for fuels production and often improves their quality. They can be primary fuel or used as additives for fuels of petroleum origin. Such fuels are characterized with better cleaning properties and better combustion and during combustion form less carbon monoxide and hydrocarbons.

The recommended concentration of oxygenates in gasoline is 3–15% (vol.). It is chosen to provide oxygen content in fuel within 2.7%. The reason is that this content of oxygenates, despite their lower value compared to petrol oil calorific value, will not have negative effects on engine power characteristics [1, 2].

In general, the use of alcohol as motor fuels and high-octane additives is known from the beginning of the last century, but their widespread industrial use began only in 80–90 years of the twentieth century.

Analysis of the existent literature [1–17] confirms that the addition of the aliphatic alcohol in fuels changes their properties.

The world centers of biofuel production in 2012 were the United States, Brazil, and the European Union. For example, in 2010, they concentrated 85% of the world production of biofuels; only the United States accounted for 48% global production of biofuels.

The most common type of biofuel – bioethanol – has 82% share in the total volume of fuel from biological raw materials produced in the world [5]. Its leading producers are the United States and Brazil.

Since today the use of aliphatic alcohols, methanol, ethanol, and butanol as additives to gasoline is widely distributed, we intend to analyze the prospects for aviation gasoline modification through the use of oxygenates and study the prospects of using alcohol as a part of aviation gasoline.

1.1 Methanol

Methanol is an alcohol with one carbon atom (CH_3OH). Methanol is one of the most promising fuels due to its high octane value [6]. As an additive to gasoline, methanol is used infrequently [3]. The most typical mixture used as motor fuel is M85 (85% methanol and 15% carbohydrates) and pure methanol M100 (100% methanol) [9]. In all cases, it allows to reduce the toxicity of engine exhaust. The use of absolute methanol is limited because of its high toxicity and corrosivity in relation to constructional materials, which reduce the life of the engine parts and quality of fuel, increasing the risk of emission of volatile organic compounds that can lead to depletion of ozone. The typical emissions of methanol combustion process include formaldehyde, while during the combustion of ethanol mainly acetaldehyde is emitted [10].

When using methanol, engine efficiency increases by 5–15% compared to gasoline. This is the high temperature of methanol vaporization, which reduces the temperature of mixture and increases the value of the fuel-air ratio and reduces the heat in the engine cylinders and exhaust gas temperature while maintaining capacity.

The most serious problems that complicate the use of methanol additive are its toxicity, poor solubility in hydrocarbons, and high water absorption. The tendency for mixture separation increases with decreasing temperature, leading to the accumulation of water and reduction of aromatic content in gasoline. To stabilize the gasoline-methanol blends, special additives, like propanol, isopropanol, isobutanol, and other alcohols, are used. The content of methanol in gasoline can be administered about 5% (vol.); in this case gasoline-methanol mixture is homogeneous [3].

Operating characteristics, energy performance, and starting quality of methanol fuel are improved after additional introduction of higher alcohols and esters. Such fuel is called mixed alcohol fuel. Tests of one of the mixed fuel compositions have shown an increase in engine power by 4–7% and improved fuel economy (compared to pure alcohol) by 10–15%, while the content of nitrogen oxides is reduced by 25–30% compared with the work on gasoline [3].

Great interest to the use of methanol as a fuel is observed in countries with rich resources of coal and insufficient oil resources. Methanol can be produced from natural gas, coal, and biomass.

1.2 Ethanol

Ethanol is of much greater interest as an additive for fuel, because it is more soluble in hydrocarbons and less hygroscopic. Ethanol molecule consists of two carbon atoms – C_2H_5OH.

Widely known use of "gasohol" which is a mixture of gasoline with 10–20% ethanol in the United States and Brazil, with significant resources of alcohol derived from sugar cane. Sweden has introduced the state program of oil abandonment by 2020. In general, the use of ethanol as fuel is more interesting for the countries with rich plant resources, including Ukraine.

Petroleum gasoline is the largest source of man-made carcinogens. Therefore, the main environmental benefit of using ethanol as a part of mixed gasoline is the ability to exclude the use of highly toxic antiknock additive of metal and methyl tert-butyl ether (MTBE). With the addition of even 10% ethanol, gasoline is enriched with oxygen, which promotes more complete combustion and reduces emissions of carbon monoxide by 30%. Also, it decreases toxic emissions by 30% and emissions of volatile organic compounds – by more than 25 . Using the mixture of gasoline and ethanol, E10, allows all the major car manufacturers to improve engine performance by adding 2–3 units of detonation resistance to fuel, prevents engine overheating, acts as antifreeze for fuel system, and does not cause contamination of the fuel injectors [10].

However, when using ethanol, there are a number of specific issues. Thus, bioethanol in its physical and chemical properties is significantly different from gasoline; it has higher octane number at 92 units by motor method and lower heating value (but higher corrosion activity); at the concentration of more than 12%, it may adversely affect the engine (Fig. 1.1).

The potential annual production of bioethanol from the available raw materials (molasses, corn, grain, sugar sorghum) according to the expert group "Ukrspirt" reaches 2 million tons; that will potentially replace 40% of gasoline consumed in Ukraine. The EU official documents consider Ukraine as a potential exporter of ethanol to Europe [11].

The value of efficiency of engines using alcohol gasoline in the whole range of the mixture is increased, so that the specific energy consumption per unit of power is reduced. Total fuel efficiency gradually increases with the percentage of ethanol in fuel [12].

Under real conditions, water inevitably gets into the gasoline-alcohol fuel during storage, transport, and use, which leads to phase separation. This problem does

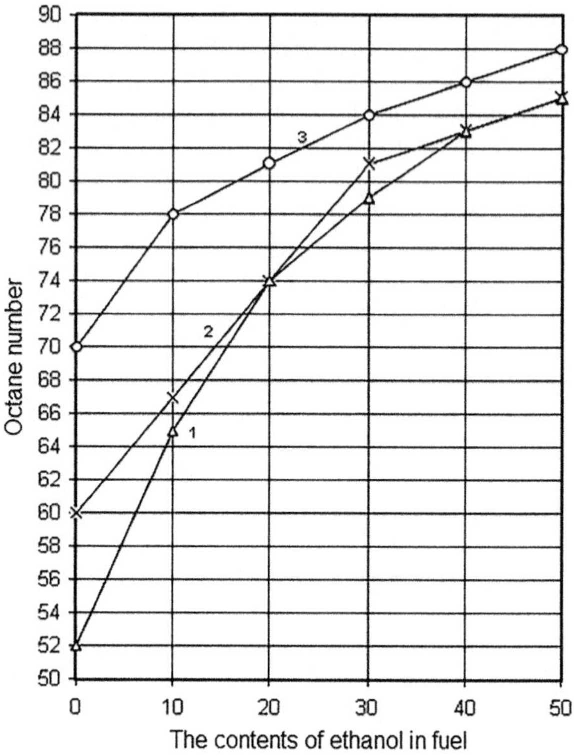

Fig. 1.1 Ethanol affects the formation of gasoline octane number (Notes: *1* ishymbaysky gasoline; *2* B-59; *3* B-70 [5, 22])

not disappear gasoline-alcohol mixtures and when using absolute ethanol. Benzyl alcohol is proposed to be used as stabilizer [14].

Today authors [14] have developed recipes of biological motor fuel E85 (ТУУ 24.6-35523958-001:2009 "motor biological fuel specifications") that meets environmental and operational requirements to motor fuels for modern cars and takes into account raw materials of Ukraine.

1.3 Butanol

In addition to methanol and ethanol, aliphatic alcohol family includes propanol (three carbon atoms, C3H7OH) and butanol (four carbon atoms, C_4H_9OH) [15].

The use of butanol fuel is the next significant stage in the development of biofuels, the use of which has to meet the growing demand for environmentally friendly, renewable transport fuels [16]. Butanol is an alcohol (colorless liquid with a characteristic smell of fusel oil). The term "biobutanol fuel" is used to refer to butyl alcohol (butanol), which is produced from plant material.

Since butanol does not have corrosive properties, it can be transported through the existing infrastructure [15].

Butanol provides cleaner energy for duty cycle than ethanol or methanol and about 10% more than gasoline. Currently derived from corn, butanol attracted increasing attention of experts for its use as a fuel, in connection with the introduction of new highly efficient butanol fuel production technologies. It is possible that in the next 10–15 years, ethanol will lose its priority. The success is due to a number of advantages of butanol over ethanol, including:

- Butanol contains 25% more energy than ethanol: 110,000 BTU per gallon of butanol to 84,000 BTU per gallon of ethanol. Gasoline also contains about 115,000 BTU per gallon.
- Butanol is much less aggressive substance than ethanol, so it can be transported through the existing fuel pipelines, while ethanol must be transported by rail or water transport.
- Butanol can be mixed with gasoline.
- Butanol can completely replace gasoline, while ethanol can only be used as an additive to gasoline with the maximum content in the mixture not more than 85% and only after significant alterations in engine structure. Currently, the mixture of 10% ethanol is dominant in the world.
- Modified butanol has higher energy output than ethanol.
- Burning butanol produces no sulfur or nitrogen oxides, which gives significant advantage in terms of ecology.

Butanol fuel improves fuel efficiency and increases vehicle mileage per unit of fuel consumed [15]. Biobutanol fuel is produced from the same raw material – corn, sugar beets, sorghum, cassava, sugarcane, cornstalks and other biomass, and ethanol – but it can replace gasoline in equal measure.

Biobutanol fuel provides significant environmental advantages over petroleum-based fuel, including lower levels of greenhouse gas emissions. Biobutanol fuel will also reduce the emission of carbon dioxide into the atmosphere [15].

Today, biobutanol fuel can be added to gasoline in concentrations of up to 10% in Europe and 11.5% in the United States without engine modifications. In the future, there is potential to increase the maximum use of biobutanol fuel in gasoline to 16% by volume.

Biobutanol fuel, despite the presence of water, is less prone to separation than ethanol/gasoline, so it allows distribution through the existing infrastructure, not requiring modifications to facilities for mixing, storage, or refueling.

According to [17], the relative cost ceiling of biobutanol fuel is 0.73 of the cost of MTBE and 0.8 of the cost of bioethanol. Values are benchmarks in assessing the competitiveness of high-biobutanol fuel and as a component of motor gasoline.

The main advantages of biobutanol fuel include larger combustion heat than that of ethanol; therefore, it can be used in higher concentrations in gasoline. Obtaining biobutanol fuel from non-plant materials is an efficient way of recycling wastes of agriculture and timber industry.

Butanol is safer in exploitation, because it evaporates six times less intensively than ethanol and it is 13.5 times less volatile than gasoline. Vapor pressure of butanol by Reid is 2.3 kPa, gasoline is 31 kPa, and ethanol is 14 kPa. It makes butanol safer when used as an oxygenate and does not require any changes in the proportions of mixture for winter and summer. Now it is used as an oxygenate in the states of Arizona, California, and others.

Butanol can replace gasoline as a fuel better than ethanol due to its physical properties, economy, security, and the fact that its use does not require modernization of motor vehicles. Until recently, no one knew of butanol as an alternative fuel, and its production has never been considered economically feasible [15].

Analysis of the existing literature leads to the conclusion that the addition of aliphatic alcohol affects the properties of traditional fuels [3–15]. We have compared the physical and chemical properties of different components of aviation gasoline.

According to Table 1.3, we see that the heat of combustion of ethanol, methanol, and butanol is significantly lower than the heat of combustion of aviation gasoline, causing increase in fuel consumption when using these alcohols. However, the oxygen content in the composition of oxygenates results in a higher completeness of combustion, so the difference in combustion heat is not so noticeable.

The main advantages of alcohols include their high-efficiency workflow and high detonation resistance (octane number between 99 and 111). The value of alcohol efficiency in the engines is over that for gasoline in the whole range of mixtures, so that the specific energy consumption per unit of power is reduced [16, 17]. Ignition limits of gasoline-ethanol mixture are broader than for pure gasoline [18].

The use of alcohol reduces emissions of products of incomplete combustion, reduces the formation of soot but at the same time increases emissions of aldehydes (as a product of incomplete oxidation of alcohols), and may also increase emissions of nitrogen oxides. This problem can be eliminated by adding water (5%) or 0.8% aniline to alcohol, heating inlet air to the engine [18]. The development and implementation of catalytic converters of exhaust gases can provide oxidation of aldehydes, acids, and other products of incomplete combustion of fuel alcohol to water and carbon dioxide [16].

The main disadvantage of gasoline-alcohol fuels is their phase instability under even small amounts of water and, consequently, limited mutual solubility of the components. Introduction of special additives for corresponding modification and stabilization to alcohols cannot overcome the emerging challenges. To ensure the stability of the alcohol-containing gasoline during production, storage and use it is necessary: remove water; use stabilizing additives or cosolvents, homogenizing system gasoline-water-alcohol. We recommend adding alcohol to gasoline directly before refueling [1].

The following stabilizers are offered for gasoline-alcohol mixtures: aliphatic alcohols C3–C12 with normal and branched structure, phenols, alkylacetate, ethers and esters and their organometallic derivatives, ketones, amines, surfactants and glycols and their ethers, aldehydes, ketals, acetals, alkylcarbonates, carboxylic acids, and mixtures of these compounds.

Among aliphatic alcohols, the most researched and effective is ethyl alcohol. Its high antiknock quality is known to have widespread use in internal combustion engines with forced (spark) ignition [18]. Equally efficient performance is typical for methanol, but its high toxicity and aggressiveness are significant obstacles for its application.

Table 1.3 Characteristics of physicochemical properties of alcohols and aviation gasoline [7]

Characteristics	Avgas 100 LL		B91/115	82 UL	Methanol	Ethanol	Butanol
Density at 20 °C, kg/m^3	Report		Does not normalize		792	789.3	810
Boiling point, °C	75–170		40–180	–	64.5	78.4	117.5
Freezing point, °C	Minus 58		Minus 60	Minus 58	Minus 98	Minus 114.1	Minus 90.2
Weight fraction oxygen, %		–	–	–	50	34,7	21,6
The heat of evaporation, kJ/kg		Not specified			1115	839.3	591.2
The heat of combustion, kJ/kg		43,500	42,947	40,800	24,000	26,945	35,520
Vapor pressure, kPa		38–49	29.3–47.9	–	–	17	8.4
Solubility in water temperature 20 °C, %		Insoluble			Unlimited		7.9
Octane number Research method Motor method		Not determined 99.5	Not determined 91	Not determined 82	111 94	108 92	99,6 94
Kinematic viscosity at temperature 20 °C		Not determined			–	1.52 cSt	3.64 cSt
TEL		0.56 g/l	2.5 g/kg	0	–	–	–
Limit content in gasoline by EN 228:2000		–	–	–	3%	5%	7–10%

Today growing interest is paid to butyl alcohol as an antiknock additive. Its advantage is that it can be transported through the existing fuel supply system and manufactured at plants producing ethanol with minor changes in technology. Butanol is less poisonous; its maximum allowable concentration is 10 mg/cm^3, while ethanol is 5 mg/cm^3 and methanol – 1 mg/cm^3.

Foreign countries are more active in terms of producing aviation fuel; today they have patented a wide range of alternative fuels for aviation. Specifically, the patent US 7559961 B2 presents alternative aviation petrol composite forming hydrocarbon mixture of alcohols. According to US 0011765 A1, the regulated brand includes ingredients extracted from biomass, but they should be distinguished from pure chemicals. In addition, this patent states that the water content in the fuel should not exceed 2%. If the water content is higher, it will grade fuel as unacceptable. Also, amines are used, which are quite expensive chemical components, production of which in Ukraine is not yet organized. The patent US 7553404 B2 mixtures include 60% traditional gasoline, butane, isopentane, and cyclohexane. Overall, 93% are mixed composite fuel. But all of them are expensive and therefore cannot be used for economic reasons [6].

Having analyzed the existing studies, we have formed an integrated comparative characteristic of physical, chemical, and environmental properties of alcohols (Table 1.4).

Table 1.4 Comparative physicochemical and environmental properties of alcohols

Alcohols	Properties
Methanol	Highest octane number (156 units)
	Improvement in fuel economy by 10–15%
	Engine efficiency increases by 5–15%
	High toxicity and aggressiveness on construction materials
	Content in the exhaust gases of nitrogen oxides is reduced by 25–30%
Ethanol	Octane number (132 units) is significantly higher than gasoline (82.5–98 units)
	Separation in the presence of water
	Requires modification of the engine with increasing content
	Low heat of combustion
	Is less toxic
	A couple of the less flammable
	Significantly higher electrical conductivity
Butanol	Octane number (104 units)
	Energy intensity close to the energy intensity of gasoline
	The mixture does not separate in the presence of water
	Does not require engine modifications
	Has no corrosive properties
	Low vapor pressure of butanol fuel should not lead to the emergence of high levels of emissions of volatile organic compounds

References

1. Boychenko, S. V., Boychenko, M. S., Boar, S. N., & Lychmanenko, O. G. (2015). Effect of additives on properties of aliphatic alcohols gasoline: Analytical review. *Science-Intensive Technologies, 1*(25), 86–92.
2. Boychenko, S. (2015). Causal analysis of aviation gasoline modification. In S.V. Boychenko, Lady K, O.G. Lychmanenko (Eds.), *Proceedings of the National Transport University* (pp. 3–13). Series “Engineering”. Scientific and technical collection. 2(32).
3. Danilov, A.M. (2000). Application additives in fuel for cars: Ref. ed. A.M. Danilov. Chemistry,. 232:ill.
4. Asyaev, A.N. (2010). Study effect of alcohol quality and composition uhlevodorodnoy fraction on physico-chemicals and эkspluatatsyonnыe indicators byoэtanolnoho fuel E85. In A.N. Asyaev, V.E. Emelyanov, E.A. Nikitin (Eds.), *Technologies of oil and gas* (pp. 24–27). 4.
5. Bondarenko, K. (2011). Prospects for the introduction of alternative fuels in aviation. In K.V. Bondarenko, S.V. Boychenko, V.G. Semenov (Eds.), *Avyatsyonno kosmycheskaya-technics and technologies* (pp. 76–80). 9(86).
6. Onoychenko, S.M. (2000). Research and development tracks unleaded petrol containing ethanol, dis. ... .. Candidate Sc. Science.: 05.17.07/Onoychenko Svetlana. M: 168 p.
7. Vdovin, V (2014). Modification of aviation gasoline by adding aliphatic alcohols: Literature review. In V.S. Vdovin, K.V. Bondarenko (Eds.), *Open consumer information and computer yntehryrovannыe Technology* (pp. 164–171). 64.
8. Market research biomethanol and its derivatives: [analyte. Report/Research. Techart]. M.: 2008. 22 p.
9. Lotko V., Lukanin V.N., Khachiyan A.S. (2000). Use of alternative fuels in internal combustion engines. MADI (TU) 311.
10. Karpov S.A, & Kunashev L.H., et al. (2006). Demonstrate how the aliphatic alcohols as environmentally friendly additives in car. Oil and gas business 2. The magazine. Access: http:/ /ogbus.ru/.
11. Ablaev A.P. Biofuels: Thinking outside of the oil pipe, A.P. Ablaev, Ecologist. Vestnik. Russia. 2009. - № 2. 23–26.
12. Polunkyn, E. V., Kamenev, T. N., Pylyavskyy, V. S., et al. (2012). Operational properties of alternative motor fuels on the basis of oxygenates. *Catalysis and neftehymyya, 20*, 70–74.
13. Pylyavskyy, V. S., Gaidai, O. O., Kyrpach, K. O., et al. (2012). Operational properties of alternative motor fuels on the basis of oxygenates. *Catalysis and neftehymyya, 21*, 162–166.
14. Gaidai, O.O. (2011). Environmental and operational characteristics of biological motor fuel E-85 [electronic resource]. In A.A. Gaidai, Zubenko S.A., Polunkin E.V., Pylyavskyy V.S. (Eds.), *Collected articles “Third Ukrainian Congress ecologists with international participation.”* Ball. 1. 308–310. Access: http://eco.com.ua/.
15. Imankulov, N.N. (2010). Biobutanol – alternative motor fuel substitute. In N.N. Imankulov (Ed.), *Scientific Bulletin of the southern region* (pp. 3–7). 5–6 (35–36).
16. Vnukovo, N.V. (2011). Alternative fuel type yak basis resursozberezhennya i ekobezpeki road vehicles. In N.V. Vnukovo, M.V. Barun (Eds.), *Alternative energy sources* (pp. 45–55). 9 (91).
17. Ershov, M.A. (2012). Research biobutanol as a high octane gasoline component: avtopef.dis. for the degree of CAD. those. Sciences: spec. 05.17.07, “Chemical technology of fuel and high-energy substances” MA Ershov. 27.
18. Boychenko, S.V., Kuchma, N.M., Efimenko, V., Titov, A.S., Chernyak, L.M. (2006). Chemmotology: Teach method. manual. K: portrait water-in NAU, 160 pp.
19. Levinter, M.E., Akhmetov, S. (1992). Deep processing of oil: A manual for schools. M.: Chemistry. 224 p.: silt.
20. Smyshlyaeva, Y. A., Ivanchina, E. D., Kravtsov, A. V., Zuong, C. H. T., & Fan, F. (2011). Development of a database on the octane number of a mathematical model of the process of compounding of commercial gasoline. *Bulletin of the Tomsk Polytechnic University, 318*(3), 75–80.

21. Gureev, A.A. (1981). Production of high octane gasoline/. In A.A. Gora, Y.M. Zhorov, E.V. Smidovich. Moscow: Chemistry.
22. Boichenko, S., Lejda, K., Lychmanenko, O., Boichenko, M., & Reshetilowski, V. (2015). Modification of aviation gasoline with aliphatic alcohols additives: An analytical review of prospects. *Internationl Journal Sustainable Aviation, 1*(4), 324–332.
23. Lychmanenko, O.G. (2015). Pespektyvy reformulovanyh aviation gasoline. In O. Lychmanenko, S. Boychenko, K. Lady (Eds.), *Systemy i srodki transportu samochodowego: Wibrane zagadnienia* (pp. 251–256). Monografia № 6. Seria: Transport. Rzesow (Poland).

Chapter 2
Ignition Performances of Alternative Fuels Under Altitude Conditions According to Their Aromatic Content

Renaud Lecourt

Nomenclature

ER	Equivalence ratio
MERCATO	Moyen Experimental de Recherche en Combustion Aérobie par Techniques Optiques
P	pressure, MPa
T	temperature, K

Subscripts

ch	chamber

2.1 Introduction

During the last years, the introduction of alternative fuels, produced from other sources than oil, has been quickly developed because of environmental constraints and anticipated economical issues. For their use in aircrafts, the fuels must comply with accurate international specifications (ASTM, DefStan). The first produced alternative fuels, XtL fuels from the Fischer-Tropsch process and hydroprocessed

R. Lecourt (✉)
Aerodynamics and Energetics Models Department, ONERA, Mauzac, France

Centre du Fauga-Mauzac, 31410, Mauzac, France
e-mail: lecourt@onera.fr

T.H. Karakoç et al. (eds.), *Advances in Sustainable Aviation*,
DOI 10.1007/978-3-319-67134-5_2

Table 2.1 Fuel compositions

FUEL	AROMATICS (%)	ACRONYM
Jet A-1	20.6	JETA
SPK	2.5	SPK
Jet A-1 + SPK	8	SPK08A
SPK + aromatic cut	20.6	SPK20AC
SPK + aromatic cut	8	SPK08AC
Aromatic cut	100	AC

esters and fatty acids (HEFA), do not contain any aromatic compound and have to be mixed up to 50% with Jet A-1 to comply with the above quoted specifications which require at least 8% aromatics in the aviation fuel composition. Nevertheless, this specification of 8% aromatics has no absolute technical justification. Therefore, to widen the spectrum of the aviation fuels, it is important to understand the influence of the fuel composition and determine the true acceptable limit.

In the frame of an in-house program, EXPLIC (exploration of limits of hydrocarbon fuels), ONERA is investigating the influence of the fuel composition, aromatic content and carbon molecule distribution on the fuel behaviour under high and low temperature conditions. The present paper deals with the most critical issues for aeronautical engines and fuels operability, altitude relight and cold start, which has been rarely described in the open literature, especially for actual turbojet injection systems and combustion chambers [1].

In a first step, it was decided to investigate the influence of the content of aromatics in the fuel composition by measuring the ignition performances of the six fuels described in Table 2.1, under altitude conditions. The content of aromatics of the samples was measured by high-performance liquid chromatography according to ASTM D 6379-04 specifications.

Jet A-1 is used as a reference. The sample used contained 20.6% in mass of aromatic compounds. The SPK (synthetic paraffinic kerosene) fuel was purchased from SkyNRG. The fuel aromatic rate is 2.52%, exclusively mono-aromatics. The SPK was mixed with Jet A-1 to produce a fuel with 8% aromatics. In addition, the petroleum company TOTAL Marketing Services provided aromatic streams, representative of the aromatic composition of Jet A-1, named aromatic cut (AC). This aromatic cut was blended with SPK to form two mixtures with, respectively, 20.6% and 8% aromatics. Finally, the pure aromatic cut was also tested. Some physical properties of the main fuels tested in this work are presented in Table 2.2.

2.2 Methodology

2.2.1 Test Bench

The experiments were carried out at the MERCATO (Experimental Mean for Research in Air-breathing Combustion by Optical Techniques) test facility at

Table 2.2 Fuel properties at fuel temperature 20 °C

Fuel designation	JETA	SPK	SPK08A	SPK08AC
Density (kg/m^3)	799.6	756.9	775.1	766.0
Viscosity (mm^2/s)	1.70	1.65	1.67	1.60
Surface tension (mN/m)	28.3	27.5	27.5	25.8
H/C ratio (−)	1.9	2.1	2.0	2.1
Aromatics (vol%)	15.6	<1.0	8.2	8.2
Naphthalenes (vol%)	1.72	0.08	0.65	1.06
Sulphur dioxide (wt%)	0.066	0.003	0.0316	0.0024

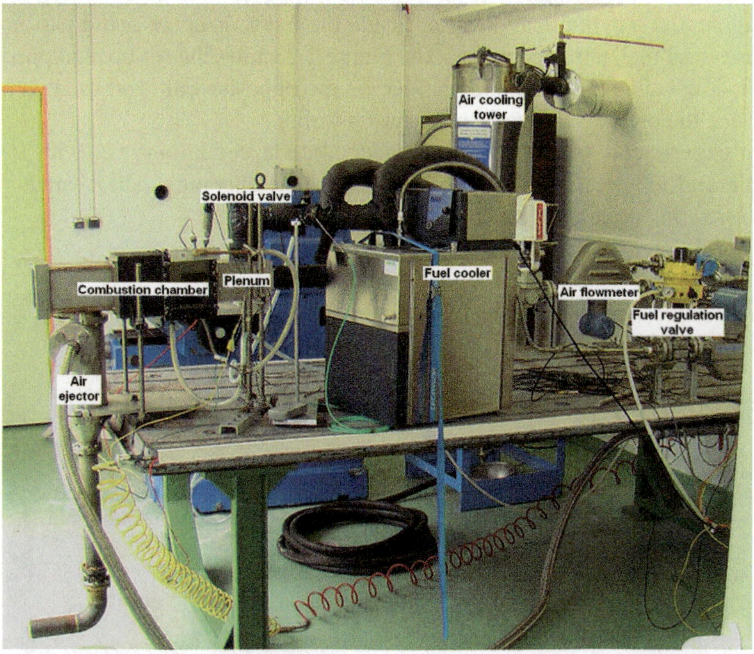

Fig. 2.1 MERCATO test bench

ONERA Fauga-Mauzac (Fig. 2.1). It is a small air-breathing propulsion research facility. Air and fuel are supplied at low flow rates, up to 100 and 10 g/s, respectively, but at a wide range of temperature from 233 to 473 K for air and from 233 K to ambient for fuel. Air is cooled through an air/LN2 cooling tower, fuel through a cooling bath. The pressure in the test chamber can be varied from 0.05 to about 0.4 MPa. This facility was extensively used to investigate the ignition phenomenon [2], especially under altitude conditions.

2.2.2 Test Set-Up and Conditions

We used the same experimental set-up as the one described in Lecourt et al. (2009) (Fig. 2.1). It is composed of a square section combustion chamber which is 129 × 129 × 235 mm^3 large. The injection system is implemented in a plenum chamber, upstream from the combustion chamber. Downstream from the combustion chamber, an interface part supports an air ejector, which is used to lower the pressure in the test chamber. The test chamber can receive four lateral windows for optical diagnostics. The injection system was provided by Turbomeca. Its geometry is fairly simple. The air circuit is composed of a radial swirler, an axial convergent and a cylindrical diffuser. Inside the convergent, an ovoid part holds a pressure swirl fuel atomizer on the axis. Figure 2.2 shows the set-up, equipped with one window facing the spark plug holder, on the opposite wall, and used with cold air (notice the presence of frost on the set-up walls).

The experiments were performed under altitude conditions, i.e. with low air pressure and temperature, but with fuel at room temperature. The characteristics of the operating point are described in Table 2.3.

Fig. 2.2 MERCATO set-up under altitude conditions

Table 2.3 Operating point

Air flow rate (kg/s)	0.0192
Air pressure (MPa)	0.06
Air temperature (K)	233
Fuel temperature (K)	290

A Vibro-Meter ET64A1 spark plug was used to ignite the air/fuel mixtures. It is a high-voltage spark plug using air as dielectric. It is supplied with an ignition exciter HEP502T2 from Vibro-Meter and operated at a frequency of 6 Hz. It is located 56 mm downstream from the injection face and flush mounted on one of the lateral walls. The capacitive discharge energy of the ignition device is 2 J. The spark energy was evaluated with tension and intensity measurements at electrodes by the manufacturer. This evaluation gave a spark energy value of 0.4 J.

2.2.3 *Test Procedure*

The objective of the present ignition experiments was to determine the ignition boundary for each fuel (Table 2.1) at the chosen operating point (Table 2.3). In an engine, the ignition procedure is a short one, and then the test duration was only 6 s. The procedure was designed as follows:

- Air was flowed continuously at the right flow rate (0.0192 kg/s) and temperature (233 K).
- At $t = 0$ s, a fuel solenoid valve, close upstream from the injection system, was opened.
- In parallel, the fuel regulation valve was set from 0% to n % opening to inject the required fuel flow rate.
- Then, at $t = 0.5$ s, the time at which the fuel flow rate has reached steady state, the spark plug was triggered and flashed at about 6 Hz, during 4.5 s, producing about 28 sparks during the tests.
- At $t = 4.9$ s, the kerosene regulation valve was set down to 0% opening.
- At $t = 5$ s, the kerosene solenoid valve was closed.
- Finally, at $t = 6$ s, the data recording was stopped.

An example of an ignition test is illustrated through Figs. 2.3, 2.4 and 2.5.

The flow rate curves (Fig. 2.3) show that the air flow rate is nearly constant during the experiment. A steady measurement of the fuel flow rate is obtained only after 2 s, because of the long response time of the Coriolis flowmeter. It must be pointed out that the steadiness of the flow rate curves allowed an accurate calculation of the global equivalence ratio of each experiment. As air was flowed continuously, its temperature was perfectly constant during the test (Fig.2.4). The fuel temperature, measured in the fuel feeding line, where the feeding line enters the plenum chamber, shows an increase from the line temperature surrounded by the cold air flow to the fuel temperature flowing around the thermocouple after the start of the test.

As shown in Fig. 2.5, ignition is easily observed on the chamber pressure curve thanks to the steep pressure increase of about 150 hPa. Then, pressure goes down to a level of about 50 hPa higher than the one before ignition until full combustion of the fuel injected in the combustion chamber during the 5 s test. In parallel, a signal extracted from the ignition exciter is recorded and gives the sparking times (rising edges, except the last one) (Fig. 2.5). In this experiment, ignition was obtained with the 17th spark.

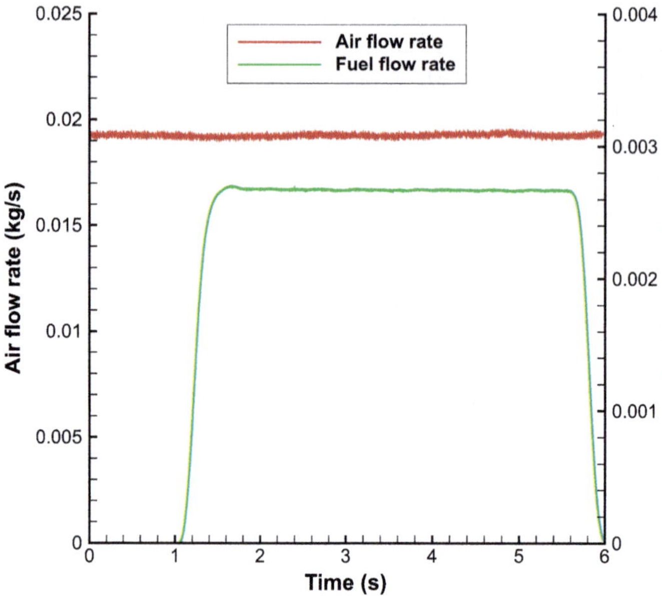

Fig. 2.3 Flow rates during an experiment

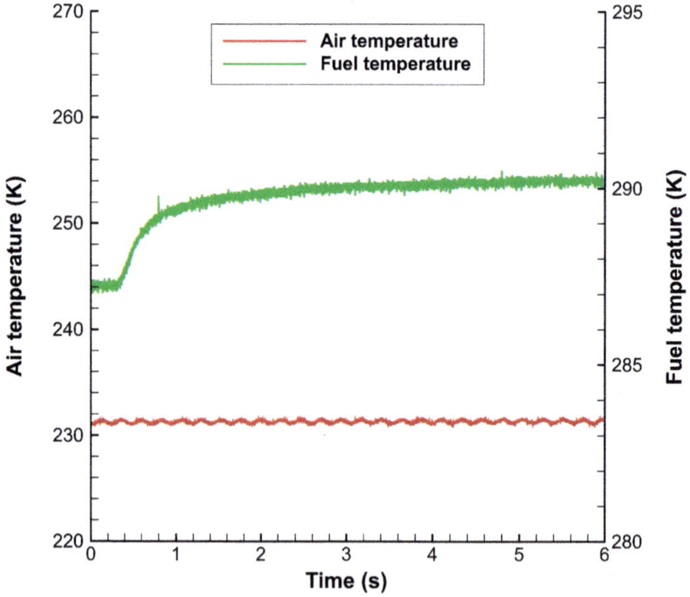

Fig. 2.4 Temperatures during an experiment

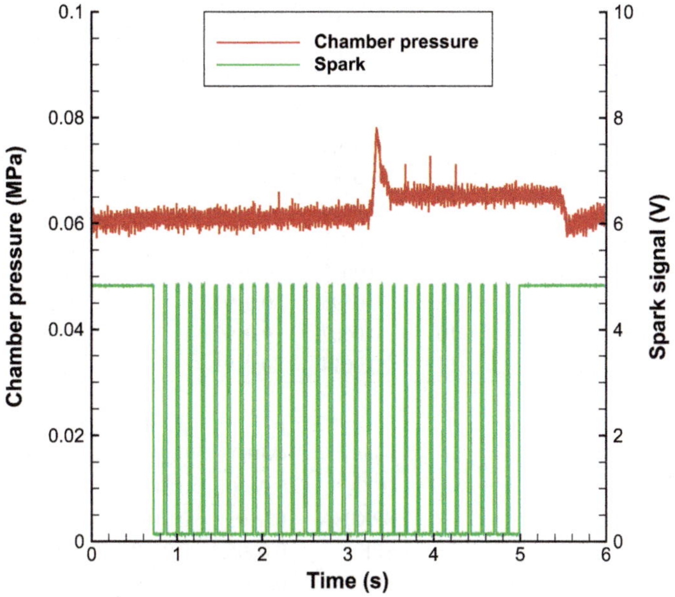

Fig. 2.5 Chamber pressure and spark plug operation during a successful ignition experiment

Finally, the steadiness of the flow rates, temperatures and chamber pressure during the experiments shows that the altitude conditions were well mimicked.

2.3 Results and Discussion

For every fuel, about 10 to 20 experiments were performed with different settings of the fuel regulation valve to obtain different global equivalence ratios from test to test. As ignitions were obtained with the highest values of equivalence ratio and misfires with the lowest values, an ignition boundary could be determined, as an equivalence ratio value, between these two sets of experiments:

$$ER_{\text{boundary}} = \left[\max(ER_{\text{misfire}}) + \min(ER_{\text{ignition}})\right]/2 \tag{2.1}$$

It must be pointed out that this equivalence ratio value was computed with one unique C-H composition, close to the one of an "average" Jet A-1, and not with the actual C-H composition of the tested fuels.

Some experiments were discarded because the targeted chamber pressure value, 0.06 Mpa, was not well reached. Finally, 10 to 12 experiments were kept by fuel. Table 2.4 gathers the characteristics of the experiments kept for the measurement of the ignition performances of the SPK08A fuel. From test to test, the discrepancies are within ±2.5% at most.

Table 2.4 Characteristics of the experiments for SPK08A fuel

Test #	P_{ch}(MPa)	Air flow rate (kg/s)	T_{air}(K)	T_{fuel}(K)
1	0.0600	0.0196	233	290
2	0.0607	0.0192	233	291
3	0.0613	0.0190	232	291
4	0.0604	0.0197	232	291
5	0.0586	0.0193	233	292
6	0.0600	0.0193	233	292
7	0.0597	0.0191	232	292
8	0.0608	0.0197	232	292
9	0.0595	0.0191	232	292
10	0.0607	0.0192	234	292
11	0.0614	0.0191	234	292

Table 2.5 Average characteristics of the experiments

Fuel	P_{ch} (MPa)	Air flow rate (kg/s)	T_{air} (K)	T_{fuel} (K)
JETA	0.0600	0.0194	233	289
SPK08A	0.0603	0.0193	233	292
SPK	0.0605	0.0193	232	290
SPK08AC	0.0601	0.0193	233	291
SPK20AC	0.0601	0.0193	233	290
AC	0.0601	0.0191	233	291

The average characteristics of the operating point of these experiments for every fuel are gathered in Table 2.5. Because of the good reproducibility from test to test, the average characteristics of the operating point are close to each other for all fuels. Therefore, their ignition performances will be able to be well compared.

Figure 2.6 gathers all the results, ignition or misfire, as the global equivalence ratio for each test and each fuel. First, on an overall, the results show that, under altitude conditions, a high equivalence ratio, about 2 and more, is needed to obtain ignition. Then, around the ignition boundary, we can observe misfires with equivalence ratio slightly higher than for some ignitions. This demonstrates the existence of a fuzzy zone around the ignition boundary because of the stochastic nature of the ignition phenomenon as it has been shown by [3] for one-phase flows and, which is more representative of aeronautical issues [4], for two-phase flows.

Concerning the influence of aromatics on the ignition performances (Table 2.6), a clear trend is observed with the SPK/AC blends. The fuels are easier and easier to ignite, for ER 2.03–1.91, when the aromatics content is increased from 0% to 20%. But the pure aromatic cut is the most difficult fuel to ignite with an ignition boundary of ER 2.3. Another surprising result is that SPK/Jet A-1 blend (SPK08A) is easier to ignite than each of its components. Nevertheless, we can observe that both the fuels with 8% aromatics (SPK08A and SPK08AC) have similar ignition boundaries, respectively, ER 1.93 and ER 1.96.

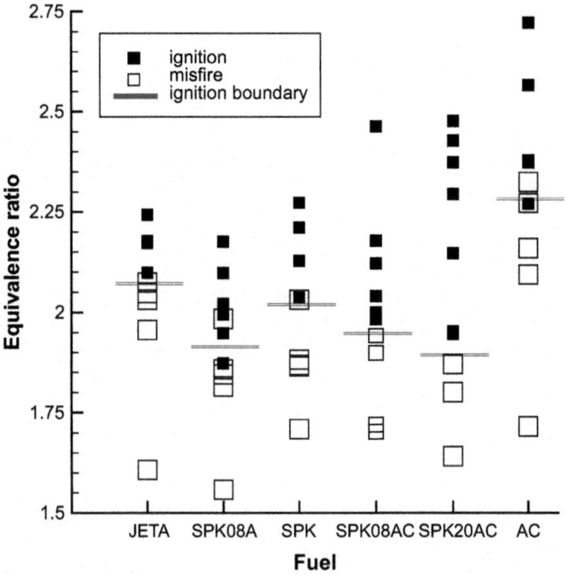

Fig. 2.6 Ignition results

Table 2.6 Ignition boundaries

Fuel	Fuel flow rate (kg/s)	ER (boundary)
JETA	0.00270	2.09
SPK08A	0.00253	1.93
SPK	0.00267	2.03
SPK08AC	0.00257	1.96
SPK20AC	0.00250	1.91
AC	0.00298	2.30

2.4 Conclusions

Successful measurements of the ignition performances of different fuel compositions have been performed under altitude conditions. In this first step, the main parameter was the content in aromatics. The results show that, to some extent, the ignition performances increase with the concentration in aromatics. Nevertheless, a part of the results show that other parameters play a role in ignition, which can work against the positive effect of the concentration in aromatics. Concerning the fuel specifications and with regard to the ignition performances, the results show that, because of the low influence of the aromatic content on them, 1.91 at most against 2.03 ER, it could be acceptable to lower the specification under 8%. Moreover, as the performances are similar for the SPK/Jet A-1 and SPK/AC mixtures, alternative fuels without Jet A-1 could be certified according to ignition properties.

References

1. Jarymowycz, T. A., & Mellorf, A. M. (1987). Effects of alternative fuels on ignition limits of the J85 annular combustor. *Journal of Propulsion and Power, 3*(3), 283–288.
2. Lecourt, R., Bismes, F., & G. Heid. (2009). Experimental investigation of ignition of an air/kerosene spray under altitude conditions, ISABE2009–1235, 7–11 September 2009, Montreal, Canada.
3. Ahmed, S. F., Balachandran, R., Marchione, T., & Mastorakos, E. (2007). Spark ignition of turbulent nonpremixed bluff-body flames. *Combustion and Flame, 151*(1–2), 366–385.
4. Lang, A., Lecourt, R., & Giuliani, F.. (2010). Statistical evaluation of ignition phenomena in turbojet engines, ASME2010–23229, ASME Turbo Expo 2010: Power for Land, Sea and Air, June 14–18, 2010, Glasgow, UK.

Chapter 3
Review of Renewable Biofuels in the Aviation Sector

Derya Unlu and Nilufer Durmaz Hilmioglu

Nomenclature

CH_4	Methane
CO	Carbon monoxide
CO_2	Carbon dioxide
FT	Fischer-Tropsch
H_2	Hydrogen
HC	Hydrocarbon
LPG	Liquefied petroleum gas
NOx	Nitrogen oxides
PM	Particulate matter

3.1 Introduction

The aviation is one of the few sectors that are the fastest growing in the world. Aviation mostly uses liquid fuels such as jet fuels which are mostly obtained from fossil source. Greenhouse gas emissions from aviation are a major problem for this industry [1]. The aviation industry is responsible for the 2% of the CO_2 emissions and the 10% of the fuel consumption [2, 3]. The aviation traffic is expected to grow by 5% per year to 2026 and fuel demand by about 3% per year, so fuel consumption and CO_2 emissions will increase in next years [4]. The aviation industry aims to have carbon neutral rise by 2020 and a 50% decrement in CO_2 emission in 2050. The actions to reach these targets include improving fuel efficiency by 2% per year

D. Unlu (✉) • N.D. Hilmioglu
Kocaeli University, Chemical Engineering Department, İzmit, Kocaeli, Turkey
e-mail: derya.unlu@kocaeli.edu.tr; deryaunluu@gmail.com

T.H. Karakoç et al. (eds.), *Advances in Sustainable Aviation*,
DOI 10.1007/978-3-319-67134-5_3

and the use of renewable fuels [5]. To reach these targets, the petroleum-derived fuels displace with biofuels [6, 7]. Therefore, large amounts of biofuel and biomass have to be produced in order to meet the 2020 and 2050 aviation sector goals.

Biofuels are gaining attention due to having the low greenhouse gas emissions. Use of biofuels significantly decreased fossil fuel dependence and increased the utilization of renewable resources. Bio based jet fuels can reduce greenhouse gas emissions up to 89% accordingly petroleum jet. Fuels with a high energy density are preferred for the aviation transport. Alternative biofuels have some properties such as good pour point, cloud point, and thermal endurance. Also, requirements of these fuels are rapid evaporation and atomization, low risk of explosion, appropriate viscosity, physical and chemical stability, and low freezing point. The developed biofuels can be suitable for the current airplane engine [8–10]. The raw materials which are utilized in the manufacturing of aviation biofuels have a biologic source and they are renewable. Nonedible oil plants, municipalities, sewage and forest wastes, etc. are the main resources for biofuels. These biomasses convert to valuable fuels by thermochemical and biochemical production technologies such as gasification, pyrolysis, enzymatic hydrolysis, and fermentation [5, 10]. This study deals with the natural raw material sources for the production of biofuel.

In addition to the decrement in the emission of greenhouse gasses, alternative aviation biofuels have a number of great advantages over conventional fuel. The production costs of fossil fuels are high. Biomass-derived fuels are renewable and cheaper than others. Therefore, production of aviation biofuel is profitable. One of the other problems is that fossil fuels are not satisfying the demand. Convenient supply is ensured easily for the aviation biofuels worldwide, and also they can be produced easily. Aviation biofuels have good properties against conventional aviation fuels such as low flash point, good thermal endurance, etc. Concisely the aviation biofuels are environmental friendly and economical, and their production is sustainable [10–12].

In addition to the all these, the alternative fuels such as Fischer-Tropsch fuels and hydroprocessed esters and fatty acids are purely paraffinic fuels. It means that these fuels have only alkane molecules. However, conventional jet fuels have aromatic molecules. Aromatics play an important role in the formation of particulate materials. Therefore, alternative biofuels are used to blend with conventional jet fuels for reducing levels of aromatics [13, 14].

In literature, there are researches about the potential feedstock for renewable aviation fuel production and the impact of biofuels on aviation emissions. Zhang (2015) produced jet and diesel fuels from sawdust. They also studied some important fuel properties of the biofuels synthesized in that work, and the obtained results showed that the jet fuel can be directionally synthesized using lignocellulosic biomass sawdust. Also, it was said that biomass-derived product can met the main requirements of jet fuels [15]. Winchester (2015) studied the fuel properties of butyl butyrate. It was found as a compatible fuel with petrol, and it has similar properties. The properties of these fuels were compared to other oxygenated biofuels, and relevant hydrocarbons [16]. Nine potential biofuels derived from sustainable sources were investigated and tested for their compatibility with Jet A-1 aviation kerosene. The main fuel properties of the biofuels were determined. Esters were found as compatible with aviation kerosene. Biodiesel was also found as suitable for aviation [17]. The biomass-derived jet fuel blends were examined. It was produced via thermochemical methods [8].

This study deals with natural raw material sources, biofuel and production methods, and occasions and difficulties in the synthesis of biofuel for aviation. Properties of raw material, the techniques used in the production process, and a comparison of alternative fuels in this field are also included.

3.2 Natural Raw Material Sources

Natural raw material sources are important for the production of aviation biofuel. The advantages of these resources are environmental sustainability, renewability, friendly to the earth, and independence on fossil fuels. The raw material resources such as sugar- and starch-based groups, oil-based group, cellulosic materials, and wastes are shown in Fig. 3.1.

3.2.1 Sugar and Starch Group

Starch is a semicrystalline polysaccharide. Starch is an energy storage used as an energy source by plants.

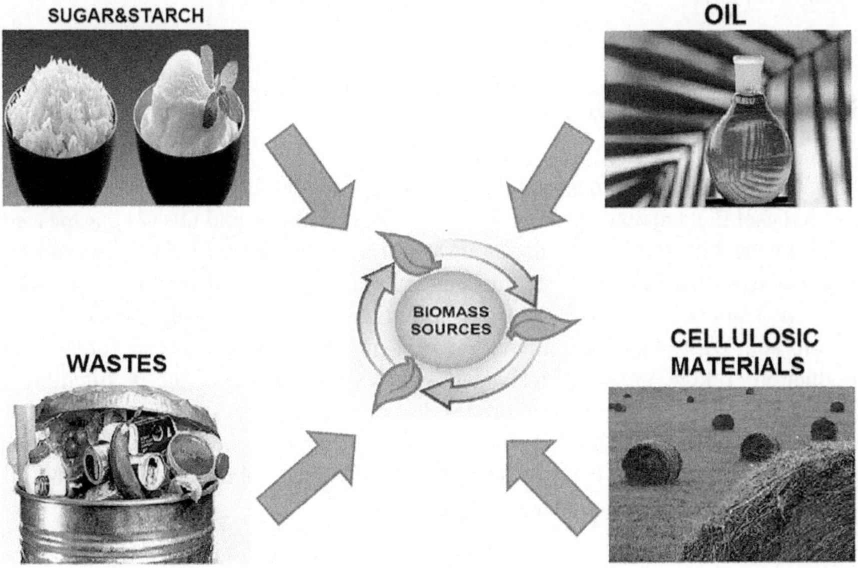

Fig. 3.1 Natural raw material sources

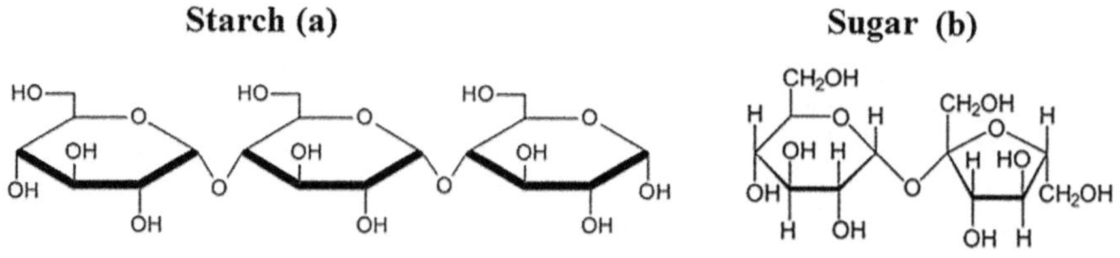

Fig. 3.2 Molecular structure of starch (**a**) and sugar (**b**)

Microorganisms used the six-carbon sugars in the synthesis of bioethanol and other bioproducts. Biomass-derived materials have a high amount of glucose, and they can be easily converted to biofuels such as bioethanol. Compared to starch and lignocellulosic biomass, the conversion of sugar into ethanol is a simple process, because there is no need for pretreatment operation. It can be decomposed by using the microorganisms [18]. Though, the sugar amount in human foods is too expensive for producing biofuel. Also, sugarcane, sugar beet, sweet sorghum, and various fruits could be feedstock for biofuel production, but these materials are also human food, and they are the most important raw materials in tropical countries. Molecular structure of starch and sugar are shown in Fig. 3.2.

The other alternative biofuel feedstock is starch. Starch as a polysaccharide is formed from chains of glucose molecules [19]. Starch is a high-yield feedstock for biofuel production. Grains, potato, corn, and rice are starchy materials. Grains are usually used for biofuel production. Starchy materials react with water for obtaining sugar. Enzymes are used to break the chemical bonds in the stirred and heated reactor [18, 20].

3.2.2 Oil-Based Group

A variety of oils as feedstocks are used to produce the biofuel. The main feedstocks for biodiesel fuel are unused oil, waste vegetable oil, and animal fats [21]. Rapeseed and soybean oils are usually used raw material for biofuel. Soybean oil can be obtained from field pennycress, jatropha, sunflower, palm oil, and coconut. The waste vegetable oil thrown away from restaurants is getting popular as feedstocks for biofuel. Waste vegetable oil is the best alternative raw material for biofuel production. Tallow, grease, chicken fat, and the by-products of the production of Omega-3 fatty acids from fish oil are used as biofuel feedstocks. Also algae which are grown in sewage are a good source of biofuel feedstocks. Other feedstocks for biofuel are halophytes. They can be grown in saltwater. Table 3.1 shows some oil sources for biofuel production and oil yield [22].

Table 3.1 Some oil sources for biofuel production and oil yield (%) [22]

Raw material	Oil yield (L/ha)
Corn	172
Soybean	446
Canola	1190
Jatropha	1892
Coconut	2689
Oil palm	5950
Microalgae	136,900

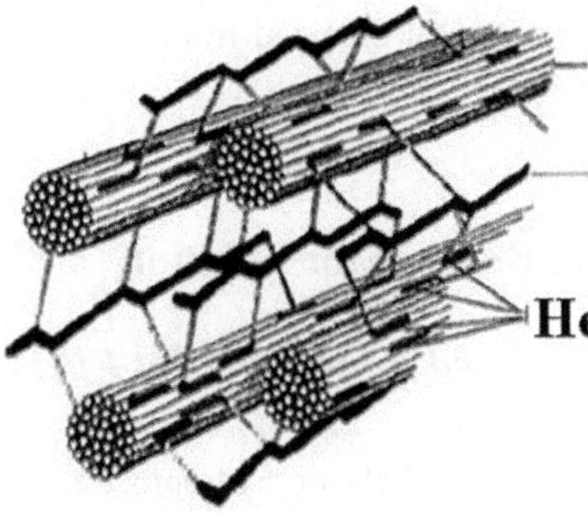

Fig. 3.3 Cellulosic materials [25]

3.2.3 Cellulosic Materials

Sugar and starch groups are in the human food chain and they are expensive. Cellulosic materials are alternative raw materials [23]. Paper, cardboard, wood, and other fibrous plant materials are examples of cellulosic materials [24].

Cellulosic resources are extensive and plentiful. For example, forests formed the great majority of the world's biomass. Cellulosic biomass comprises about 50% of world biomass. Therefore, cellulosic biomass is an inexpensive feedstock due to sugar and starch groups.

Cellulosic biomass consists of polysaccharides. The structure of cellulosic material is shown in Fig. 3.3. Cellulosic materials consist of lignin, hemicellulose, and cellulose. These materials are called as lignocellulosic materials. Lignin provides the structural support for the plant. Trees have higher lignin contents than grasses. Lignin doesn't contain sugars. It is enclosed in the cellulose and hemicellulose molecules; this property makes it difficult to reach lignin structure. Therefore, lignocellulosic materials need a pretreatment stage. Biofuel production from lignocellulosic forms five stages: biomass pretreatment, hydrolysis, fermentation, separation, and effluent treatment [18].

Cellulose molecules comprise of chains of glucose molecules. They have a different structure according to starch molecules. Therefore, lignocellulosic material is more difficult to hydrolyze than starchy materials [20].

3.2.4 Wastes

Wastes from different sources are raw materials for the manufacturing of biofuels. Wastes such as forest products, paper residues, and agricultural residues can be converted to biofuel. These resources are of low price and simply accessible. The utilization of waste resources for the manufacturing of biofuels can get over many challenges such as the demand of fertilizer and watering. Also, the use of waste feedstock for the production of biofuel can decrease waste management problems, environmental, air and soil pollution, greenhouse emissions, and the utilization of fossil fuels. Usage of biofuel obtained from waste causes to decrease the global warming [26, 27].

3.3 Alternative Biofuels for Aviation

3.3.1 Hydroprocessed Renewable Jet Fuels

Hydroprocessed renewable fuels or hydroprocessed esters are paraffinic liquid fuels. They are manufactured by the hydrodeoxygenation of vegetable and animal oils, waste grease, algal oil, and seed oils from oil-seed plants like *Jatropha* and *Camelina*. Hydroprocessed renewable fuels have similar properties of Fischer-Tropsch jet fuel. These renewable fuels, which have high energy, can be utilized directly as fuel. One of the main advantages of these fuels is a decrease in the emission of carbon monoxide (CO), hydrocarbons (HC), nitrogen oxides (NOx), aromatic content, and particulate matter (PM). The hydroprocessed renewable jet fuels have a high cetane number and high thermal endurance.

The hydroprocessed renewable jet fuels can be used without modification in conventional aircraft engines. Also, it can be used as an additive or blended with conventional jet fuel. Several aircraft companies have tested hydroprocessed renewable jet fuels [28]. These fuels avoid the motor corrosion. Due to the good cold flow properties, these fuels are suitable for higher flights. The lack of oxygen and sulfur in the fuel reduces its lubricity [10].

3.4 Fischer-Tropsch Fuels

Fischer-Tropsch fuels are hydrocarbon fuels. Synthesis process of these fuels is the catalytic conversion of syngas [29, 30]. The Fischer-Tropsch process is shown in Fig. 3.4. Gasification is the first step for the heating biomass by partial oxidation or in the presence of steam. This step is created a syngas from biomass. The mixture of CO and H_2 called as syngas is obtained, and then syngas is processed by Fischer-Tropsch catalysts into hydrocarbon chains. The Fischer-Tropsch process

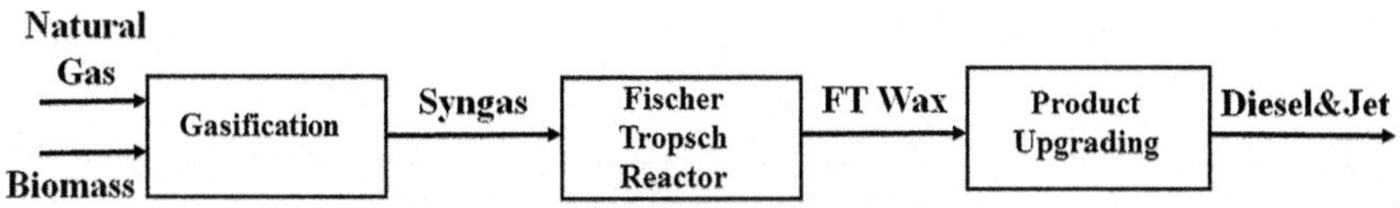

Fig. 3.4 Liquid fuels from biomass through Fischer-Tropsch synthesis [32]

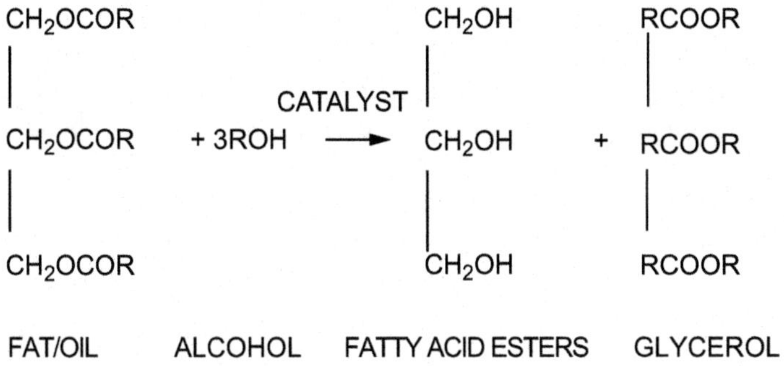

Fig. 3.5 Transesterification reaction [33]

can produce a cleaner diesel oil fraction with a high cetane number. Raw materials of Fischer-Tropsch can comprise of any matter containing carbon. There are lots of biomass raw materials such as woody crops, wood waste, grass crops, and agricultural residues for the production of syngas. Fischer-Tropsch fuels have clean burning properties such as low emission of nitrogen oxides and particulate and low sulfur and aromatic content. However, the production process of Fischer-Tropsch fuels is expensive and has low efficiency [10, 31].

3.5 Biodiesel

Biodiesel is produced by the process of transesterification using vegetable oils and animal fats. Transesterification reaction is shown in Fig. 3.5. Biodiesel is a biodegradable fuel and shows superior lubricity. Biodiesel has advantages according to other fuels such as high flash point and good miscibility with fossil-derived fuels. Biodiesel shows lower greenhouse gas and sulfur emissions. Biodiesel fuels do not cause any water, soil, or air pollution because of the nontoxic properties. However, the biodiesel fuel is not appropriate to utilize as an aviation fuel. The energy density of biodiesel is lower than conventional jet fuels. It has low efficiency. The cloud point of biodiesel is higher than petroleum-based aviation fuel. This property has negative effects for high flights. Biodiesel contains unsaturated fatty acids. These structures cause the decrement stability of the biodiesel. Therefore, biodiesel is not currently attractive as an alternative aviation fuel [10].

3.6 Biohydrogen and Biomethane

Biohydrogen and biomethane can be produced by fermentation and anaerobic digestion. Biomass is converted to biogas in anaerobic digestion, as a biological conversion process. Biohydrogen can be biologically produced by bacterial fermentation [34]. Liquid hydrogen is an alternative jet fuel. Biohydrogen is manufactured from biomass resources by thermal and biochemical procedures. Liquid hydrogen provides more energy per weight due to conventional aviation fuel. The liquid hydrogen fuels cause low harmful emission [35–37]. However, hydrogen fuel has some disadvantages. It cannot be used as jet fuel without engine machine modification, and it also needs high storage volume.

Liquid methane can be utilized as a fuel in cryogenic aircrafts. The use of liquid methane fuel causes the decrement of carbon dioxide emission about 25%. Engine modification is required for the using of the liquid methane fuels [10].

3.7 Bioalcohols

Ethanol and butanol are generally utilized as biofuels. Bioethanol is not applicable for aviation sectors due to its high volatility, low flash point, and energy density. Biobutanol has also similar properties. Alcohol as an aviation fuel needs extra engine design and storage. Mixing of bioalcohols with conventional aviation fuels is not practical because of its inadequate fuel properties. Figure 3.6 shows bioethanol production from sugar and starch [10].

Bioethanol is produced from biomass mostly via a fermentation process. Firstly, the raw material (sugar, starch, etc.) is converted into glucose. Sugar is easily converted to glucose by dissolving in the water, but starch requires pretreatment to transform it to glucose such as saccharification and liquefaction. After the glucose obtainment, the enzymes add and the glucose is available for fermenting into alcohol. While the ethanol production from sugar requires the use of yeast, the starch materials need enzymes.

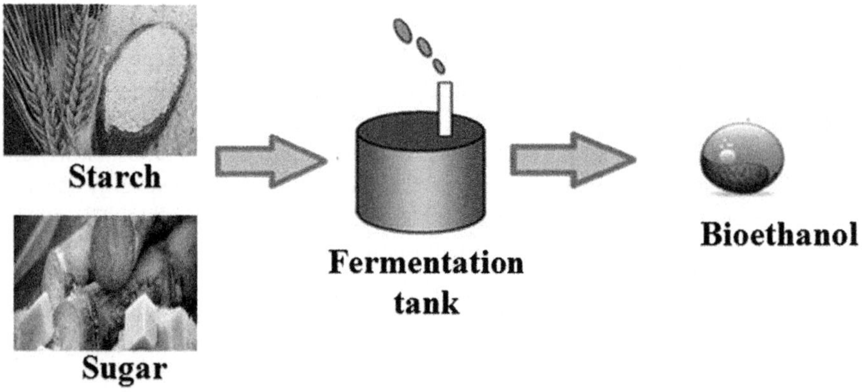

Fig. 3.6 Bioethanol production from sugar and starch

3.8 Production Methods

3.8.1 Thermochemical Process

Thermochemical process is defined as the conversion of biomass to fuel by pyrolysis and gasification. Pyrolysis is the thermal decay of biomass. According to this process, bio-oil and methane are occurred. Gasification process is defined as pyrolysis products that are treated with air or steam to obtain synthesis gas. Syngas occurs from mixing hydrogen and carbon monoxide [10]. Thermochemical process is shown in Fig. 3.7.

3.8.2 Hydroprocessing

The oxygen is removed from the raw materials (fats and oils) by using hydrogen at hydroprocessing. After the hydrodeoxygenation, isomerization and cracking steps have taken place. After these steps, desired specifications of fuels are obtained such as low temperature properties. As a result of hydroprocessing, clean paraffinic fuels are obtained. The scheme of hydroprocessing is shown in Fig. 3.8. These fuels have high thermal stability and have no aroma and sulfur. These fuels are called as hydroprocessed renewable jet fuels. In this method, oxygen is usually taken away

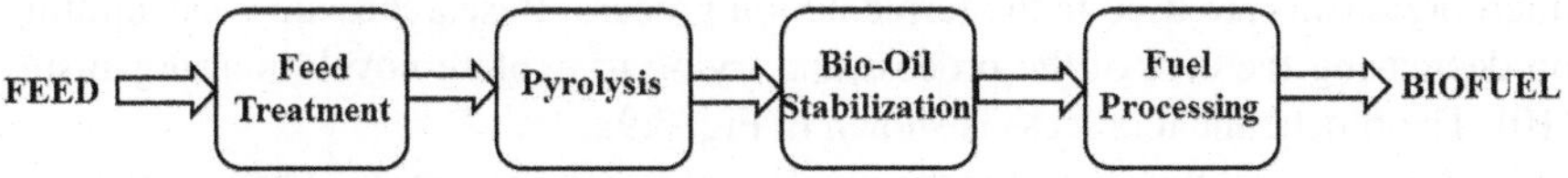

Fig. 3.7 Thermochemical process – pyrolysis [38]

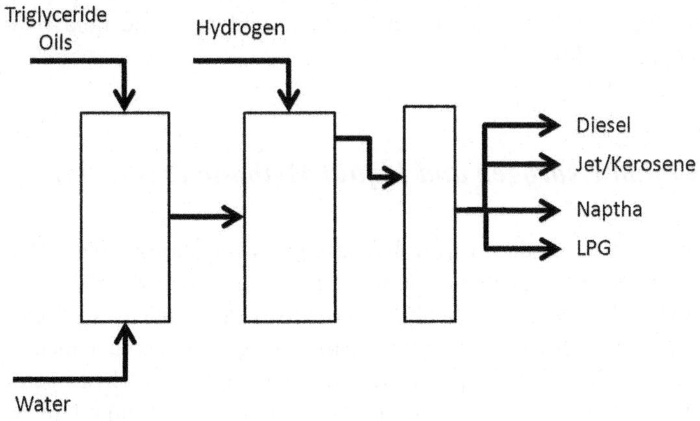

Fig. 3.8 Hydroprocessing process

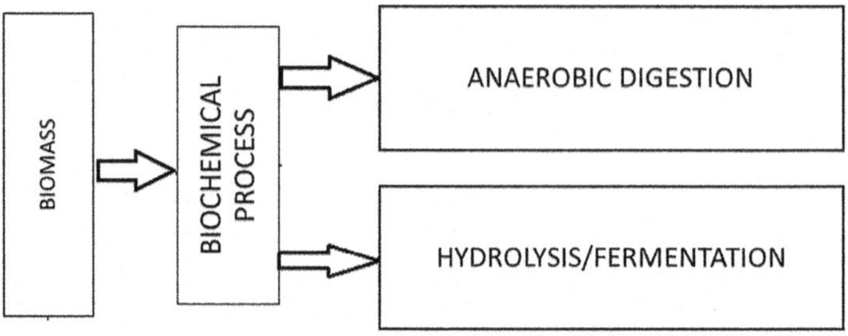

Fig. 3.9 Biochemical processes [41]

as water, and one of the by-products is propane. The commonly used raw materials for the manufacturing of hydroprocessed renewable jet fuels are plant oils like jatropha, camelina, and algal, bio-oil, animal fats, and waste grease. The method is effective. This type of fuel has good cold flow properties and high cetane number. Hydroprocessed renewable fuels are more economical than other conventional fuels [10, 39, 40].

3.8.3 Biochemical Process

Biochemical process is the conversion of biomass to carbohydrates. Enzymes or microorganisms are used in the fermentation process. Researches are concentrating on decreasing the cost of the process and trying to explore novel microorganisms [10]. The biochemical process is shown in Fig. 3.9.

There are two types of biochemical conversion processes. One of them is anaerobic degradation for biogas ($CH_4 + CO_2$) synthesis and the other one is ethanol (C_2H_5OH) fermentation. These biochemical conversion processes need to use substrates, and also these substrates should meet the kind of microbial population preferred [41].

3.8.4 Liquid Hydrogen and Liquid Methane Production

Biohydrogen can be produced from biomass by gasification process. Hydrogen is obtained from biomass at higher temperatures and pressures, and CO and CO_2 are obtained as by-products. Membrane processes or chemical ways can be used for separation of hydrogen. Catalytic steam reforming is another method for the production of hydrogen from pyrolysed oil. The synthesis of hydrogen by using algae and bacteria is a biological method. The usage of liquid hydrogen for aviation reduces the harmful gas emissions [42]. Biomethane is manufactured by anaerobic decomposition or fermentation of renewable feedstocks. The use of liquid

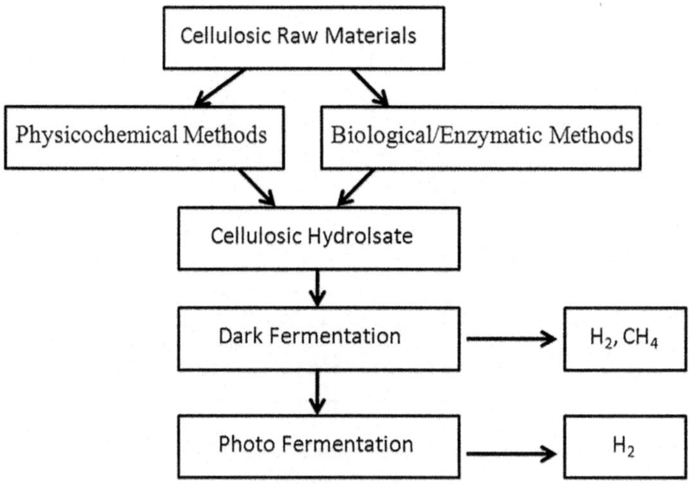

Fig. 3.10 Scheme of biohydrogen and biomethane production [43]

methane fuels is needed for the modification of engines [10]. Figure 3.10 shows the biohydrogen and biomethane production from cellulosic raw material.

In the biohydrogen and biomethane production, microbial and enzymatic pre-treatment is used to convert biomass to fermentable sugars. When the biohydrogen is produced from cellulosic feedstock by dark fermentation, cellulose is hydrolyzed and transformed into H_2 gas as simultaneously [43].

3.8.5 Transesterification of Oils/Fats

Biodiesel is synthesized by transesterification process which is shown in Fig. 3.11. Fats and oils convert as catalysts into alkyl esters of fatty acids. The biodiesel production process is inexpensive. Nonedible oils are used as raw materials; therefore, it is not for human food consumption. The biodiesel fuel is nontoxic and biodegradable. Biodiesel is not appropriate for aviation fuels, because of the low energy density and poor cold flow properties of the fuel [10].

3.9 Conclusion

The synthesis of alternative aviation fuels from renewable feedstocks attracts considerable attention because of its energy security and its ability in reducing the dependency on fossil fuels. Hydroprocessed renewable jet fuel and Fisher-Tropsch fuels as alternative fuels have advantages according to the conventional jet fuels.

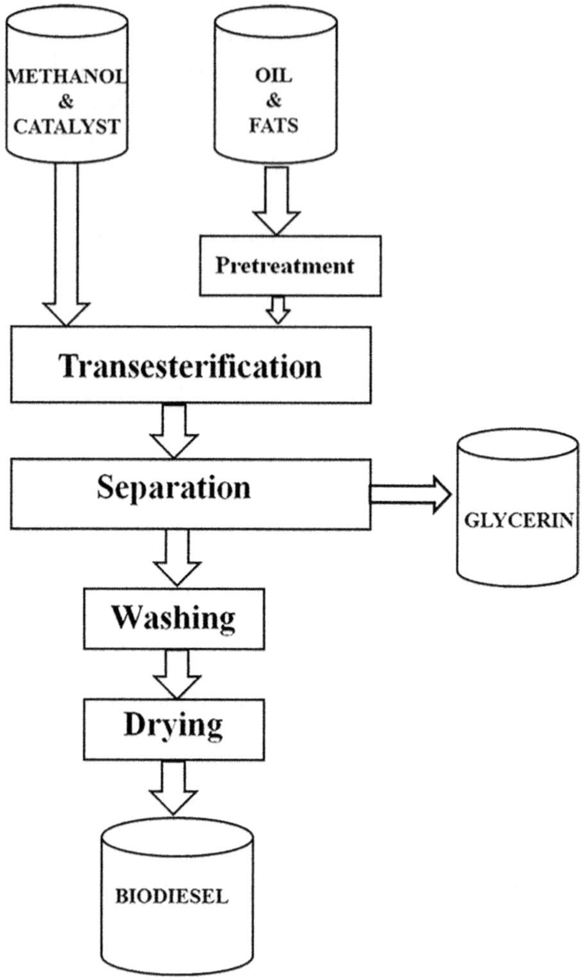

Fig. 3.11 Transesterification process of oils and fats [44]

Biodiesel and bioalcohols as aviation fuels have poor fuel properties. The liquid hydrogen and liquid methane fuels are other alternative fuels, but they have high production cost and they need to modify for the usage in aircraft engine. Some difficulties come up such as accessibility of raw material, blending of alternative fuels with conventional fuels, environmental worry, and manufacturing and delivery problems. The government and organizations can help the enhancement of the sector and commercialization.

References

1. Köhler, J., Walz, R., Marscheder-Weidemann, F., & Thedieck, B. (2014). Lead markets in 2nd generation biofuels for aviation: A comparison of Germany, Brazil and the USA. *Environmental Innovation and Social Transitions, 10*, 59–76.
2. Blakey, S., Rye, L., & Wilson, C. W. (2011). Aviation gas turbine alternative fuels: A review. *Proceedings of the Combustion Institute, 33*, 2863–2885.
3. Upham, P., Tomei, J., & Boucher, P. (2009). Biofuels, aviation and sustainability: Prospects and limits. In S. Gössling & P. Upham (Eds.), *Climate change and aviation: Issues, challenges and solutions* (1st ed., pp. 309–320). London: Earthscan.
4. Chiaramonti, D., Prussi, M., Buffi, M., & Tacconi, D. (2014). Sustainable bio kerosene: Process routes and industrial demonstration activities in aviation biofuels. *Applied Energy, 136*, 767–774.
5. Cantarella, H., Nassar, A. M., Cortez, L. A. B., & Junior, R. B. (2015). Potential feedstock for renewable aviation fuel in Brazil. *Environmental Development, 15*, 52–63.
6. Sandquist, J., & Güell, B. M. (2012). Overview of biofuels for aviation. *Chemical Engineering Transactions, 29*, 1147–1152.
7. Han, J., Elgowainy, A., Cai, H., & Wang, M. Q. (2013). Life-cycle analysis of bio-based aviation fuels. *Bioresource Technology, 150*, 447–456.
8. Atsonios, K., Kougioumtzis, M. A., Panopoulos, K. D., & Kakaras, E. (2015). Alternative thermochemical routes for aviation biofuels via alcohols synthesis: Process modeling, techno-economic assessment and comparison. *Applied Energy, 138*, 346–366.
9. Cremonez, P. A., Feroldi, M., Araújo, A. V., Borges, M. N., Meier, T. W., Feiden, A., & Teleken, J. G. (2015). Biofuels in Brazilian aviation: Current scenario and prospects. *Renewable and Sustainable Energy Reviews, 43*, 1063–1072.
10. Hari, T. K., Yaakob, Z., & Binitha, N. N. (2015). Aviation biofuel from renewable resources: Routes, opportunities and challenges. *Renewable and Sustainable Energy Reviews, 42*, 1234–1244.
11. Kivits, R., Charles, M. B., & Ryan, N. (2010). A post-carbon aviation future: Airports and the transition to a cleaner aviation sector. *Futures, 42*, 199–211.
12. Gegg, P., Budd, L., & Ison, S. (2014). The market development of aviation biofuel: Drivers and constraints. *Journal of Air Transport Management, 39*, 34–40.
13. Pągowski, Z. T. (2003). New perspectives for biofuels in aviation. *Journal of KONES Internal Combustion Engines, 10*, 3–4.
14. Zhang, C., Hui, X., Lin, Y., & Sung, C. J. (2016). Recent development in studies of alternative jet fuel combustion: Progress, challenges, and opportunities. *Renewable and Sustainable Energy Reviews, 54*, 120–138.
15. Zhang, Y., Bi, P., Wang, J., Jiang, P., Wu, X., Xue, H., Liu, J., Zhou, X., & Li, Q. (2015). Production of jet and diesel biofuels from renewable lignocellulosic biomass. *Applied Energy, 150*, 128–137.
16. Winchester, N., Malina, R., Staples, M. D., & Barrett, S. R. H. (2015). The impact of advanced biofuels on aviation emissions and operations in the U.S. *Energy Economics, 49*, 482–491.
17. Chuck, C. J., & Donnelly, J. (2014). The compatibility of potential bioderived fuels with Jet A-1 aviation kerosene. *Applied Energy, 118*, 83–91.
18. Cardona, C. A., Orrego, C. E., & Paz, I. C. (2009). The potential for production of bioethanol and bioplastic from potato starch in Colombia. *Fruit, Vegetable and Cereal Science and Biotechnology, 3*, 102–114.
19. Cherubini, F. (2010). The biorefinery concept: Using biomass instead of oil for producing energy and chemicals. *Energy Conversion and Management, 51*, 1412–1421.
20. Badger, P. C. (2002). Ethanol from cellulose: A general review. In J. Janick & A. Whipkey (Eds.), *Trends in new crops and new uses* (pp. 17–21). Alexandria: ASHS Press.

21. Ahmia, A. C., Danane, F., Bessah, R., & Boumesbah, I. (2014). Raw material for biodiesel production. Valorization of used edible oil. *Revue des Energies Renouvelables, 17*(2), 335–343.
22. Burns, N. A. (2010). Biomass—the next revolution in surfactants. *International News on Fats, Oils and Related Materials, 21*(12), 717–780.
23. Escobar, J. C., Lora, E. S., Venturini, O. J., Yanez, E. E., Castillo, E. F., & Almazan, O. (2009). Biofuels: Environment, technology and food security. *Renewable & Sustainable Energy Reviews, 13*, 1275–1287.
24. Demirbaş, A. (2005). Bioethanol from cellulosic materials: A renewable motor fuel from biomass. *Energy Source Part A, 27*, 327–337.
25. Almarsdóttir A.R., & Gunnarsson I.B. (2009). Framleiðsla etanóls úr flóknum lífmassa með hitakærum bakteríum. University of Akureyri.
26. Demirbaş, A. (2007). Progress and recent trends in biofuels. *Progress in Energy and Combustion Science, 33*, 1–18.
27. Naik, S. N., Goud, V. V., Rout, P. K., & Dalai, A. K. (2010). Production of first and second generation biofuels: A comprehensive review. *Renewable and Sustainable Energy Reviews, 14*, 578–597.
28. Hileman, J.I., Ortiz, D.S., Bartis, J.T., Wong, H.M., Donohoo, P.E., Weiss, M.A., Waitz, I.A. (2009). Near-term feasibility of alternative jet fuels. Technical Report.
29. Corporan, E., DeWitt, M. J., Belovich, V., & Pawlik, R. (2007). Emissions characteristics of a turbine engine and research combustor burning a Fischer-Tropsch jet fuel. *Energy Fuel, 21*, 2615–2626.
30. Lobo, P., Hagen, D. E., & Whitefield, P. D. (2011). Comparison of PM emissions from a commercial jet engine burning conventional, biomass, and Fischer Tropsch fuels. *Environmental Science & Technology, 45*, 10744–10749.
31. DeWitt, M. J., Corporan, E., Graham, J., & Donald Minus, D. (2008). Effects of aromatic type and concentration in Fischer-Tropsch fuel on emissions production and material compatibility. *Energy Fuel, 22*, 2411–2418.
32. *Coal, petroleum and other form of fuels for energy conversion*. [online] http://coalandfuel.blogspot.com.tr. Accessed 22 Mar 2017.
33. Garba, M. U., Alhassan, M., & Kovo, A. S. (2006). A review of advances and quality assessment of biofuels. *Leonardo Journal of Sciences, 9*, 167–178.
34. Monlau, F., Barakat, A., Trably, E., Dumas, C., Steyer, J. P., & Carrère, H. (2013). Lignocellulosic materials into biohydrogen and biomethane: Impact of structural features and pretreatment. *Critical Reviews in Environmental Science and Technology, 43*, 260–322.
35. Contreras, A., Yigit, S., Ozay, K., & Veziroglu, T. N. (1997). Hydrogen as aviation fuel: A comparison with hydrocarbon fuels. *International Journal of Hydrogen Energy, 22*, 1053–1060.
36. Koroneos, C., Dompros, A., Roumbas, G., & Moussiopoulos, N. (2005). Advantages of the use of hydrogen fuel as compared to kerosene. *Resources, Conservation and Recycling, 44*, 99–113.
37. Pohl, H. W., & Malychevc, V. V. (1997). Hydrogen in future civil aviation. *International Journal of Hydrogen Energy, 22*, 1061–1069.
38. *Thermochemical conversion processes*. [online] http://www.energy.gov/ processes. Accessed 23 Apr 2016.
39. Hui, X., Kumar, K., Sung, C. J., Edwards, T., & Gardner, D. (2012). Experimental studies on the combustion characteristics of alternative jet fuels. *Fuel, 98*, 176–182.
40. Pearlson, M., Wollersheim, C., & Hileman, J. (2013). A techno-economic review of hydroprocessed renewable esters and fatty acids for jet fuel production. *Biofuel Bioproducts and Biorefining, 7*, 89–96.
41. *Texas renewable energy resource assessment.* Chapter 5 Biomass Energy [online] http://www.seco.cpa.state.tx.us/publications/renewenergy/. Accessed 24 Apr 2016.
42. Yılmaz, İ., İlbaş, M., Taştan, M., & Tarhan, C. (2012). Investigation of hydrogen usage in aviation industry. *Energy Conversion and Management, 63*, 63–69.

43. Saratale, G. D., Chen, S. D., Lo, Y. C., Saratale, R. J., & Chang, J. S. (2008). Outlook of biohydrogen production from lignocellulosic feedstock using dark fermentation – a review. *Journal of Scientific and Industrial Research, 67*, 962–979.
44. *Ultrasonics improve biodiesel process efficiency*. [online] https://www.hielscher.com. Accessed 23 Apr 2016.

Chapter 4
Case Study of Alternative Jet Fuel Production with Bio-additives from Plant Oils in Ukraine and Poland

Anna Yakovlieva, Sergii Boichenko, Oksana Vovk, Kazimierz Lejda, and Oleksandr Gryshchenko

4.1 Introduction

One of the features of the modern world is the increased attention of the international community to the problems of rational and efficient use of energy resources, introduction of energy-saving technologies, and searching of renewable energy.

Today, renewable energy development in the world took an accelerated character that is associated with the growing of global multifactorial crisis phenomena. On the one hand, there is limitation of geological reserves of the main types of fuel resources – oil and gas – which leads to an inevitable rise in prices for them. On the other hand, there is the obvious growth of the negative environmental impacts, which are caused by the effects of human activity.

Today application of alternative energy sources in various spheres of human activity becomes more and more popular all over the world and in Eastern European countries as well. Along with the development and implementation of alternative motor fuels, scientists around the world study perspectives of partial or total substitution of traditional jet fuels with their alternative analogues. Following the world tendencies, we see the need to consider the potential in development and application of alternative jet fuels produced from plant oils. The analysis was done for Eastern Europe countries, such as Ukraine and Poland.

A. Yakovlieva (✉) • S. Boichenko • O. Vovk • O. Gryshchenko
National Aviation University, Kyiv, Ukraine
e-mail: pinchhuk_anya@ukr.net

K. Lejda
Rzeszow University of Technology, Rzeszow, Poland

T.H. Karakoç et al. (eds.), *Advances in Sustainable Aviation*,
DOI 10.1007/978-3-319-67134-5_4

4.2 Literature Overview

Modern civil aviation is developing constantly. The world volume of aircraft transportation increases 4–5 % annually. As a result during the last decade, consumption of fuels for air-jet engines has increased to 21 %. As it is stated in [7], about 5.5 thousand barrels of jet fuel are produced and consumed in the world daily. Dynamics of jet fuel consumption in Ukraine and Poland are shown in Fig. 4.1.

Thus, today aviation is one of the most significant consumers of oil-derived jet fuels. Oil and other fossil fuels used for production of both jet and other kinds of fuels are exhausted irreversible. Oil deposits are estimated to last for 40 years, natural gas for 70 years, and coal for 230 years. CO_2 emissions accompanying processes of fuel production and use stipulate the increase of greenhouse effect and therefore global warming [6]. The Intergovernmental Panel on Climate Change and International Energy Agency (IEA) state that modern aviation is a source of about 2% of the world's total CO_2 emissions (Fig. 4.2).

Such rapid increase of CO_2 in the atmosphere is connected to active development of the aviation industry. According to the forecasts, the number of aircrafts will increase two times till 2031 and will total to about 31 thousand units. Enlargement of air fleet and number of flights consequently leads to increasing jet fuel consumption. During the 10-year period (1992–2002), the level of jet fuel consumption has risen by 21%. Besides CO_2 aircraft exhaust gasses contain a number of other components that negatively influence the state of the environment: SO_x, NO_x, CO, CH_4, soot, and others [7].

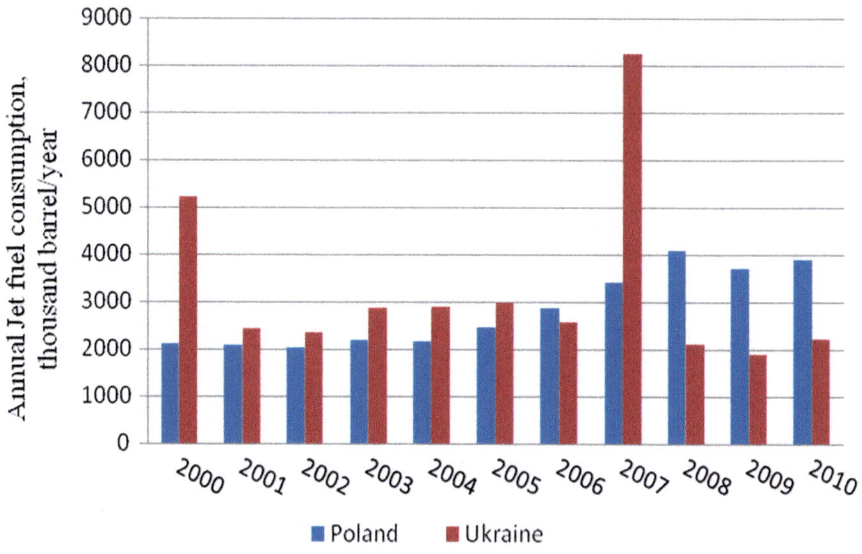

Fig. 4.1 Statistics of jet fuel consumption in Poland and Ukraine

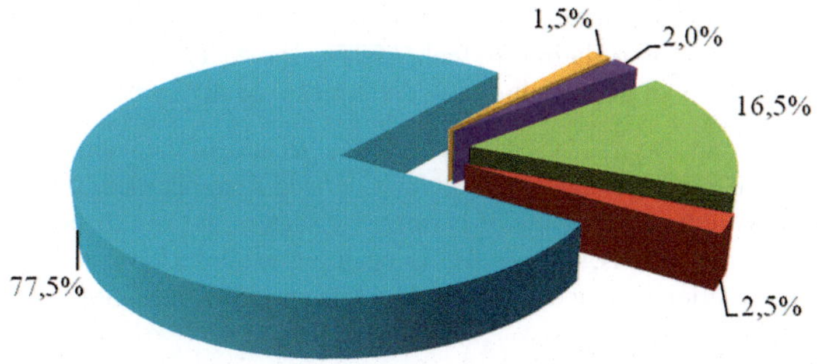

Fig. 4.2 The share of CO_2 emissions from air transport compared to other kinds of transport: 1.5%, international aviation; 2%, international marine transport; 16.5%, motor transport; 2.5%, other kinds of transport; 77.5%, other energy sources

Quantitative and qualitative composition of aircraft emissions determines ecological properties of jet fuels. Mostly they depend on the content of heteroatom compounds and aromatics in fuel. In other words, ecological properties of jet fuels and aircraft's emissions are determined by the quality of crude oil [2]. We should mention that the amount of greenhouse gas formation during jet fuel production from natural gas is 1.8 times lower and from coal is 2.4 times higher compared to crude oil. Processing of bituminous coal is accompanied with CH_4 emission that is also one of the greenhouse gasses. At the same time SO_x emissions are absent during the combustion of fuel made of natural gas [2].

The main environmental damage associated with global climate change of the Earth is greenhouse effect, caused mainly from mining, processing, and burning of fossil fuels – coal, oil, and gas. The greenhouse effect is up to 75% share of the anthropogenic environmental damage. In this regard, the satisfaction of growing needs of the world's population in fuel, electricity, and heat simultaneously with the environmental safety necessitates the development of renewable energy, because oil is not single raw material for getting of high-efficient organic fuels for engine.

In terms of global pollution, air transport is responsible for direct and indirect emissions of several greenhouse gasses: CO_2 (2% of global emissions by IATA – International Air Transport Agency), tropospheric ozone, methane, etc. [2]. More than once the Committee on Environment Impact of Aviation (CAEP) raised the issue of reducing emissions including emissions of NO_x and CO_2. With the support of ICAO, the strategic document of the 38th Assembly of ICAO has presented strategic plans in the field of environmental protection regarding emissions of aircraft engines that affect local air quality, the impact of international aviation, and climate change, including the progress on the part of new aviation standards on emissions of CO_2, plans of action of countries and ASSIST countries, environmental alternative fuels for the aviation, market activities, and global goals [9, 22].

The initiator of the development and implementation of alternative (including biological) fuels for aviation is the International Civil Aviation Organization (ICAO). Back in 1983, the ICAO established the structure of the technical committee on environmental protection in aviation (CAEP), which is the active target group (subcommittee) on alternative fuel (AFTF). The AFTF is concerned with the economic feasibility of aviation biofuels, determined mainly by two criteria:

- The need to improve fuel efficiency in aviation because of the well-known global problem of depletion of hydrocarbon fuels derived from petroleum.
- The need to reduce emissions of harmful substances in the exhaust gas of gas turbine engines, that is, an increase in environmental performance of aviation fuel. This criterion is now the dominant and determining the actions of scientists and practitioners.

Following 6 years of negotiations, governments meeting at the International Civil Aviation Organization (ICAO) are finalizing the design elements of a global market-based measure for international aviation. It is part of a series of actions the aviation industry is taking to reduce its carbon emissions which includes investing in new technology, scaling up the use of sustainable alternative fuels, improving operational performance of aircraft already in the fleet, and using more efficient infrastructure.

The carbon offsetting and reduction scheme for international aviation (CORSIA) has been agreed by government negotiators at the 39th ICAO Assembly. The CORSIA will be the world's first market mechanism for dealing with climate change from any industrial sector. The global aviation industry, coordinated by the Air Transport Action Group (ATAG), has been instrumental in proposing the plan and is now encouraging states to support it.

Such a situation promoted strengthening of ecological requirements of quality of jet fuels. A number of influential organizations took measures on the prevention of negative impacts of aviation on the environment [1]. The International Air Transport Association (IATA) has set up the task to reduce the level of CO_2 emissions from air transport by 50% by 2050. In addition, the European Commission in 2011 has set a policy target of achieving a 60% reduction of CO_2 by 2050. Low-carbon fuels in aviation should reach 40% by 2050. Besides, the European Committee insists on using about 4% of fuel produced from plant feedstock in aviation. In the future this may help to step forward in achieving the goal to cut down CO_2 emissions. According to the experts' forecast, the use of jet biofuels can guarantee an 80% reduction of CO_2 emissions compared to conventional jet fuel [3].

Thus, we can see that the search in methods and technologies of alternative jet fuel production from renewable feedstocks was acknowledged as a key direction for solving such questions about energy and resource saving and environmental protection in aviation.

4.3 Modern Situation in Sphere of Jet Fuel Production

Traditionally fuels for jet engines are produced from crude oil, coal, natural gas, oil sands, and oil shales. However, technologies are quite energy intensive and difficult to realize. The limitation of the world's oil deposits and other fossil fuels promotes the development of alternative technologies for jet fuel production. Today numbers of organizations pay much attention to the investigation of existing and newly appeared technologies for alternative jet fuel production. Scientists define jet fuels derived from conventional oil, unconventional oil (oil sands and oil shale), natural gas, coal, or biomass via the FT process, renewable oils (biodiesel, biokerosene, hydroprocessed renewable jet or hydrotreated vegetable oil – HVO), and alcohols (ethanol and butanol). However, alcohol-derived fuels are suitable for aviation piston engines only [6]. The most perspective are fuels derived from plant feedstock – the so-called biofuels. IEA forecasts that the share of biofuels in the total balance of fuel in the transport sector will reach 4–6% till 2030.

The development and use of jet biofuels contribute to the appropriate legal regulation. Implementation guidelines strictly define certain types of biofuels in the consumer market. Today areas used for biofuel crop cultivation in Europe are more than 2500 ha. Basically these crops are cereals and rape seed. At present, the 13.2 ha of EU lands are available for cultivation of biofuel cultures. Till 2020 this area may increase up to 20.5 ha and till 2030 to 26.2 ha. According to the European Commission, in order to achieve the 2020 goal for energy crops, it is necessary to use 17.5 ha or about 10% of agricultural lands used in the EU countries.

Biofuel holds a special place in the structure of renewable energy sources. Being one of the few alternative fuels in the transport sector, biofuels are seen as an important resource in the choice of energy sources, in ensuring energy security, in the development of agriculture and rural areas, as well as in the mitigation of climate change by reducing greenhouse gas emissions.

The formal classification of energy crops to the appropriate categories (Fig. 4.3) has been published in a report titled Biofuels in the European Vision, Vision 2030 and Beyond. According to these reports, biofuels are divided into the first- and second-generation biofuel. The Department of Transport and Energy of the European Commission has figured out a third-generation biofuel. The third-generation fuel, known as a future fuel, may be obtained from oils produced from micro- and macroalgae.

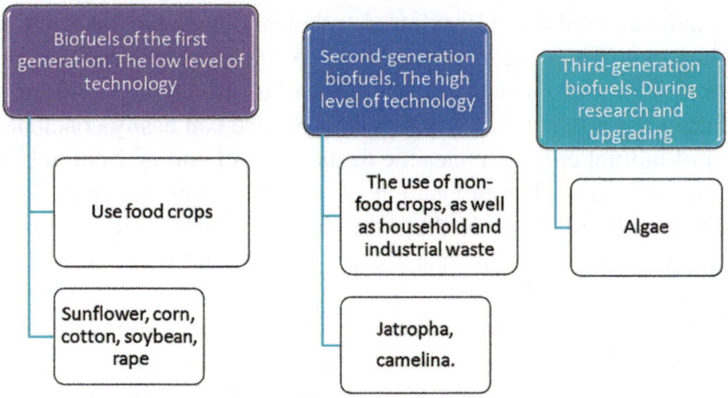

Fig. 4.3 Classification of biofuels according to generations

Conditionally biomass as feedstock for biofuel production can be divided into three generations. The following are the current generations of biofuels:

- Edible oil and sacchariferous terrestrial plants
- Inedible cellulosic plants
- Inedible aquatic plants, i.e., algae

First-generation biofuel is made from sugar, starch, vegetable oil, and animal fat using conventional technology. The main sources of raw materials are the seeds or grains, for example, from rapeseed extract vegetable oil, which can then be used as biodiesel, and from wheat-obtained starch, after fermentation – bioethanol.

Deforestation, negative impacts on traditional agriculture, the imbalance of agricultural land use in the direction of industrial crops, and the threat of the food security are some of the problems humanity is facing during the production of biofuels. The main problem in the production of fuel from biomass is food security, because first-generation biofuels made from agricultural crops enter the food chain of humans and animals (corn, soybeans, palm oil, rape, sugarcane, wheat, rye). The public has thought suddenly that large areas where food was produced by commercially oriented farmers was given to the technical culture. Because the world population grows and requires more and more food, the use of these areas for the production of biofuels reduces the amount of available food and increases their prime cost.

Second-generation biofuels are produced from non-food raw materials. Sources of raw materials are lignocellulosic compounds the remaining after, as suitable for use in the food industry part of vegetable raw materials are removed. For this purpose, fast-growing trees and grasses (poplar, willow, *Miscanthus*, *Jatropha*, etc.) can also be used. They are also called energy forests or plantations. Twenty different species of plants, including arboreal, shrubs, and herbaceous, are tested.

In September 2013, the European Parliament decided to reduce the share of first-generation biofuels in total biofuel balance because of its environmental load. It is easy to consider, when the yield of a crop like rape is 30 kg/ha, then it is possible to obtain 1.0–1.3 tons of oil per hectare. It means that about nine million hectares of arable land was used for cultivation of technical culture. At the same time 45–80 kg of nitrogen, 18–40 kg of phosphorus, 25–100 kg of potassium, 30–150 kg of calcium, 5–15 kg of magnesium, and 30–45 kg of sulfur were taken out from the soil during the production 1 ton of seeds. So, we can easily conclude that the production of biofuel crops depletes the fertile layer of soil of 2 cm thick that was formed over 100 years. The production of renewable raw materials for biofuels without scientific justification leads to the depletion of the soil and reduction of crop yields, deterioration of raw material for biofuel, and water and soil pollution by agrochemicals and fertilizers.

During the last years, we observe criticism of first-generation biofuels because the fuel produced from rapeseed and similar crops is not effective for reducing CO_2 emissions. The proportion of first-generation biofuels in transport fuel balance according to the decision of the European Parliament should be reduced to 5.5%. In addition, after 2020, the EU may completely abandon the first-generation biofuels.

This fact stimulates the search for new, cost-effective technologies, which will use the second-generation feedstock, which resources are almost unlimited. These sources should be crops of non-food purposes, perennial grasses, and agricultural waste of food and forestry industries. The main advantages of the second-generation biofuels compared to the first generation are the following:

- Variety of biomass, suitable for processing.
- Higher efficiency – on average by 30–40%.
- Reduction of greenhouse gas emissions in case of applying second-generation biofuels can reach up to 90% (first-generation biofuel – 50%).

If we talk about the state of biofuel production in Ukraine, we should conclude that the level of biofuel branch development is significantly lower than in EU countries. According to [11] the total supply of primary energy sources in Ukraine is more than 130 M ton of oil equivalent, but the amount of biofuel is just 1.48 M ton, that is, 1.13% in structure of energy supply. At the same time in EU countries, this index reaches 6.72%. The total supply of primary energy in EU countries is estimated at 1759 M ton of oil equivalent with the share of biofuels of 118 M ton (Fig. 4.4a–c) [12].

It should be stated that the existing potential of Ukraine in the production of biofuels is not used completely, even concerning the first-generation biofuels. This is due to the lack of modern plants for its production. Transition to the use of the second-generation feedstock is possible only through the development of modern industry for the processing of raw materials of the first generation, which can create the technological basis for a gradual use of the second-generation feedstock, as well as the expansion of the resource base by selection of alternative crops, making them productive varieties and their implementation into production.

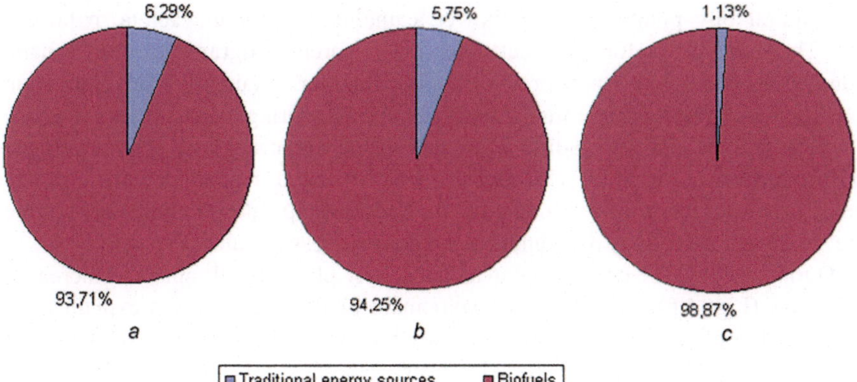

Fig. 4.4 Share of biofuels in structure of energy supply (**a**) in EU countries, (**b**) in Poland, (**c**) in Ukraine

4.4 State and Perspectives of Jet Biofuel Production in Ukraine and Poland

Poland and Ukraine as well as other European states are characterized by the deficiency of oil deposits. Ukraine consumes about 130 M ton of fuel-energy resources annually, satisfying its need in energy resources on around 53%. In order to satisfy needs of civil aviation, Ukraine requires about 400,000 ton of aviation kerosene. At the same time, sphere of jet fuel production is characterized by the opposite situation. The production of aviation kerosene at Ukrainian enterprises in 2011 decreased at 16.6% compared to the previous year up to 282,000 ton. The final energy consumption for transportation purposes in Poland in 2005 was 10.27 million ton of energy equivalent (Mtoe).

Generalizing the world's experience in alternative jet fuel development, we have concluded that the most optimal kind of aviation biofuels is biokerosene, which is derived from renewable oils [8]. The technology of biokerosene manufacturing seemed to be the most rational for both Ukrainian and Polish conditions. This kind of biofuel is obtained via processing of oil plants. Biokerosene is a mixture of traditional kerosene, produced from crude oil and biocomponents in certain concentrations. The percentage of biocomponent can reach 50%. Advantages are explained by the availability of feedstock, maturity of technology, presence of necessary equipment, satisfactory physical-chemical and exploitation properties of such kinds of biofuel, and also its ecological safety.

4.4.1 Feedstock Base for Jet Biofuel Production

The feedstock for biocomponent production is oils, obtained from seeds of various agricultural oily plants: rape, sunflower, camellina, *Jatropha*, canola, palm oil, etc. The main factor for feedstock selection is surely geographical and climatic conditions typical for country-producer and optimal for certain oily plant cultivation [3]. The base of European biofuel industry is rape and during the last years camelina. Ukraine and Poland are traditionally agricultural countries with well-developed oil production branch [17, 18]. Due to this they have a great potential for the development of jet biofuel manufacturing. The most typical oil cultures in Ukraine are sunflower, rape, and soy; camelina and corn are less popular (Fig. 4.5).

During 1990–2011 cultivating areas under oily plants in Ukraine had increased 3.7 times (Fig. 4.6). Such strong development of the industry was explained by stably high profitability of cultures, annually growing demand for feedstock in food industry, animal farming, and biofuel industry.

Due to the large area of agricultural terrains, Poland has a great potential to use renewable energy. The main renewable energy source in Poland is biomass. Today Poland is one of the most promising countries in EU concerning biofuel feedstock production. The most typical oil cultures in Poland are sunflower and rape (Fig. 4.7) [17].

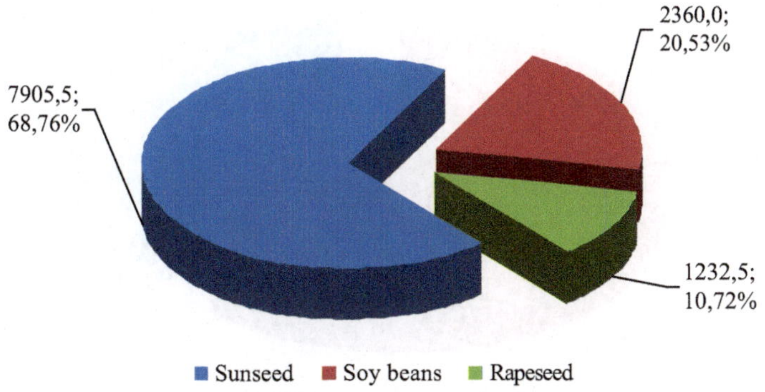

Fig. 4.5 Production of main oily plants in Ukraine in 2012 (thousand ton)

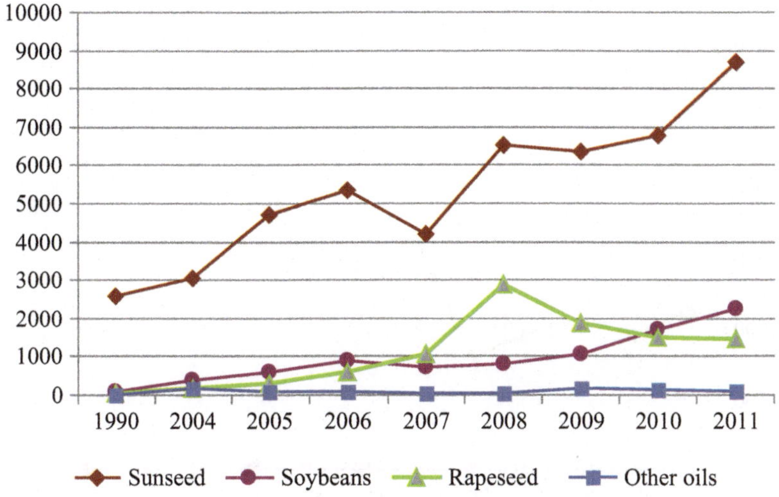

Fig. 4.6 Dynamics of main oily plant production in Ukraine during 1990–2011 (thousand ton)

Such plants as soy, camelina, etc. are still not cultivated in industrial scales. However, compared to Ukraine, only rape oil is considered to be used as a feedstock for biofuel production (Fig. 4.8).

4.4.1.1 Review of the Main Oily Plant Potential for Biofuel Production in Ukraine

Sunflower is the main oily plant cultivated in Ukraine. Its part in total yield is about 70%. Seeds of ordinary breeds contain 50–52% of oil, while selected breeds contain up to 60%. Compared to other oily plants, sunflower allows production of quite a high volume of oil – 952 l per ha (890 – l per ha on average in Ukraine). The last

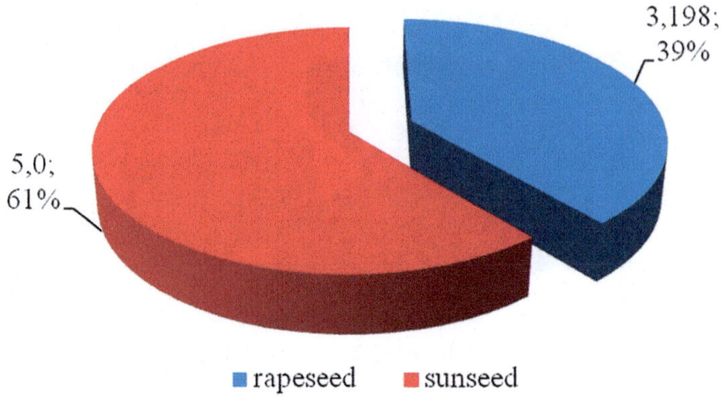

Fig. 4.7 Production of main oily plants in Poland in 2014 (thousand ton)

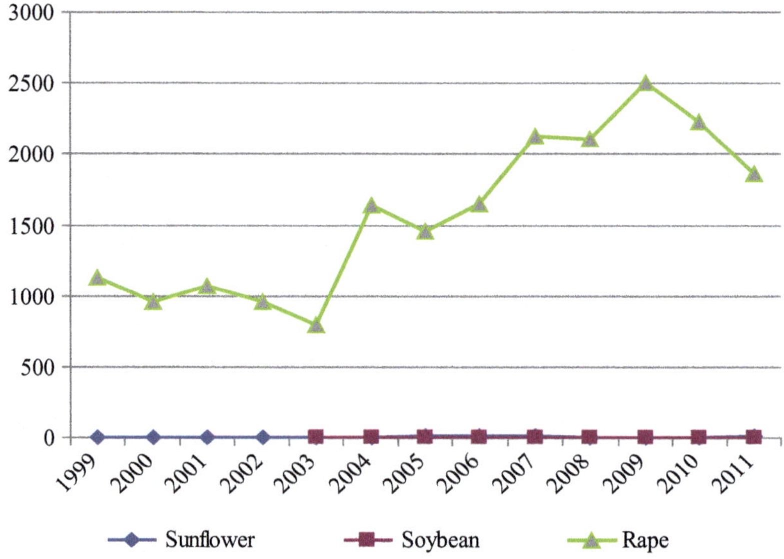

Fig. 4.8 Dynamics of main oily plant production in Poland during 1990–2011 (thousand ton)

years are characterized by significant growth of sunflower production [18]. Today sunflower fertility is about 1840 kg per ha that is much higher compared to years 2002–2004 (about 1100–1300 kg per ha) (Fig. 4.9). The main part of sunflower is used in the food industry and in the production of high-protein forage, oil meal, and other technical needs. A significant amount is exported that makes a valuable income for the country. Despite so huge production volumes, it is not rational to consider sunflower as a feedstock for biofuel industry, mainly manufacturing of jet biofuel.

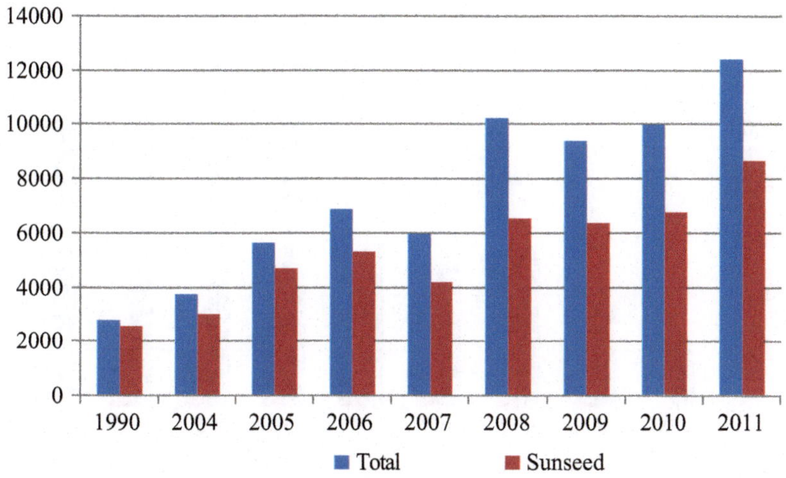

Fig. 4.9 Dynamics of sunflower production in Ukraine and its share in total volume of oily plant cultivation during 1990–2004 (thousand ton)

Another oily plant – rapeseed – is being cultivated in Ukraine in industrial volumes only during the last 8–10 years. It is explained by strong world demand on rapeseed oil for biofuel production. The most active rapeseed production in Ukraine was observed during years 2004–2008. Rapeseeds contain 45–50% of oil, 24–31% of proteins, and 6–12% of fiber. According to [4], today fertility of rape is about 1500–1700 kg per ha and can reach up to 2400 kg per ha. Average yield of rapeseed oil is around 1190 l per ha. Certain rapeseed breeds with low content of erucic acid have been already selected that influence positively biofuel properties. Compared to sunflower oil, rape oil doesn't need preparatory purification from various related components. That is why it is considered to be more appropriate for biofuel production. Several years ago rapeseed was cultivated in Ukraine with the aim of export only. Its internal processing was almost absent and didn't exceed 10%. However, now a significant increase in volumes of internal processing is observed (Fig. 4.10).

Another popular oil culture in Ukraine is soy. Compared to rapeseed, during the recent years, high rates of soy production are observed in Ukraine (Fig. 4.11). Soy average fertility is about 1500–2000 kg per ha with comparatively low oil yield – 446 l per ha [4, 18]. The content of oil in soy seeds is about 16–27%. Soy oil is characterized by high content of unsaturated fatty acids (86–87%); this makes a certain impact on its physical-chemical properties. At the same time, it contains 40–50% of high-quality proteins [12]. Due to this soy culture is widely used for production of food oil, protein concentrates, and various food products. These factors made it irrational for biofuel production. It should be mentioned that soy production requires significant resources and material-technical base for its cultivation.

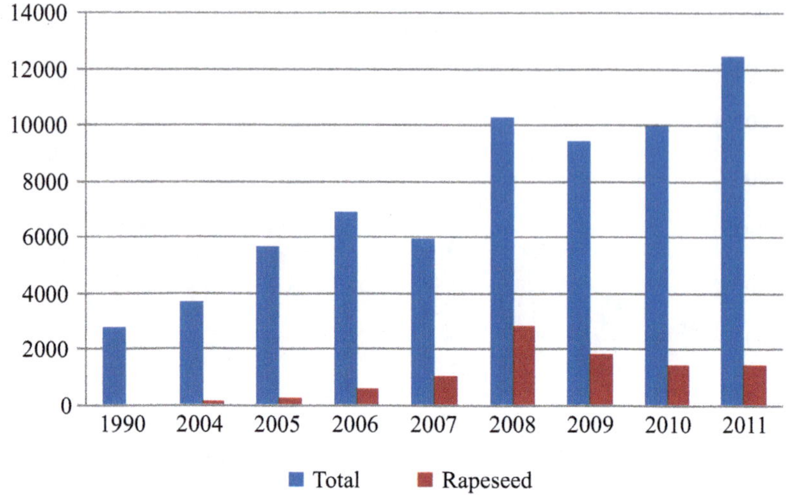

Fig. 4.10 Dynamics of rapeseed production in Ukraine and its share in total volume of oily plant cultivation during 1990–2011 (thousand ton)

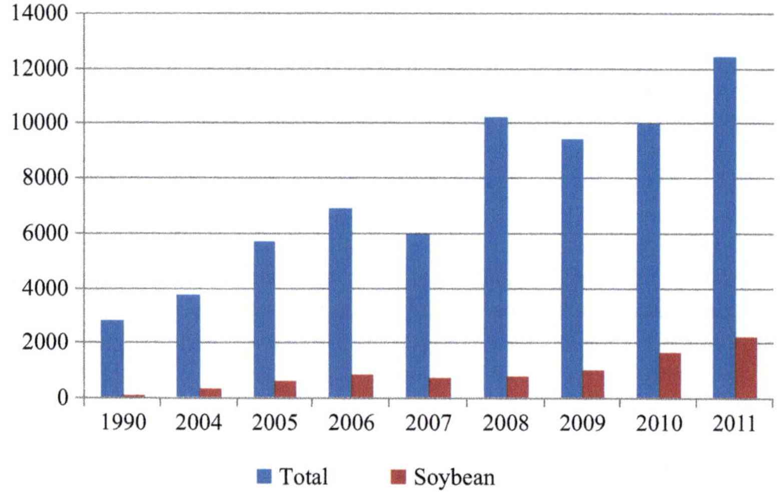

Fig. 4.11 Dynamics of soy production in Ukraine and its share in total volume of oily plant cultivation during 1990–2011 (thousand ton)

4.4.1.2 Review of the Main Oily Plant Potential for Biofuel Production in Poland

During the last years, Poland has become one of the leading countries within the EU, with regard to biofuel production. Having an area of 18.3 million ha devoted to agricultural production, Poland presents great biomass opportunities in energy crop

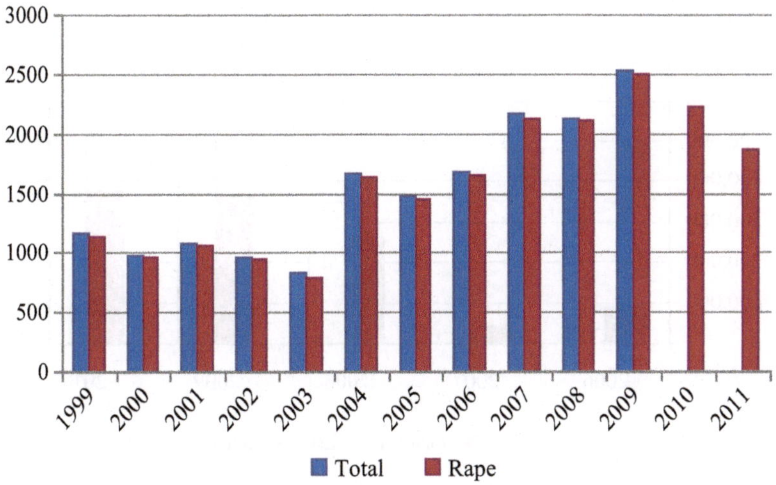

Fig. 4.12 Dynamics of rapeseed production in Poland and its share in total volume of oily plant cultivation during 1999–2011 (thousand ton)

production. Today Poland is treated as one of the main rapeseed-producing countries [23] (Fig. 4.12). As a result, the new capacities in the country are being installed, mainly encouraged by the incipient growth of domestic consumption.

The biofuel sector in Poland is mainly driven by the European incentives and directives aiming at considerably increasing biofuel production and use. Along with other European states according to recent European regulations, the mandatory target for 2020 has been set at a minimum level of 10% biofuel component in transportation fuels in Poland. Considering the Poland case, the potential of rape cultivation should be determined. More specifically, in Poland, the production of rapeseed increased sharply from 0.7 in 2003 to 1.25 M tons in 2004 [17]. In recent years an increasing trend in biofuel production and consumption can be seen (Fig. 4.13). In 2011, in Poland there were registered 20 producers of biofuels made of rape oil [13].

However, from the graph below (Fig. 4.14), it may be easily seen that volumes of rape production in Poland are generally higher than in Ukraine. It may be explained by the presence of high level of governmental support to biofuel-producing industry. This support is regulated and based on European biofuel implementation programs.

4.4.1.3 Review of *Camelina sativa* Potential for Second-Generation Biofuel Production

A popular plant for the production of the second-generation biofuels is *Camelina sativa*. It is an oleifera annual grass. The genus of *Camelina* includes ten species, which are grown in Europe and Asia in the areas of cool climate of the steppe

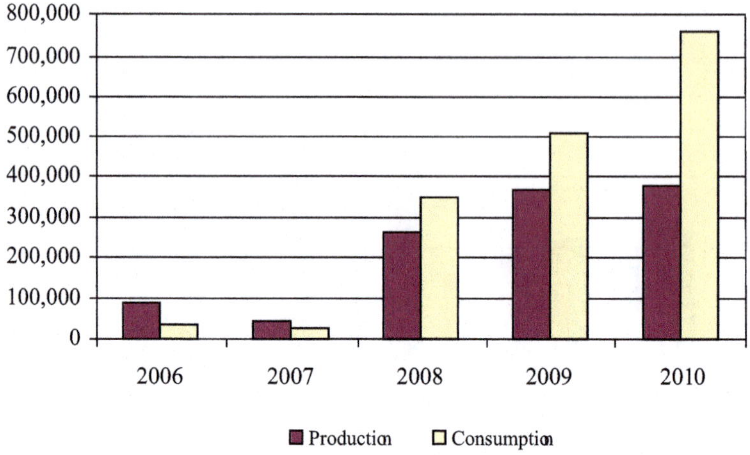

Fig. 4.13 Production and consumption of rape oil-derived biofuels in Poland [13]

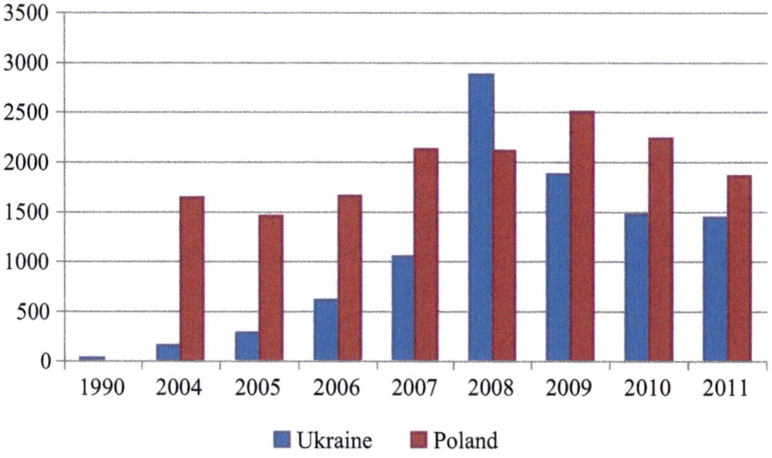

Fig. 4.14 Dynamics of rape production in Poland and in Ukraine (thousand ton)

and forest areas. It can be grown on the fields and fallow lands and along roads. *Camelina sativa* comes from the countries of Asia Minor, where it was cultivated for centuries. In the second half of the nineteenth century, *Camelina* was firstly cultivated in Russia and France. In the 1940s–1950s, about 400 ha of the former Soviet Union was covered with *Camelina* culture. Camelina oil was used primarily in the aerospace industry [20]. Nowadays interest to camelina oil is explained by its potential for use as biofuel feedstock in Europe and the United States. Today camelina is cultivated on some territories of Northwest and Central Europe and not for a long time in Ukraine. Camelina is easily grown on unfertile soils, undemanding to external conditions and resistible to low temperatures and dry weather. It can be

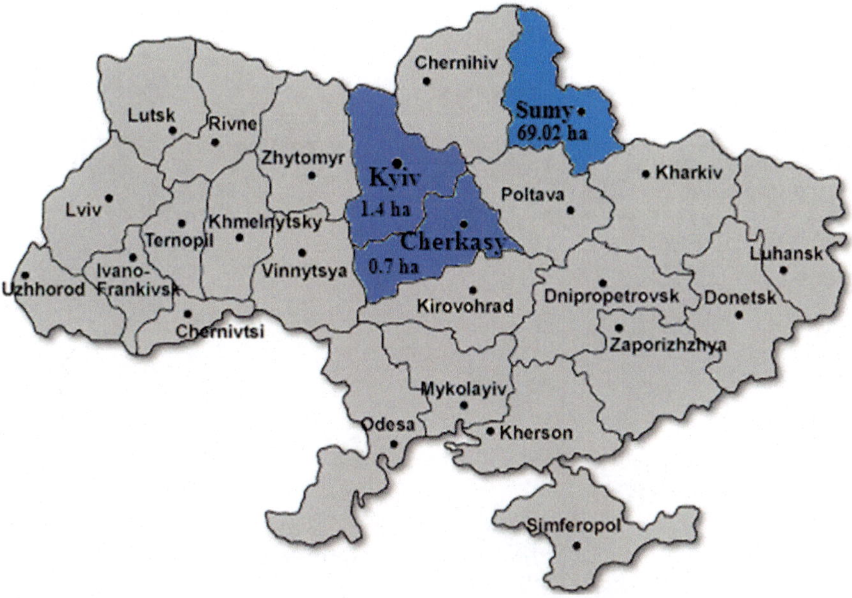

Fig. 4.15 Ukrainian regions where camelina is cultivated [4]

sown when the soil resting from wheat and other grains as part of a crop rotation. Plant has short vegetating period; this allows growing other cultures after harvesting camelina. When it grows naturally its yield is 1000–1200 kg per ha on average and can reach 2000 and even 3000 kg per ha. Camelina seeds contain 40–50% of oil, and oil yield is about 1250 l per ha [16]. During recent years application of camelina oil as a feedstock for biofuel production is highly popular [15].

In Ukraine camelina oil production is still developing. During several decades camelina was cultivated in Ukraine as agricultural crop of secondary importance. Due to the increase in the area under sunflower, rapeseed, and other crop cultivation, camelina was almost expelled. Only small areas are being used continuously for growing up camelina to manufacture soaps and cosmetics, as well as a component of birdseed. Nowadays within Ukraine *Camelina* culture is grown on small areas in the Polissia region and in the northern forest steppe (Fig. 4.15), although there are prerequisites for the expansion of the areas under this crop. This culture is technical, so biofuel production will not compete with food industry.

In Poland camelina is still not considered as a biofuel feedstock. Taking into account world and European tendencies in particular, we may predict that camelina production in Poland and Ukraine will be increasing rapidly [11, 18]. Camelina is grown in Poland in regions of Podlasie, Wielkopolsce, and Opolszczynys (Fig. 4.16).

If to speak in general, *Camelina sativa* practically isn't cultivated in Ukraine [9, 10]. It can be easily seen from the diagram (Fig. 4.17) that describes areas taken for the cultivation of *Camelina sativa* and rapeseed during the period of 2005–

Fig. 4.16 Polish regions where camelina is cultivated [13]

2015. However, taking into account requirements of leading world organizations concerning the sustainability of biofuels, *Camelina sativa* seemed to be a potential alternative crop which may replace rapeseed in the nearest future.

In order to understand the general characteristics of both biofuel crops camelina and rapeseed, let's compare these crops from the point of view of agriculture (Table 4.1) [19].

As you can see from Table 4.1, *Camelina sativa* has a larger potential compared with rape that is closely connected with its stability in unfavorable soil and climate conditions. As a result it makes it possible to grow it up in soil of bad quality which is not good for other crops, and it means that we have a chance to get vegetable oil good for producing biofuel without reducing the production of food. Besides, it gives us a real chance to develop agricultural areas with low-quality soils [24]. Scientific research work devoted to technological methods of growing up *Camelina sativa* and increasing its seed productivity taking into consideration soil and climate conditions in Ukraine, paying special attention to the determination of optimal thickness and regulation of the optimal terms of sowing it, has been done,

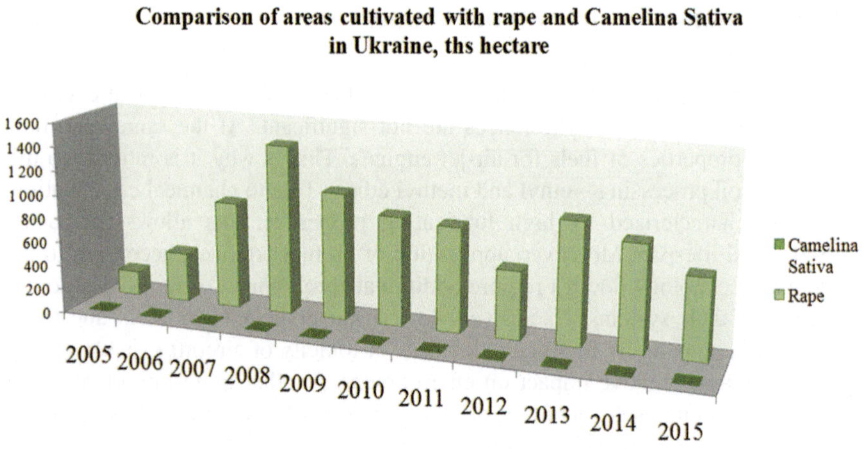

Fig. 4.17 Areas cultivated with *Camelina sativa* and rapeseed in Ukraine during 2005–2015

Table 4.1 The comparative characteristics of rapeseed and *Camelina sativa* cultures

Characteristics	*Camelina sativa*	Rapeseed
General characteristic	Annual crop	Annual forage crop
Drought resistance	Low water demand	Moisture-loving crop
Potential of germination	All types of soil are good	Highly demanding to soil
Threat of soil fertility decreasing	It is often cultivated along with dead crops and is used as an intermediating crop; after harvesting it's possible to cultivate some other crops	Emaciate soil, it's possible to cultivate again rape on the same area only in 3–4 years
Fertilizers	Are not needed	Nitrogen-containing fertilizer is necessary
Presence of weeds	It produces ethereal oil which puts down growth and evolution of weeds from the phase of stalk formation to full ripe seeds	A lot of weeds
The period of vegetation	60–75 days from the moment when crops are already standing	90–100 days and nights
Loss of seeds	High firmness of pods from chapping	Low firmness of pods from chapping
Vulnerability of crops to vermin	Vermin and diseases are not found	Badly damaged by vermin

and standards of mineral fertilization have been chosen. In addition, growing up the seeds of *Camelina sativa* and sorghum is ecologically safe because both cultures are characterized as extremely conditions of growing up; they don't need any fertilizers, pesticide and fungicide.

4.5 Technology of Jet Biofuel Production

Oils derived from the abovementioned plants have different physical-chemical properties; however, these differences are not significant. At the same time they differ from properties of fuels for air-jet engines. That is why it is rational to use products of oil processing – ethyl and methyl ethers. Due to chemical composition, biofuel is characterized by high lubricating properties; this allows increasing aircraft detail lifespan. Moreover, application of biofuel produced according to the mentioned technology doesn't require additional reequipment or modernization of engines and their systems. Exclusively low sulfur content in biofuels due to its natural origin may allow significant decrease of toxicity of aircraft exhaust gasses, thus minimizing negative impact on environment. Another quite important factor that determines interest to jet biofuels is its ability to be decomposed easily in natural environment without doing harm to living objects [9].

Considering the suitability of various oils for biofuel production, it is necessary to analyze fatty acid composition of oils (Table 4.2) [14]. It is a well-known fact that fatty acid composition of oils determines physical-chemical properties of oils and thus biofuels. The degree of unsaturation (amount of acids with single or multiple double bonds) determines chemical stability of oil. Under chemical stability we assume resistance of fatty acids to oxidation under various factors, i.e. temperature, light, presence of other compounds, long-term storage, etc. In general, the higher the content of unsaturated fatty acids, the less chemical stability. At the same time the degree of unsaturation influences such important properties as viscosity and melting (or cloud) point of oils. In this case the higher the content of unsaturated fatty acids, the less viscosity and better low-temperature properties. Usually it may be explained by the curved form of unsaturated fatty acids and less compact mutual displacement [5].

From the Table 4.3 we may see that both studied plant oils are reach in unsaturated fatty acids. Camelina oil contains high proportion of linoleic acid (with two double bonds) with considerable content of oleic (one double bond) but lower content of palmitic (no double bonds) acid. The analysis shows that camelina oil contains high amounts of linolenic (three double bonds) acid and eicosenoic acid

Table 4.2 Average fatty acid composition of rapeseed and camelina oils

	Plant oil	
Fatty acid	Camelina oil	Rape oil
Myristic acid	–	0–1.5
Palmitic acid	5.0–6.3	1.0–4.7
Stearic acid	2.5–3.0	1.0–3.5
Oleic acid	14.5–16.0	13–38
Linoleic acid	15.0–17.0	9.5–22
Linolenic acid	36.0–38	1–10
Erucic acid	1–4,0	40–64
Eicosenoic acid	14.8–16.2	0.3–1.0

Table 4.3 Comparative analysis of properties of oils and esters of rapeseed and camelina oils

Characteristics	Camelina oil	Methyl ethers of camelina oil	Rapeseed oil	Methyl ethers of rapeseed oil
Density at $t = 20\ ^\circ C$, kg/m^3	923	885	920	880
Kinematic viscosity at $t = 20\ ^\circ C$, mm^2/s	80	5,18	76	6,9–8,2
Cetane number	39	55	34	49–56
Flash point, °C	250	180	285	170
Freezing point, °C	−18	−5	10	−7
Heat value kJ/kg	37,400	37,000	37,400	37,000–39,000

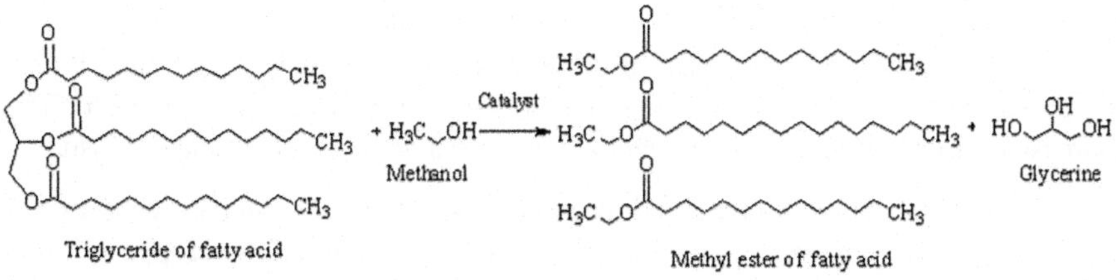

Fig. 4.18 Chemical reaction of plant oil esterification

that has the longest carbon chain. The typical characteristic of rapeseed oil is high content of erucic acid – long-chain acid with one double bond.

Thus, the fatty acid composition of oils determines their physical and chemical properties (Table 4.3). Camelina oil possesses slightly higher viscosity compared to rapeseed oil similarly to density values. It means that the energy value of camelina oil is comparatively higher than rapeseed oil.

The technology of ethyl (methyl) ethers production is quite simple and similar to the process of biodiesel manufacturing. The process is based on the reaction of esterification of fatty acids containing in plant oils using methanol or ethanol and basic catalyst (Fig. 4.18). In a result of this process, esters of plant oil and glycerine are obtained [21]. After that esters of fatty acids undergo the process of vacuum distillation. Vacuum distillation provides a better level of ester purification [10]. This allows using them as components of jet fuels.

The comparative analysis of physical-chemical characteristics of *Camelina sativa* and rapeseed oils can demonstrate which of the mentioned crops are more suitable for producing biofuels (Table 4.3). For the beginning we introduced some basic physical-chemical parameters of oil and biocomponents made of oils (Table 4.3).

Significant differences in physical-chemical properties of camelina and rapeseed oils, and thus its esters, could be explained by quantitative composition of fatty acids of oils that is presented on the diagram (Fig. 4.19) [19].

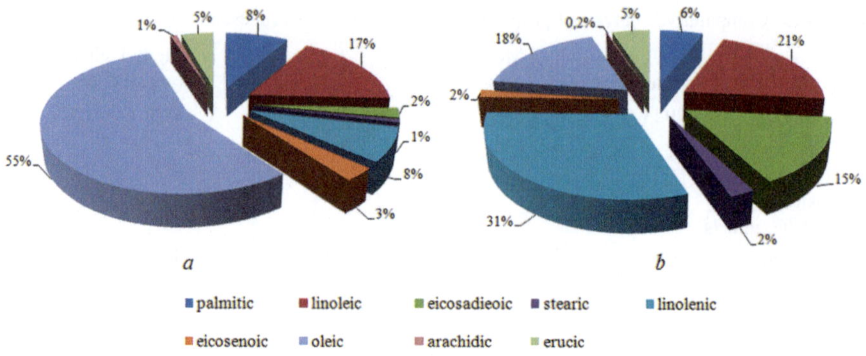

Fig. 4.19 Composition of fatty acids of (**a**) rapeseed and (**b**) camelina oils, %

In order to conclude, camelina oil raises interest as a potential raw material for the production of aviation fuel components (biofuel of the second generation). The increase of area cultivated with *Camelina sativa* allows to resolve range of issues.

Social issue: in contrast to rapeseed, crop cultivation of *Camelina sativa* does not require utilization of rich soils which could be occupied by food crops.

Agricultural issue: Camelina sativa could be easily grown in poor soils and does not require usage of pesticides and fungicides.

Environmental issue: usage of plant-based fuel components decreases volume of emission of greenhouse gas. For Ukraine this becomes especially important considering its scarce energy resources that could be used for aviation fuel production, while at the same time, the country possesses a significant area of agriculture lands. Additionally it is worth mentioning that the existing potential in the area of biofuel production is not realized in Ukraine due to the absence of modern production facilities.

Today the technology of biofuel production is realized successfully both in laboratories and industrial enterprises. Taking into account experiences of European countries, biofuel manufacturing in Ukraine can be organized at the low-tonnage installations (local level for certain enterprises) with productivity 300–3000 ton/year, regional factories, producing 10,000–30,000 ton/year, and large-scale industrial factories [8]. Technological process of jet biofuel production from plant oils is significantly simpler compared to crude oil refining and profitability of the process in two to three times higher. Except that, plant feedstock is a renewable resource, and thus it is comparatively unlimited.

4.6 Conclusion

In general Ukraine and Poland among other countries have considerable resource potential for jet biofuel production from plant oils. The main advantage of feedstock cultivation in Poland is the presence of high level of governmental support to biofuel-producing industry that is based on European biofuel implementation programs. However, Poland's potential in biofuel feedstock production is not realized completely. Today only rapeseed is cultivated for providing feedstock for biofuel production; at the same time, the biofuel potential of sunflower, soy, or camelina oils is not realized. On the contrary with no or almost absent governmental support, Ukrainian producers develop oil production from traditional oily plants for satisfying biofuel industry needs. Being one of the leading countries in Europe by volumes of rape and sunflower production, Ukraine is still one of the top rapeseed exporters to the EU. At the same time, Ukrainian oil producers promote growing of new biofuel cultures, such as camelina, soy, and some others. These countries have all possibilities for jet biofuel manufacturing for satisfaction of its own needs and for export also.

Implementation of biofuels derived from plant feedstock surely has potential and perspective for its future development. Application of biofuels should totally be the scope of sustainable development principles and doesn't do harm to food industry. In the future application of alternative jet fuels will have the following positive results: availability of feedstock that is important for countries, which don't possess its own deposits of fossil fuels; saving of exhaustible energy resources; decreasing volumes and toxicity of aircraft exhaust gasses, thus protecting air quality in lower atmosphere layers; comparative simplicity and low cost of biofuel production process; low price for biofuel compared to traditional kerosene; and stimulation of agricultural complex development in countries-producers of biofuels.

In general Ukraine and Poland have considerable resource potential for jet biofuel production from plant oils due to its favorable natural climatic conditions, availability of feedstock base, and relatively cheap human resources.

The main advantage of feedstock cultivation in Poland is the presence of high level of governmental support to biofuel-producing industry. However, Poland's potential in biofuel feedstock production may be increased due to sunflower, soy, or camelina oil production.

The opposite situation is observed now in Ukraine. With no or almost absent governmental support, Ukrainian producers develop oil production from traditional oily plants for satisfying biofuel industry needs.

At the same time Ukrainian oil producers promote growing of new biofuel cultures, such as camelina, soy, and some others. If to speak in general, these countries have all possibilities for jet biofuel manufacturing for satisfaction of its own needs and for export also.

References

1. Bauen, A., Howes, J., Bertuccioli, L., et al. (2009). Review of the potential for biofuels in aviation. [online] Biofuels in aviation E4tech Final report (for CCC), available via http://citeseerx.ist.psu.edu/viewdoc/download;jsessionid=E296A64BF3E834001B8281496DA49C7D?doi=10.1.1.170.8750&rep=rep1&type=pdf. Accessed 13 Jan 2014.
2. Boichenko, S., & Zakharchuk, M. (2012). *Aviation fuels and lubricants*. Kyiv: NAU.
3. Devianin, S., Markov, V., & Semenov, V. (2007). *Plant oil and fuels for diesel engines*. Kharkov: Novoe Slovo.
4. Dubel, A. (2010). Peculiarities and economic efficiency of rape cultivation. *Innovative Economics, 2*, 88–91.
5. Giakoumis, E. (2013). A statistical investigation of biodiesel physical and chemical properties, and their correlation with the degree of unsaturation. *Renewable Energy, 50*, 858–878.
6. Hemighaus, G., Boval, T., Bosley, C., et al. (2006). Alternative jet fuels. Addendum 1 to Aviation Fuels Technical Review Chevron Corporation, FTR-3/A1.
7. Iakovlieva, A., & Boichenko, S. (2012a). Cause-effect analysis of jet fuel production and state of environment. *Systems and Means of Motor Transport, 3*, 239–246.
8. Iakovlieva, A., & Boichenko, S. (2012b). Application of biokerosene for improvement of jet engines' ecological characteristics. *Aviation and Cosmic Technics and Technology, 7*(94), 60–64.
9. Iakovlieva, A., Boichenko, S., & Vovk, O. (2013). Investigation of properties of biocomponents used in mixture fuels for jet engines – chemmotological view. *Systems and Means of Motor Transport, 4*, 351–356.
10. Iakovlieva, A., Vovk, O., Lejda, K., et al. (2014). Improvement of technological scheme of fatty acids ethyl esters production for use as jet fuels biocomponents. *International Journal of Theoretical and Applied Science, 11*(19), 44–50. DOI: http://dx.doi.org/10.15863/TAS.2014.11.19.9.
11. Kaletnik, G. (2008). Economic efficiency of biofuel market development in Ukraine. *Problems of Science, 12*, 38–43.
12. Kondili, E., & Kaldellis, J. (2007). Biofuel implementation in East Europe: Current status and future prospects. *Renewable and Sustainable Energy Reviews, 11*, 2137–2151.
13. Królczyk, J., Latawiec, A., & Kuboń, M. (2014). Sustainable agriculture – the potential to increase wheat and rapeseed yields in Poland. *Polish Journal of Environmental Studies, 3*, 663–672.
14. Kumar, V., & Kant, P. (2013). Study of physical and chemical properties of biodiesel from sorghum oil. *Research Journal of Chemical Sciences, 9*, 64–68.
15. Nikolaeva, N., Stepycheva, N., & Kozlov, V. (2005). Change of physical-chemical properties of rapessed oil depending on term of their cultivation. *Chemistry of Plant Feedstock, 2*, 32–34.
16. Petcu, A., Carlanescu, R., & Berbente, C. (2014). Straight and blended Camelina oil properties. *Recent Advances in Mechanical Engineering, 4*, 160–167.
17. Pietak, A., & Radkowski, S. (2011). Biofuels – opportunities and challenges. *Journal of KONES Powertrain and Transport, 3*, 25–31.
18. Proskurina, O. (2011). Perspectives of production and application of biofuel in Ukraine. *Theoretical and practical aspects of Economics and Intellectual Property, 1*, 12–15.
19. Prahova, T. (2013). Camellina: Biology, productivity, technology. *Proceedings of Altai State Agrarian University, 9*(107), 24–28.
20. Rachmetov, D., Blume, Y., Yemets, A., et al. (2014). Camellina sativa – precious oil feedstock. *Institute of Food Biotechnology and Genomics, 2*, 16–21.
21. Sagar, P., Kadu, R., & Sarda, H. (2011). Use of vegetable oils by transetherification method as C.I. Engine fuels: A technical review. *Journal of Engineering Research and Studies, 2*(3), 19–26.
22. Walker, D. (2009). Biofuels, facts, fantasy, and feasibility. *Journal of Applied Phycology, 21*, 509–517.

23. Wisniewski, G., Kupczyk, A., Rogulska, M., et al. (2005). Development of biofuels market in Poland. *Proceedings of the International Conference International Conference 'Eastern Biofuels', 4*, 176–184.
24. Zelenina, O., & Prahova, T. (2009). Fatty acids composition of camellina oil seeds. *Oily Cultures, 2*, 54–57.

Chapter 5
Development of a 3D Grain Burnback Simulation Tool for Solid Rocket Motors

Yusuf Ata, D. Funda Kurtulus, and O. Ugur Arkun

Nomenclature

∀	Volume
A_b	Burning surface area
C	Corner
CAD	Computer-aided design
d	Distance value
d_b	Burning distance
dx, dy, dz	Structured grid dimension
MDF	Minimum distance function
MDV	Minimum distance value
n	Number of points in any axis
N	Number of slices
P	Intersection point
Per	Perimeter
SLP	Rendering file format
STL	Standard Template Library
V′	Extruded vertex
w_{inc}	Web increment
$w_{\max}$	Maximum web thickness
x, y, z	Cartesian space coordinates
α	Angle between two vectors

Y. Ata • O.U. Arkun
Roketsan Missiles Industries Inc., 06780, Ankara, Turkey

D.F. Kurtulus (✉)
METU Aerospace Engineering Department, 06800, Ankara, Turkey
e-mail: kurtulus@metu.edu.tr

T.H. Karakoç et al. (eds.), *Advances in Sustainable Aviation*,
DOI 10.1007/978-3-319-67134-5_5

5.1 Introduction

The accurate prediction of the grain geometry during the burnback is a critical problem in the design process of the solid rocket motors [1–4]. During the operation of the motors, the burning surface of the propellant grain is assumed to regress in a normal direction [2, 5]. This regression modifies both the configuration of the propellant grain and its burning surface area. In the current study, a new fast burnback simulation tool, namely, *Fast 3 Dimensional Burnback Tool* (F3DBT), has been developed to predict the grain geometry at every burn step, and the capabilities of the developed code are described. It is known that the geometrical changes of the solid propellant grains have direct effect on the performance of a solid rocket motors.

The developed F3DBT code has been inspired by Willcox's [6] study, and additional new capabilities were added to decrease user intervention and computation time. An important improvement which decreases the computational time is the modification of the code to be used for symmetrical models. Secondly, the code does not need any sign definition for the minimum distance calculations which simplifies mathematical operations in all volumetric regions. F3DBT method can be used for propagation of different geometries which was limited by the use of sign corrections such as sharp-angled, hull-type geometries [6]. Thirdly, an initial distance calculation is introduced to the program to decrease more the computational cost. This initial w_{max} input simplifies the search algorithm close to the grid point by eliminating the triangles outside of this initial input distance.

The developed code can take as input both the STL and SLP file formats. STL file format requires two input files, one for the propellant geometry and one for the mandrel geometry which is the port volume of propellant grain. However, SLP format output provides an easy burnback model setup. The user can initially define non-burning faces, outer boundary, and symmetry boundary conditions in a CAD program with different color codes which defines directly the propellant and the mandrel geometries in a single input file. By the use of these formats, the burned and non-burned information of the generated triangles could be distinguished.

An auto mesh interface is developed to detect the minimum and the maximum coordinates of the inner and outer geometries and to create automatically the grid points in the solution domain. Those innovations offer less user interaction in the initialization phase of the burnback simulation.

The geometry offsetting methods are generally used for rapid prototyping (RP). In literature, there are several example algorithms [7–9] which have been developed for offsetting the surfaces using STL file formats. An example of an offset methodology, which can directly be used with STL file formats, consists of extruding triangular elements in the normal direction and trimming or extending them to connect with the new geometry. The offset method is called Offset Weighted by Angle (OWA) [7].

Commercial softwares (CATIA, SOLIDWORKS, I-DEAS, AutoCAD, etc.) can perform offsets of 2D and 3D geometries. By defining grain geometries as a parametrical model, the burn area is obtained at every burn step. Constraints of the parameters should be chosen properly. In drafting techniques, user interaction is more dominant and more critical compared to other burn back techniques. The manual interactions of the model such as trimming and gap filling usually increase the errors.

Different numerical methods are also used for burnback calculations. Hartfield et al. [10] developed an interface capturing algorithm, namely, *Fast Marching Method*, which uses the upwinding nature of the propellant interface motion and which solves the eikonal-type equations (nonlinear partial differential equations) on a fixed three-dimensional tetrahedron mesh.

A computational method for simulating the evolution of the burning surface of 3D solid rocket motor propellants is developed by Willcox [6]. Using stereolithography surface information from the CAD file, the minimum distance function (MDF) is calculated, and the propellant surface burnback is simulated by use of the initial MDF.

Current study focus on developing and testing an algorithm based on minimum distance calculation for complex grain geometries to simulate burnback of propellant.

5.2 Burnback Model Initialization

5.2.1 File Formats

STL file formats are triangular surface meshes of 3D objects [11]. The file consists of a list of triangles which are uniquely defined by their normal vector and the three vertices [12]. The surface information of the geometry is defined by the use of three vertices (*V1, V2, V3*) and a unit normal vector $\vec{n}$ of the triangles.

SLP file format is almost same as STL format; however, it also includes the color code of faces which differs the burning face of the propellant, the burnout interface, and the symmetric boundary condition.

5.2.2 Solution Domain

In order to define the solution domain borders, the first input from the CAD program to F3DBT code is the STL output of the propellant geometry (Fig. 5.1a). Thereafter, the mandrel geometry of the propellant grain (Fig 5.1b) is read to distinguish the initial burning surface of the propellant and the propellant–empty volume interface. The minimum distance calculations are performed by use of the mandrel geometry.

With the SLP file output of the grain geometry, burnback analysis can be performed by using only the propellant geometry. SLP file format is typically exported from Pro/Engineer program. SLP file format output is very similar to STL format except it also includes color information of faces. Color information is used

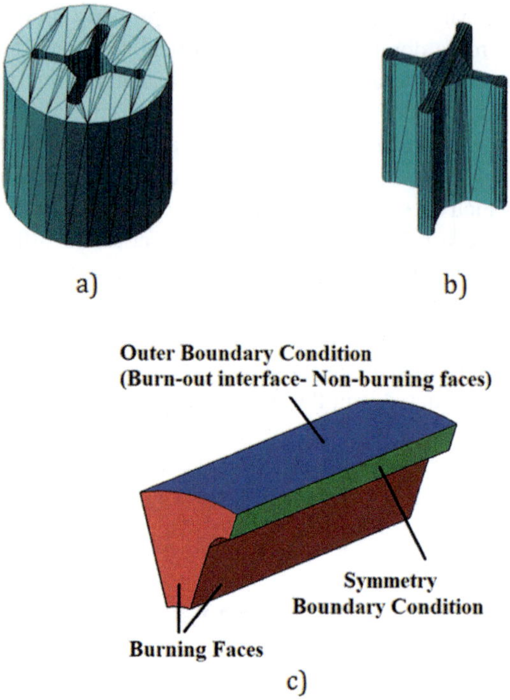

Fig. 5.1 Propellant and mandrel geometries for STL and SLP file formats. (**a**) Geometry of the propellant (STL file format). (**b**) Geometry of mandrel (STL file format). (**c**) Propellant geometry with color code (SLP file format)

to determine whether these faces are burning faces or not (Fig 5.1c). With this property, propellant geometry alone is sufficient for burnback analysis, and there is no need to eliminate the non-burning faces from the analyses. In addition, the user can easily define inhibited faces or unburned faces with the color definition. The symmetry faces and outer boundary faces are also represented with different color codes.

5.2.3 *Auto Mesh Interface*

Grid points are created in the domain such that the grid intensity and their locations can be both defined manually by the user or automatically by the developed auto mesh interface code. It is noted that the grid size and the number of grid slices along the grain length are important for an accurate burning area calculation.

Due to the complexities of both the inner and the outer geometries of the propellant grain, the auto mesh interface option can be used to determine the location of grid slices along the propellant length [6]. The number of triangles, which defines propellant grain, increases with the geometrical details of the grain. If the inner geometry of grain is varying along the z axis, the code will require many grid slices along z direction.

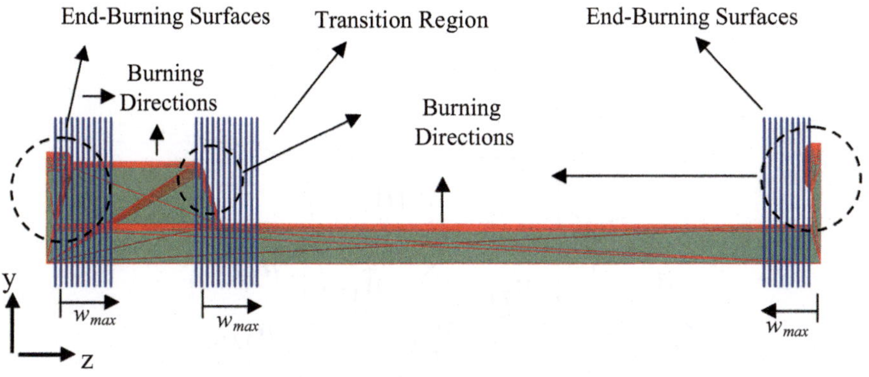

Fig. 5.2 Inner geometry of the grain and grid slices

Generally, the grain geometries have an end-burning surface and no any geometrical changes along the z direction. The program requires more than two grid slices for true burnback simulation for these cases. The program uses the user-defined $w_{\max}$ value and the distance between the grid slices to determine the number of grid slices in case that the grain has end-burning surfaces. In this way, the full propagation of the end-burning surfaces is obtained (Fig. 5.2).

In case that complex geometries are present, the number of grid slices should be sufficient to simulate the exact shape of the burning surface (Fig. 5.2). In the burnback simulation, grid slices can have different resolutions. Different resolutions of grid slices enable low computational time without additional computational error. The grid slices are grouped as block data and solved independently. The total burning area is evaluated by combining each grid slice perimeter and burned area contributions. In addition to create grid slices by using the auto mesh interface program, the code also allows the user to manually add extra z locations in case that there could be insufficient number of grid slices inside the solution domain for complex shapes.

5.3 Minimum Distance Calculation Method

The minimum distance calculation method is presented by Willcox [6] which calculates the minimum distance values between grid points to any geometrical shape in 3D space. Three volumetric regions are used for a triangle to perform the minimum distance calculation.

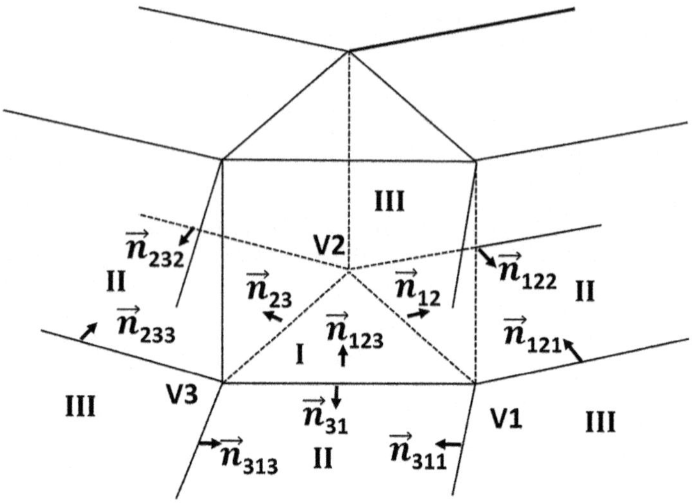

Fig. 5.3 Volumetric region definitions

5.3.1 Volumetric Region Definition

In minimum distance calculations, there exist three volumetric regions (I, II, and III) independently generated for each triangle. Three different volumetric regions are obtained by defining nine different unit normal vectors [6] with the use of the vertices of the triangle and its normal vector. The only known unit normal vector is $\vec{n}_{123}$ from STL file pointing out of the plane of the triangle. The other ones are obtained from this normal vector and three vertices of the triangle. As shown in Fig. 5.3, the volumetric region I is a semi-infinite triangular prism obtained by extruding the triangle defined by these three vertices along the normal vector $\vec{n}_{123}$. There exist three different volumetric regions of II, and they are formed by a semi-infinite rectangular prism (Fig. 5.3). They are obtained by extruding surfaces of triangular prism of volumetric region I along normal vectors of the surfaces ($\vec{n}_{12}$, $\vec{n}_{23}$, and $\vec{n}_{31}$). Lastly, three different volumetric regions of III are also defined for each vertices of the triangle, and they are located between two volumetric regions of II (Fig. 5.3).

5.3.2 Distance Calculation Methods

To reduce the computational time, initial proximity calculations are performed by introducing an initial distance (d_{int}) which is the minimum distance between the grid point and the triangle's plane (Fig. 5.4). This parameter results in a computational time reduction of 18.1% for the cases investigated in the result section.

Fig. 5.4 Initial distance calculation

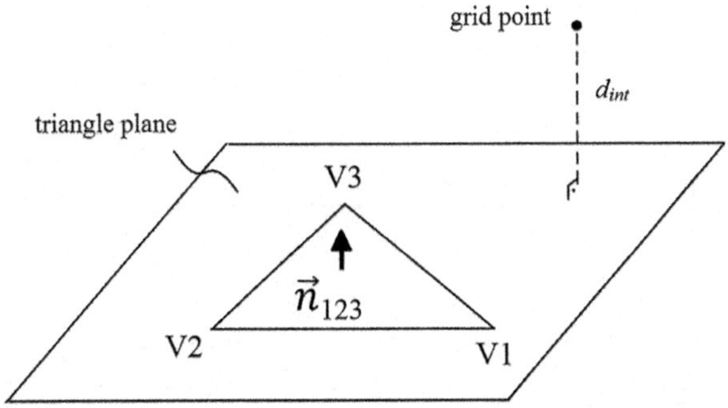

However, MDV is the minimum distance between the grid point and the triangle's vertices, edge, or inner region according to the volumetric region of the grid point with respect to the triangle. When d_{int} is bigger than w_{max}, the distance calculation will not be performed. If it is less than w_{max}, the program will perform the normal distance calculation procedure (Fig. 5.4). The code skip lots of calculations for all three volumetric regions with this initial distance check which provides computational time advantages. On the other hand, if the check is satisfied, the point distance value is obtained according to where the point lies in one of the three volumetric regions. Minimum distance calculation iteration starts with the grid point. At the end of the triangle iteration, the code use the minimum distance value among the distance values of all triangles (Fig. 5.5).

There are three checks which are used to determine whether the grid point is inside of the volumetric region I or not. The check methodology depends on finding the angles between the surface normal vector and the vector which is created by any point of the surface and the grid point. The newly developed method is quite simple and gives accurate results in lower computational time relative to other methods like barycentric coordinate system method and parametric equations system method.

One of the three angular calculations of volumetric region I is demonstrated in Fig. 5.6.

For case 1, the grid point lies on counter side of the plane created by edge of the triangle (*V2, V3*) relative to direction of normal vector $\vec{n}_{23}$. In that condition, the angle between the normal vector and the vector which is created by the grid point and *V3* will be less than 90°. If the point is exactly on the plane (case 2), the angle will be equal to 90°. Otherwise, the angle will be greater than 90° for case 3. Three checks are enough for semi-infinite triangular prism (volumetric region I) to determine whether the point is inside the region or not.

The code obtains the dot product of these two vectors for all three surfaces (Eq. 5.1) of the triangular prism.

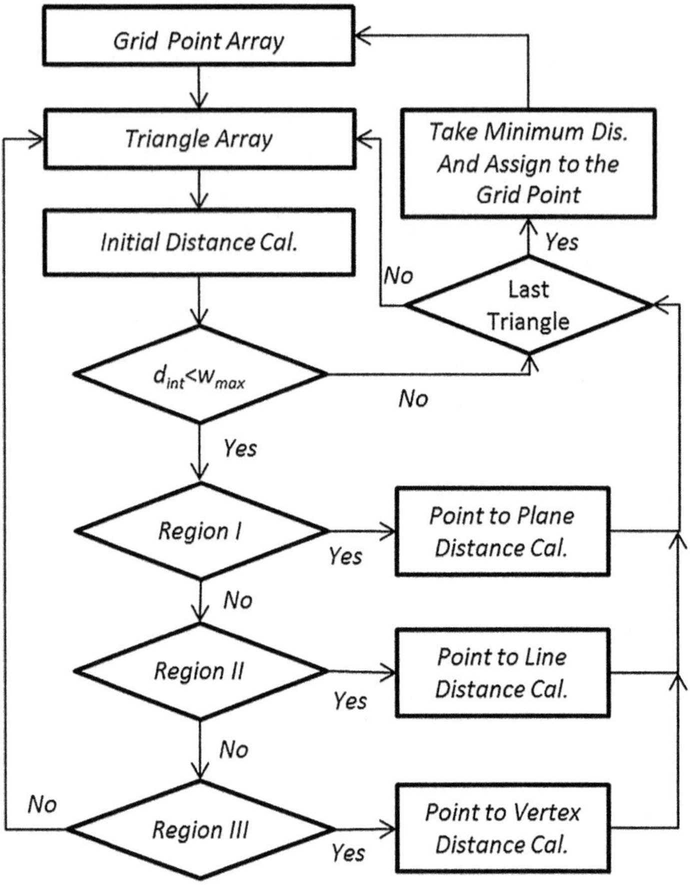

Fig. 5.5 Flowchart of the minimum distance calculation

$$\cos\ \alpha_{I1} = \frac{\overrightarrow{(V2-P)}\cdot\overrightarrow{n_{12}}}{\left\|\overrightarrow{(V2-P)}\right\|\ \left\|\overrightarrow{n_{12}}\right\|};\quad \cos\ \alpha_{I2} = \frac{\overrightarrow{(V3-P)}\cdot\overrightarrow{n_{23}}}{\left\|\overrightarrow{(V3-P)}\right\|\ \left\|\overrightarrow{n_{23}}\right\|};\quad \cos\ \alpha_{I3} = \frac{\overrightarrow{(V1-P)}\cdot\overrightarrow{n_{13}}}{\left\|\overrightarrow{(V1-P)}\right\|\ \left\|\overrightarrow{n_{13}}\right\|} \tag{5.1}$$

If the dot products of all angular values are greater and equal to zero, the point will be inside the semi-infinite triangular prism (including the surfaces). In this case, the distance is calculated between the point and the surface using Eq. 5.2.

$$d_i = \left\|\overrightarrow{(P-V1)}\cdot\overrightarrow{n_{123}}\right\|\ (i = 1, 2, 3, \ldots., \text{number of triangles}) \tag{5.2}$$

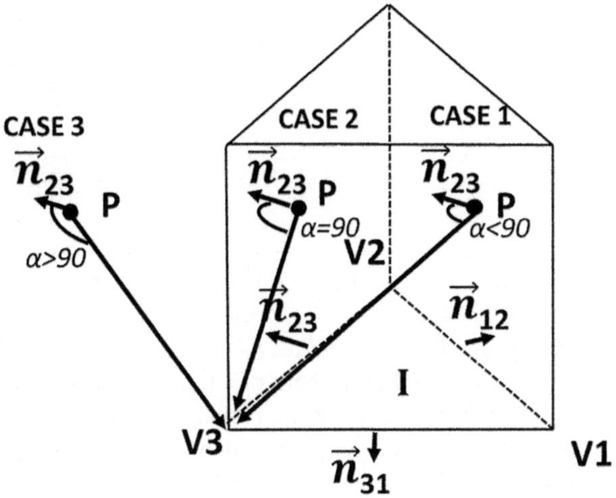

Fig. 5.6 Angular calculation method for volumetric region I

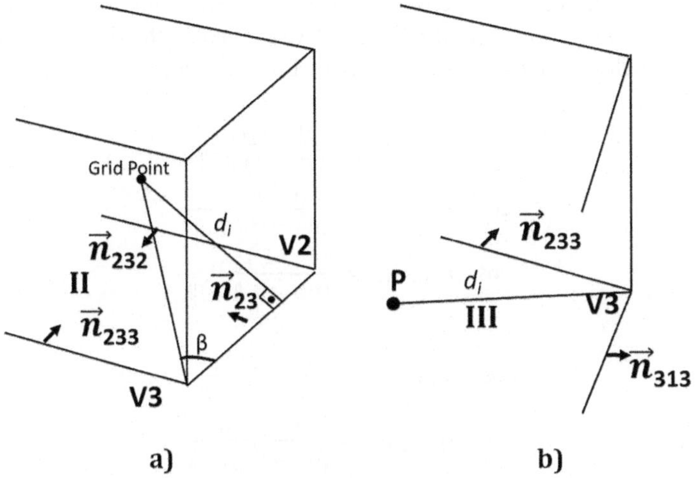

Fig. 5.7 Example distance calculation methods. (**a**) Volumetric region II. (**b**) Volumetric region III

The generated nine unit normal vectors are used to determine whether the grid point is inside or not of any of the three volumetric regions of II. For one of the volumetric regions of II (Fig. 5.3), which is the side of edge formed by vertices *V3* and *V2* of the triangle (Fig. 5.7a), similarly volumetric region I, three angular calculations will be performed by using $\vec{n}_{23}$, $\vec{n}_{233}$, and $\vec{n}_{232}$ unit normal vectors and two new vectors which are defined by grid point to *V3* and to *V2*, respectively (Eq. 5.3).

$$\begin{aligned}\cos\ \alpha_{\mathrm{II},V3,V2} &= \frac{\overrightarrow{(V2-P)}\cdot\overrightarrow{n_{23}}}{\left\|\overrightarrow{(V2-P)}\right\|\ \left\|\overrightarrow{n_{23}}\right\|}\\ \cos\ \alpha_{\mathrm{II}2,V3,V2} &= \frac{\overrightarrow{(V2-P)}\cdot\overrightarrow{n_{232}}}{\left\|\overrightarrow{(V2-P)}\right\|\ \left\|\overrightarrow{n_{232}}\right\|}\\ \cos\ \alpha_{\mathrm{II}3,V3,V2} &= \frac{\overrightarrow{(V3-P)}\cdot\overrightarrow{n_{233}}}{\left\|\overrightarrow{(V3-P)}\right\|\ \left\|\overrightarrow{n_{233}}\right\|}\end{aligned} \tag{5.3}$$

$$d_i = \sqrt{(P_x - V_{3x})^2 + \left(P_y - V_{3y}\right)^2 + (P_z - V_{3z})^2}\,\sin\beta \tag{5.4}$$

where β is shown in Fig. 5.7a.

If all $\alpha_{\mathrm{II}t,V3,V2}$ values (where $t = 1,2,3$) are less than and equal to zero, the point will be inside the semi-infinite rectangular prism of volumetric region II, and the distance calculation is performed by point to line distance calculation method using Eq. 5.4.

Otherwise, the code will check the other volumetric regions II. Equations 5.5 and 5.6 are used to control for the remaining two volumetric regions of II (for edge of *V1, V2* vertices and *V1, V3* vertices), respectively. Finally, the distance calculation for two volumetric regions II is done by using Eqs. 5.7 and 5.8.

$$\begin{aligned}\cos\ \alpha_{\mathrm{II}1,V1,V2} &= \frac{\overrightarrow{(V1-P)}\cdot\overrightarrow{n_{12}}}{\left\|\overrightarrow{(V1-P)}\right\|\ \left\|\overrightarrow{n_{12}}\right\|}\\ \cos\ \alpha_{\mathrm{II}2,V1,V2} &= \frac{\overrightarrow{(V1-P)}\cdot\overrightarrow{n_{121}}}{\left\|\overrightarrow{(V1-P)}\right\|\ \left\|\overrightarrow{n_{121}}\right\|}\\ \cos\ \alpha_{\mathrm{II}3,V1,V2} &= \frac{\overrightarrow{(V2-P)}\cdot\overrightarrow{n_{122}}}{\left\|\overrightarrow{(V2-P)}\right\|\ \left\|\overrightarrow{n_{122}}\right\|}\end{aligned} \tag{5.5}$$

$$\begin{aligned}\cos\ \alpha_{\mathrm{II}1,V3,V1} &= \frac{\overrightarrow{(V3-P)}\cdot\overrightarrow{n_{31}}}{\left\|\overrightarrow{(V3-P)}\right\|\ \left\|\overrightarrow{n_{31}}\right\|}\\ \cos\ \alpha_{\mathrm{II}2,V3,V1} &= \frac{\overrightarrow{(V3-P)}\cdot\overrightarrow{n_{313}}}{\left\|\overrightarrow{(V3-P)}\right\|\ \left\|\overrightarrow{n_{313}}\right\|}\\ \cos\ \alpha_{\mathrm{II}3,V3,V1} &= \frac{\overrightarrow{(V1-P)}\cdot\overrightarrow{n_{233}}}{\left\|\overrightarrow{(V1-P)}\right\|\ \left\|\overrightarrow{n_{233}}\right\|}\end{aligned} \tag{5.6}$$

$$d_i = \sqrt{(P_x - V_{2x})^2 + \left(P_y - V_{2y}\right)^2 + (P_z - V_{2z})^2}\,\sin\beta \tag{5.7}$$

$$d_i = \sqrt{(P_x - V_{1x})^2 + \left(P_y - V_{1y}\right)^2 + (P_z - V_{1z})^2}\,\sin\beta \tag{5.8}$$

If these checks (for volumetric regions I and II) are not satisfied for a point, the point will be inside of one of the three volumetric regions III (Fig. 5.3). Figure 5.7b represents a point inside the volumetric region III at the side of vertex *V3*. For that condition, $\cos\alpha_{\mathrm{III}1,V3}$ given by Eq. 5.9 is greater and equal to zero, and the minimum distance is calculated by the point to vertex calculation method using Eq. 5.10.

$$\begin{aligned} \cos\ \alpha_{\mathrm{III1},V3} &= \frac{\overrightarrow{(V3-P)}\cdot\overrightarrow{n_{233}}}{\left\|\overrightarrow{(V3-P)}\right\|\ \left\|\overrightarrow{n_{233}}\right\|} \\ \cos\ \alpha_{\mathrm{III2},V3} &= \frac{\overrightarrow{(V3-P)}\cdot\overrightarrow{n_{313}}}{\left\|\overrightarrow{(V3-P)}\right\|\ \left\|\overrightarrow{n_{313}}\right\|} \end{aligned} \tag{5.9}$$

$$d_i = \sqrt{(P_x - V_{3x})^2 + \left(P_y - V_{3y}\right)^2 + (P_z - V_{3z})^2} \tag{5.10}$$

Equations 5.11 and 5.12 are used to check the two volumetric regions III for sides of vertices *V2* and *V1*, respectively. If these checks are satisfied for any of remaining two volumetric regions III, the distance value between point and vertex is obtained by Eqs. 5.13 and 5.14.

$$\begin{aligned} \cos\alpha_{\mathrm{III1},V2} &= \frac{\overrightarrow{(V2-P)}\cdot\overrightarrow{n_{232}}}{\left\|\overrightarrow{(V2-P)}\right\|\ \left\|\overrightarrow{n_{232}}\right\|} \\ \cos\alpha_{\mathrm{III2},V2} &= \frac{\overrightarrow{(V2-P)}\cdot\overrightarrow{n_{122}}}{\left\|\overrightarrow{(V2-P)}\right\|\left\|\overrightarrow{n_{122}}\right\|} \end{aligned} \tag{5.11}$$

$$\begin{aligned} \cos\alpha_{\mathrm{III1},V1} &= \frac{\overrightarrow{(V1-P)}\cdot\overrightarrow{n_{123}}}{\left\|\overrightarrow{(V1-P)}\right\|\ \left\|\overrightarrow{n_{123}}\right\|} \\ \cos\alpha_{\mathrm{III2},V1} &= \frac{\overrightarrow{(V1-P)}\cdot\overrightarrow{n_{311}}}{\left\|\overrightarrow{(V1-P)}\right\|\left\|\overrightarrow{n_{311}}\right\|} \end{aligned} \tag{5.12}$$

$$d_i = \sqrt{(P_x - V_{2x})^2 + \left(P_y - V_{2y}\right)^2 + (P_z - V_{2z})^2} \tag{5.13}$$

$$d_i = \sqrt{(P_x - V_{1x})^2 + \left(P_y - V_{1y}\right)^2 + (P_z - V_{1z})^2} \tag{5.14}$$

5.3.3 Special Conditions

There may be some limitations for offset of concave and convex shapes. When the angle between two offset surfaces is convex, there is a gap between them. On the other hand, when two offset surfaces have a concave angle between them, in this case, there is an intersection or overlap between them. For an accurate offset calculation, all the intersections due to the concave surfaces should be trimmed to delete the unused part of them and all the gaps should be filled for convex surfaces. This procedure can be complex due to the number of triangles in the 3D models [8].

F3DBT code only uses the minimum distance at the grid points to obtain the offset of the surfaces. The burning surfaces from CAD programs give only the initial geometry, and there is no need for the creation of an additional offset geometry using F3DBT code. Thus, self-intersections and gaps due to the offset of convex and concave surfaces mentioned above are not considered. The initial geometry is used for all of the minimum distance calculations. F3DBT code uses only burning triangles in minimum distance calculations, and this operation doesn't require any sign assignment. Willcox's method [6] could result in an error in sign for non-convex hull-type and sharp-angled conical geometries. It should be noted that, in general, the propellant geometries do not include sharp angles except in some special case such as propellant cavity analysis where there exist irregular geometries inside the propellant [10].

5.3.4 Minimum Distance Calculation Test Cases

The current code capability is tested with some typical grain geometries as slot, wagon, anchor, and dendrite (Fig. 5.8). These example geometries have both large- and small-angled triangles, and the results show that the code developed can be used for propagation of all different geometries including geometries with sharp angles. F3DBT code has also a low computational time requirement for the simulation of even complex geometries.

5.4 Perimeter and Burned Area Calculations

5.4.1 Perimeter and Burned Area Calculations for Non-End-Burning Faces

If the perimeter line is inside the grid square, then the perimeter and the burned area calculations are done. For the burning distances given in Eq. 5.15, all the grid squares along the contour of the burning surface are determined, and the total perimeter and the burned area according to the burning distance are computed. This procedure is repeated until the maximum MDV is reached. The perimeter and the burned area in a grid slice are calculated, and the same procedure is applied for all grid slices.

The perimeter of non-end-burning surface is obtained by obtaining the intersection points of the perimeter line and the edges of the grid square. The grid squares for the point's MDV(i,j,k) and its eight neighbors in the solution domain are shown in Fig. 5.9. The intersection points on the grid edges are found using the corners of the squares and the minimum distance values.The corners of grid square (C1, C2, C3, C4) are defined by subtracting minimum distance values (MDV) of these grid points from the burning distance. This approach creates different sign values on the corners which enables the perimeter calculations. The perimeter of non-end-burning surfaces is calculated by obtaining the intersection points on the edges of relevant grid square. After finding perimeter line in a grid square, burned area is also

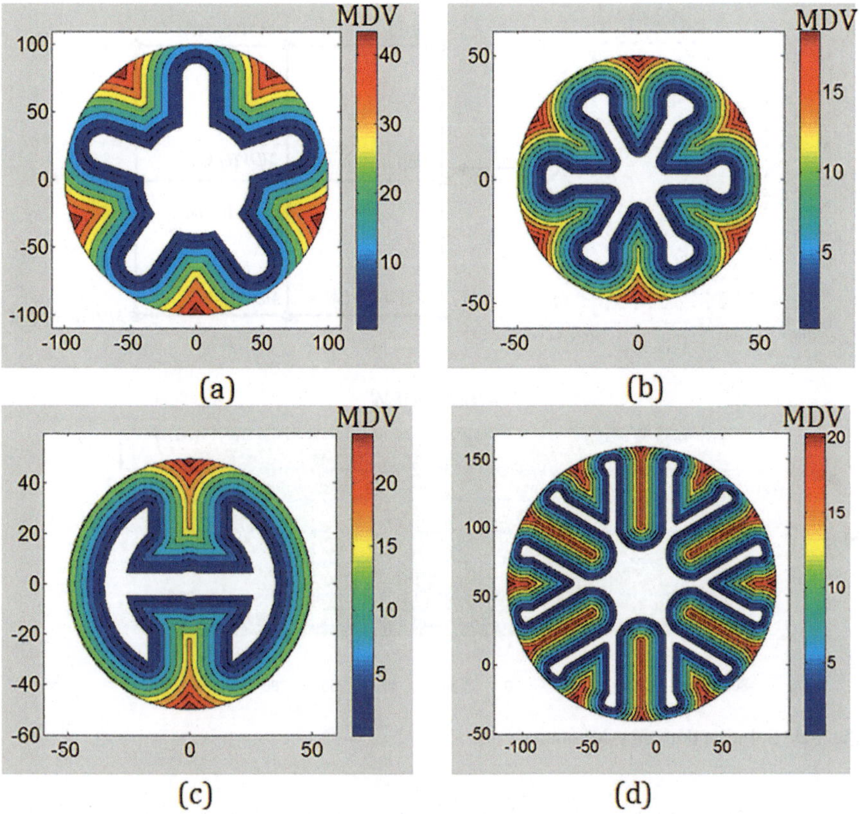

Fig. 5.8 Minimum distance solutions of different grain geometries. (**a**) Slot, (**b**) wagon, (**c**) anchor, (**d**) dendrite

calculated. Perimeter line divides the grid to burned and unburned parts. Burned area in a grid square is determined by area calculations of burned part in the grid square. There are 22 possible perimeter lines and burned area conditions for a grid square. They are solved according to sings of the corner values. The perimeter and burned area locations are obtained by Eq. 5.15.

$$d_{\mathrm{bi}} = i^* w_{\mathrm{inc}}, i = \left[1, 2, 3 \ldots, \mathrm{round}\left(\frac{\max(\mathrm{MDV})}{w_{\mathrm{inc}}}\right)\right] \tag{5.15}$$

where d_{bi} is the burning distance at i^{th} time and w_{inc} is defined by the user as an incremental burning distance value. The w_{inc} determines the number of perimeter lines in a slice, and therefore it affects the resolution of burning surface area. If it is not defined otherwise, the program takes its default value as twice of the grid dimension (*dx, dy*).

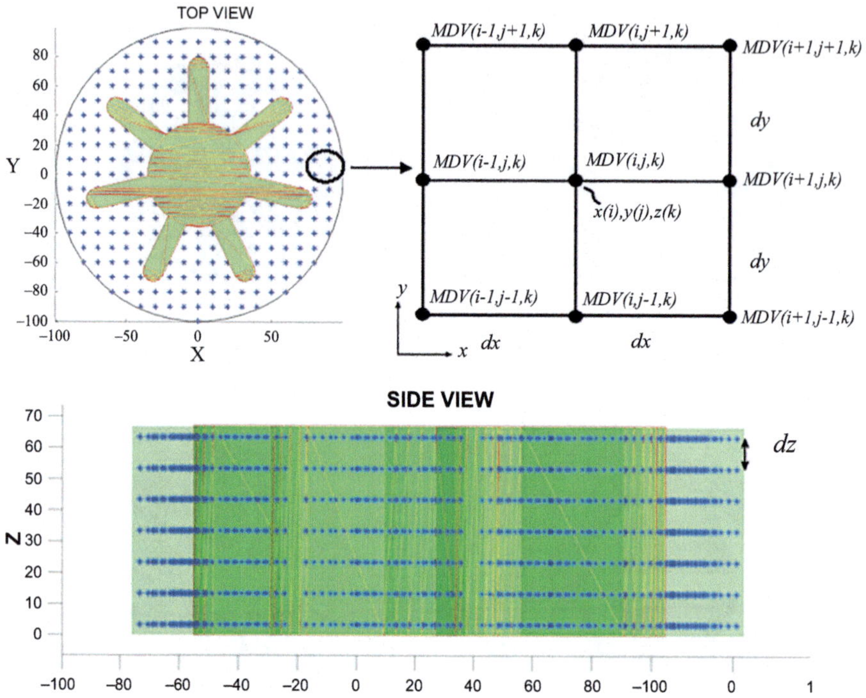

Fig. 5.9 Solution domain grid squares

Sign of the grid square corner values determines whether the perimeter and burned calculations will be performed for the grid square in consideration. If the square is in the propellant side, all corners of the square have a positive sign. Negative sign for all corners implies that the square is completely in the gas side and the perimeter calculations have already been performed for this grid square. On the other hand, if the grid square corners have different signs, this means that the square is at the solid-gas interface. One of the grid's square which has different signs at the corners in the solution domain and the perimeter line within the square is shown in Fig. 5.10.

The length of the perimeter line inside of the grid square is obtained using the coordinates of the intersection points. For the condition given in Fig. 5.10, the distances (C1, C2, C3) from points to the perimeter line are determined by using Eqs. 5.16, 5.17, and 5.18.

$$\mathrm{C1} = \mathrm{d1} == \mathrm{MDV}\,(i, j, k) - d_{\mathrm{bi}} \tag{5.16}$$

$$\mathrm{C2} = \mathrm{d2} = \mathrm{MDV}\,(i + 1, j, k) - d_{\mathrm{bi}} \tag{5.17}$$

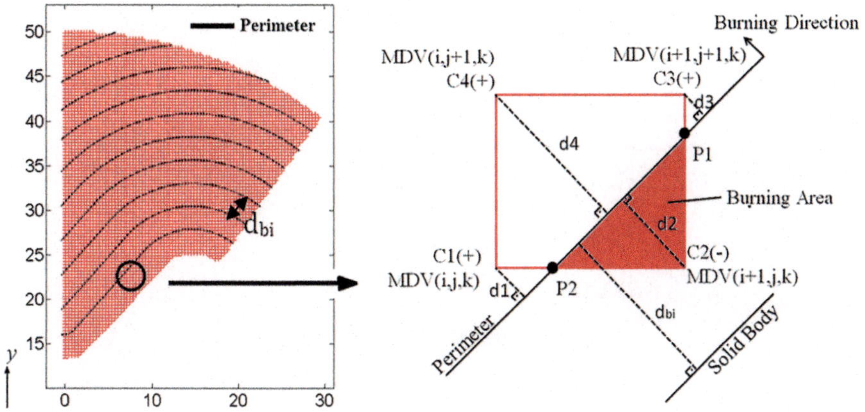

Fig. 5.10 Perimeter line and burned area inside a square

$$\text{C3} = \text{d3} = \text{MDV}\,(i+1, j+1, k) - d_{\text{bi}} \quad \left(i = 1 \ldots n_x, j = 1 \ldots n_y, k = 1 \ldots n_z\right) \tag{5.18}$$

Using the distance values, the intersection point coordinates are calculated by linear interpolation with Eqs. 5.19 and 5.20 [6]. Similar methodology is used for other possible cases.

$$\text{P1} = \left(x_i + \text{d}x, y_j + \text{d}y \times \left[\left(\frac{|\text{C2}|}{\text{C3} + |\text{C2}|}\right)\right], z_k\right) \tag{5.19}$$

$$\text{P2} = \left(\left[x_i + \text{d}x\left(\frac{\text{d1}}{\text{d1} + |\text{d2}|}\right), y_i, z_i\right]\right) \tag{5.20}$$

5.4.2 *Perimeter Calculations Around Symmetry Boundaries and Burnout Interfaces*

The geometries of almost all solid rocket propellants are symmetric. F3DBT code is developed as also being capable of performing the minimum distance calculations for symmetrical boundaries and burnout interfaces. The outer and symmetrical boundary faces are solved by the minimum distance program, and they are taken account as a perimeter line inside the grid squares. For the outer boundary faces, the program calculates the minimum distance values for the grid squares. The corners are numbered as “0” and “−2.” If the grid corner is 0, the minimum distance value of the corner is changed to negative sign. The burning distance is selected as “0” for perimeter calculations because there is no regressions for system borders. The

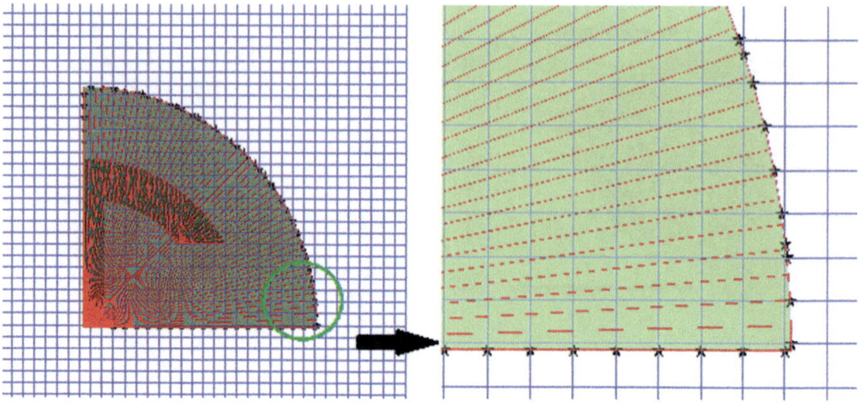

Fig. 5.11 Intersection points (P1, P2) of anchor-type grain with grid squares

Fig. 5.12 Intersection point between the boundary and the perimeter lines

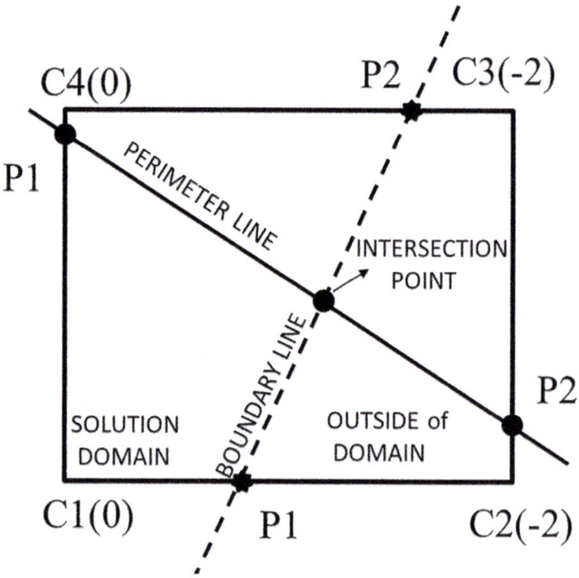

intersection points on the edges of the squares are calculated by using Eqs. 5.19 and 5.20. As an example, the intersection points for an anchor grain are shown in Fig. 5.11 with grid squares. The numbering procedure of the square corners is also presented in Fig. 5.12.

The exact borders of solution domain are determined by using these intersection points. The code calculates the minimum distances and determines the location of the perimeter line in the grid square.

The intersection point divide the perimeter line and the total perimeter calculations does not include the part of the perimeter located outside of solution domain

(Fig. 5.12). This procedure prevents any perimeter contributions from grids which are located outside of the domain.

5.4.3 Surface Area Calculations

The inner geometry of the grain can have both burned and unburned surfaces. The code uses only the burning surface while performing the minimum distance calculation and the indices of the burning faces are saved. After completing the minimum distance calculations, the initial area is obtained by summing the surface areas of triangles in burning faces.

There are two different surface area calculation methods used in F3DBT code, one is the perimeter calculation method which is used to obtain the burning area contribution of boundary cells, and the second one is the burned area calculation in the solution domain. The surface area is obtained by multiplying the average of two perimeters in the two different slices having the same burning distance and the distance between those slices as given by Eq. 5.21.

$$\mathrm{Ab}_i = \sum_{j=1}^{N-1} \frac{\left(\mathrm{Per}_{j+1} + \mathrm{Per}_j\right)}{2} \times \left(z_{j+1} - z_j\right) \tag{5.21}$$

The burning volume is obtained by multiplying the average of the burned area in the two different slices with the distance between slices as given in Eq. 5.22. The burning volume of the propellant is obtained at each step. The burned area is obtained by dividing the volume to the web increment (Eq. 5.22).

$$\forall_i = \sum_{j=1}^{N-1} \frac{\left(A_{j+1} + A_j\right)}{2} \times \left(z_{j+1} - z_j\right), A_i = \frac{\forall_i}{w_{\mathrm{inc}}} \tag{5.22}$$

5.4.4 Surface Area Calculations of End-Burning Faces

The propellant grain geometries have mostly end-burning faces. The grid points have constant distances at the end-burning faces. The perimeter and the burned area calculations cannot be performed for this case since these calculations need different distance values to the perimeter line inside the grid square. In F3DBT code, it is checked if the grid point's MDV and its neighbors have similar (within user defined tolerance) distance value or not in order to define the grid square as an end-burning surface (Fig. 5.9). In the first case, the perimeter and burned area calculations are not carried out, and this square is denoted into an array as an end-burning area. The burning distance value between the initial geometry and the grid point is used to

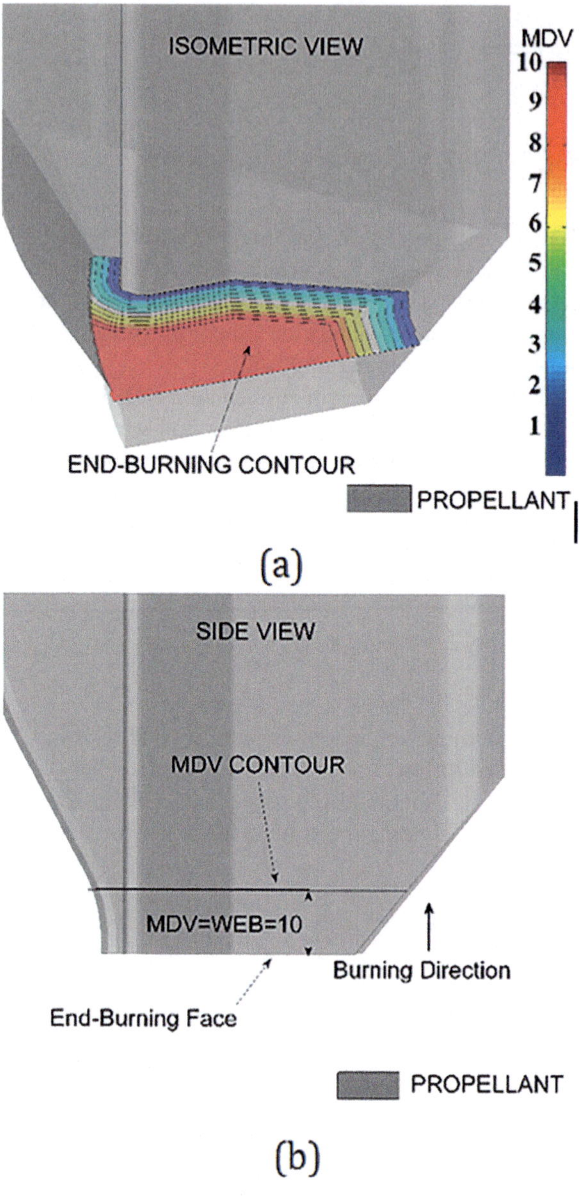

Fig. 5.13 End-burning contour and web. (**a**) Isometric view. (**b**) Side view

obtain the web of this area (Fig. 5.13). As an example, in Fig. 5.13, MDV gives the web value of 10 for the end-burning area.

In order to understand if the grid is an end-burning grid or not for symmetrical grain geometries, a different check is applied at the symmetry line. The same procedure is also applied at burnout interface. Figure 5.14 shows a symmetrical end-burning propellant geometry. By comparing the MDV of point P(i,j,k) with that

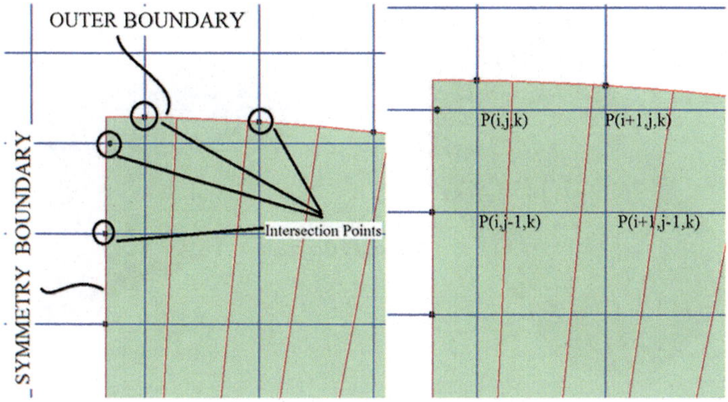

Fig. 5.14 Intersection of symmetry and outer boundary lines with propellant

of points P($i-1,j,k$), P($i+1,j,k$), P($i+1,j-1,k$) and the intersection points, if they are equal then the grid is taken as an end-burning grid. The corresponding burning areas are considered as their surface areas.

5.5 Results

5.5.1 Cases of Dendrite and Slot Grains

Two validation models have been analyzed for verification of F3DBT code. The first one is the dendrite model without end-burning faces (Fig. 5.15a), and the second one is the slot geometry with end-burning face at both ends of the grain (Fig. 5.15b). The grid dependency analysis has been performed for dendrite case since it is the most complex geometry among the investigated grains. The calculated specific distance ratio, which is the ratio of outer diameter to the grid dimension, is also used for the following burnback analysis. The total percent error of output data (burning surface area distance versus web) are compared with the reference burning area obtained from the CAD solutions.

The dendrite geometry having 12 arms and 200 mm outer diameter is a quite complex and detailed geometry. The STL output of symmetric dendrite geometry consists of 442 triangles. The burnback simulations have been performed with 2.4 GHz single processor for several different grid dimensions. The percentage of errors of surface areas relative to CAD solution and computational times of F3DBT code for each grid dimensions are given in Fig. 5.16a. The evaluated surface areas are compared with CAD simulation of the dendrite grain in Fig. 5.16b.

As the grid size becomes smaller, the total error of calculated surface area asymptotically approaches to zero. In this case, however, the computational time required for the simulation is drastically increasing. In order to obtain acceptable

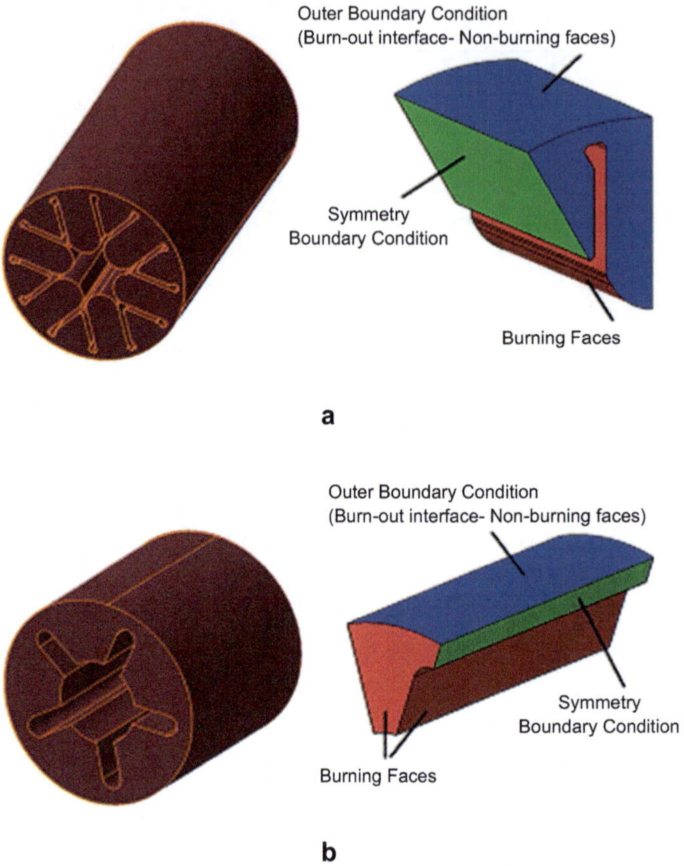

Fig. 5.15 Full and symmetric views of burnback validation models. (**a**) Dentrite model. (**b**) Slot model

results in a reasonable time interval, a compromise should be performed between the error and the computational time. For grid dimensions smaller than 0.5 mm, the burnback simulation gives quite acceptable results (errors <0.095%). It is concluded that 0.5 mm grid dimension can be used for burnback simulation of 200 mm outer diameter dendrite-type grain. In this case, the specific distance ratio will be 400. Slot geometry has 200 mm outer diameter. The 0.5 mm grid dimension is also used in this slot burnback analysis. Although the slot geometry is simpler relative to dendrite geometry, the required number of grid slice for burnback simulation is more than dendrite case due to end-burning faces at both ends of the slot geometry. Moreover, catching propagation of both inner and end-burning faces together in a grain requires more number of grids than one having no end-burning faces. The evaluated burning surface area from simulation is compared with burning surface area obtained from

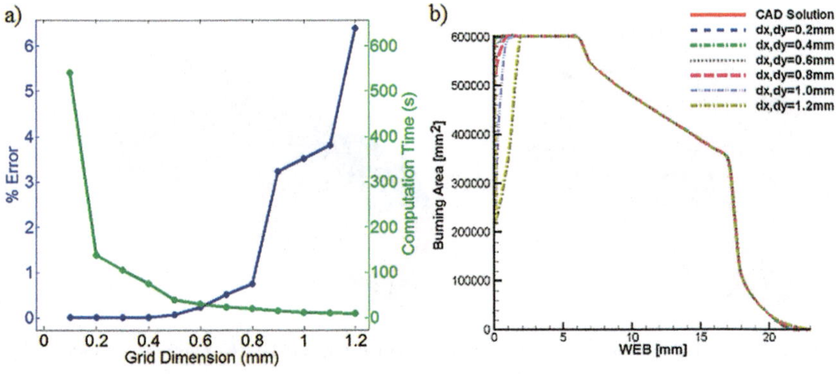

Fig. 5.16 Grid refinement study for dendrite geometry burning surface area. (**a**) % error in burning surface area and computational time. (**b**) Burning surface area grid refinement study and comparison with CAD solution

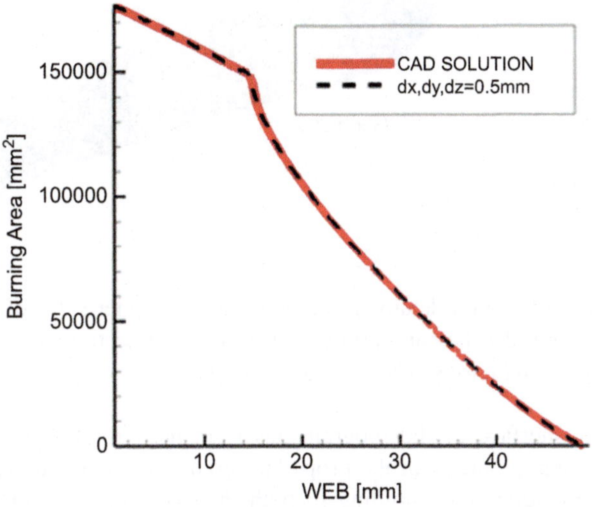

Fig. 5.17 Burning surface area comparison for slot geometry

CAD solution in Fig. 5.17. Total error of the burnback simulation is about 0.102% (with 400 specific distance ratio) which is quite acceptable computational error.

5.5.2 3D Propellant Grain

A real motor with a physical 3D propellant grain configuration has been examined. The propellant grain's full and symmetric models are given in Fig. 5.18. The

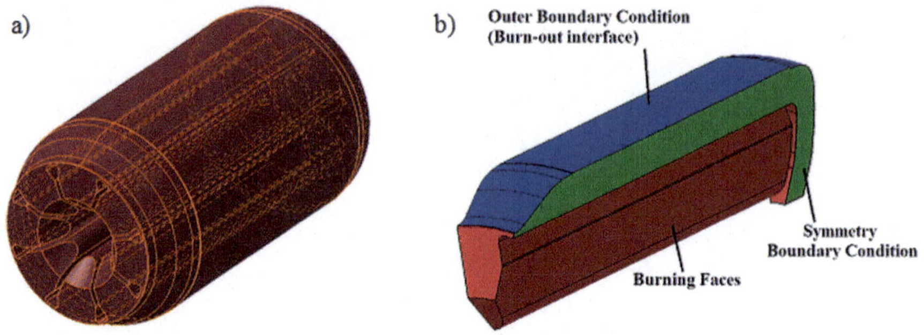

Fig. 5.18 Physical 3D propellant grain configuration. (**a**) Full model. (**b**) Symmetric model

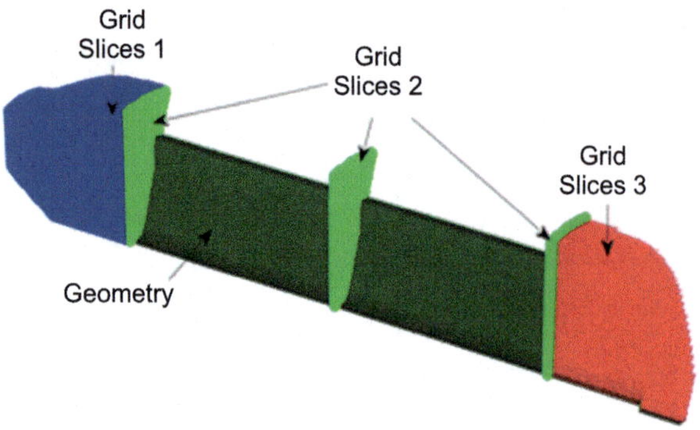

Fig. 5.19 Three different grid slices on symmetric model

geometry is complex with details at the outer and the inner sides of the grain. In addition, the grain also has an end-burning face at aft section. The propellant is wagon grain with eight arms. The simulation has been carried out with 400 specific distance ratios.

Auto mesh interface of the program has been used for the solution domain according to outer geometry of the propellant grain. The required numbers of the grid slices are determined according to both the inner and the outer geometries of the grain. Auto mesh interface compares the required number of grid slice for both inner and outer geometries and selects the larger one. In this way, the required number of grid slices for both of the geometries has been satisfied. Three different grid slices have been formed by auto mesh interface (Fig. 5.19). A constant grid dimension has been used by auto mesh interface. The specific distance ratio of 400 is used for three different grid slice groups created.

3D burning surfaces of the grain at different burning distances are given in Fig. 5.20. The evaluated burning surface areas with propellant volume and port volumes of these three grid slices are shown in Fig. 5.21.

The total burning surface area, which is evaluated by summing the burning surface area contributions of the three grid slice groups, is compared with the burning area obtained from CAD solution in Fig. 5.22. The total error of burnback simulation relative to CAD solution is about 0.48%. The propellant port volume

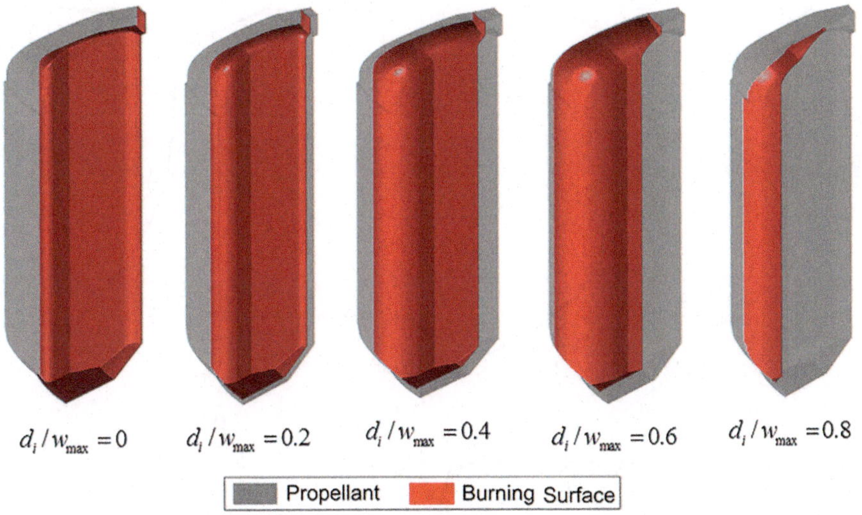

Fig. 5.20 3D burning surfaces at various d_i/w_{max} values

(which is the empty volume of propellant grain) and propellant grain volume calculations are performed by same method in the burn area calculations (Fig. 5.22).

In order to observe the computational time advantage of initial distance calculation method, the burning surface area has been evaluated with and without initial distance calculation methods. Figure 5.23 represents the comparison of the computational times for both conditions which show about 18.1% decrease in the computations for symmetrical model and about 13.2% for full model.

The full model has been analyzed in the same conditions to see the symmetry condition effect on computational time. The simulation of symmetric model has almost $2N$ ($N = 8$ for the motor with a physical 3D propellant grain) times less number of triangles than full model and also about $2N$ times less computed grid points. This means that nearly $4N^2$ times less mathematical operations are carried out in the symmetric simulation relative to full model burnback simulation. This extensively reduces the computational time. The time cost analyses show that using the initial distance method and symmetric model in burnback simulation together provides great computational time advantage ($\approx$286 times less relative to full model without initial distance method).

5.6 Conclusion

In the present chapter, a developed fast burnback tool named F3DBT is presented for internal ballistic analyses of solid rocket motors. The burnback calculation is performed by using STL and SLP file formatted geometries for propellant grain

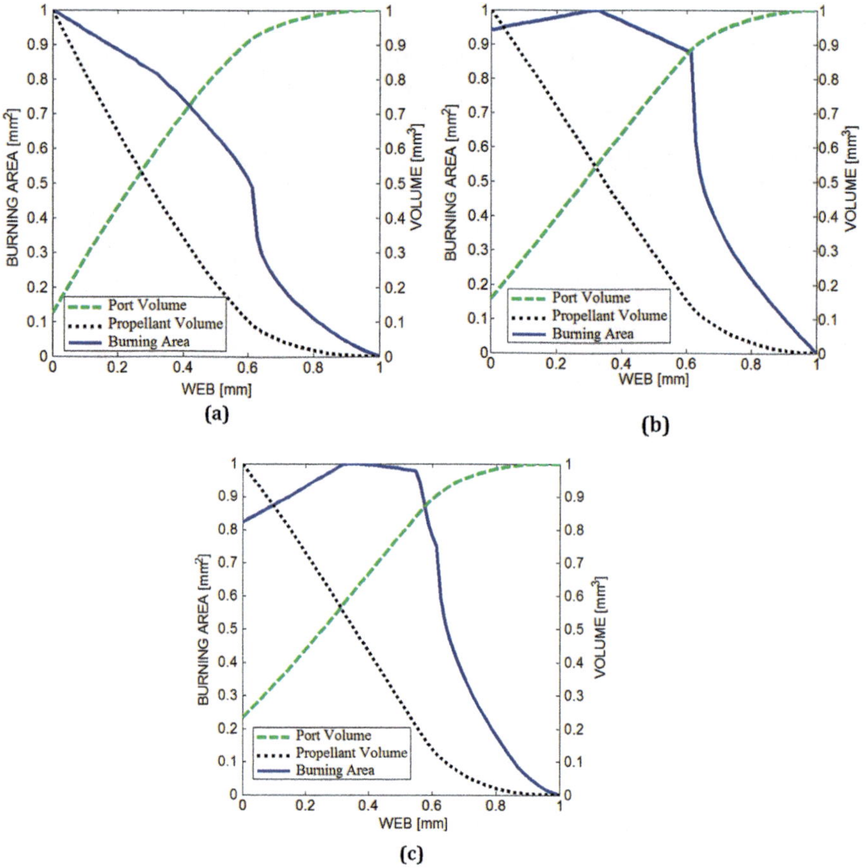

Fig. 5.21 Burning surface area, propellant volume, and port volume changes of grid slice regions. (**a**) Grid slice region 1. (**b**) Grid slice region 2. (**c**) Grid slice region 3

models. The results show that better accuracy can be achieved by using finer grids. The burnback analysis of the symmetrical model and the initial distance calculation method capabilities of the tool have been tested with a physical 3D propellant grain, and the results show that the computational time has been drastically reduced by those new capabilities. Furthermore, additional capabilities such as burnback simulation using SLP file format and the auto mesh interface of the burnback tool have been presented and implemented in the developed burnback tool to decrease the user interaction in burnback simulations. F3DBT code is validated using different grain geometries and the simulations are presented in comparison with CAD solutions. Different test cases have been examined for 3D burnback analysis, and F3DBT demonstrates its capability to perform 3D simulations of complex grain geometries. The current developed burnback tool is able to perform

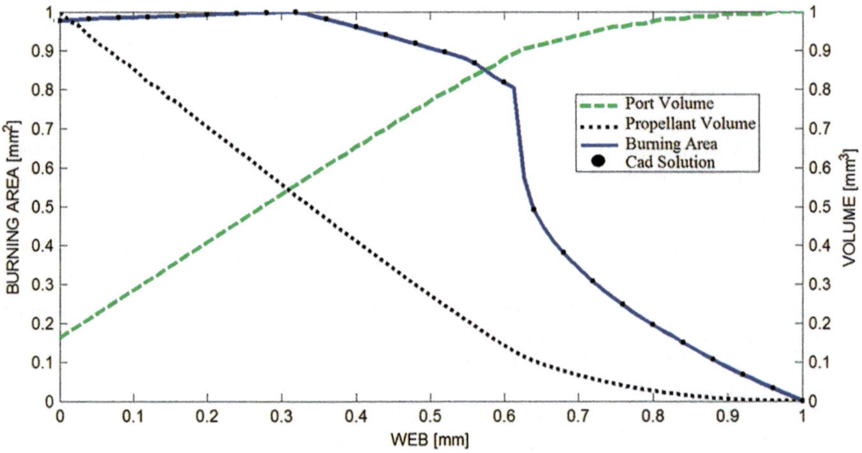

Fig. 5.22 Total burning surface area, propellant volume, and port volume versus web distance

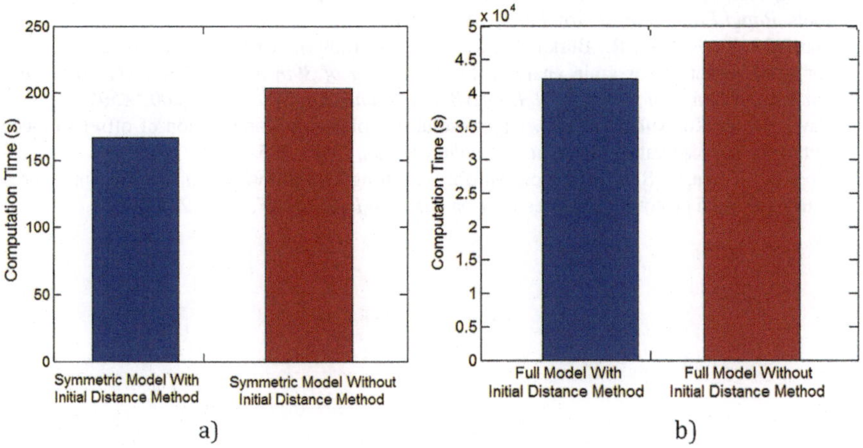

Fig. 5.23 Comparison of computational times for symmetric and full models. (**a**) Symmetric model. (**b**) Full model

accurate grain burnback analyses at low computation time and with less user interaction requirements.

References

1. Coats, D.E., Nickerson, G.R., Dang, A.L., Dunn, S.S., Kehtarnavaz, H. (1987). Solid Performance Program (SPP). In *Proceeding of 23rd AIAA/ASME/SAE/ASEE Joint Propulsion Conference San Diego, CA, U.S.A. June 29–July 2 1987, AIAA 1987-1701.*

2. Dunn, S.S., & Coats, D.E. (1997). 3-D grain design and ballistic analysis using the SPP97 code. In *Proceeding of 33rd AIAA/ASME/SAE/ASEE Joint Propulsion Conference, Seattle, WA, U.S.A., 1997, AIAA 97-3340.*
3. Zeller, B. (2005). Solid propellant grain design. In A. Davenas (Ed.), *Solid rocket propulsion technology* (1st English ed., pp. 35–84). New York: Pergamon Press.
4. Toker, K.A., Tınaztepe, H.T., Aksel, M.H., (2004). Three-dimensional propellant grain burnback calculations on tetrahedron mesh by fast marching method. In *Proceeding of 22nd Applied Aerodynamics Conference and Exhibit, Rhode Island, 16–19 August 2004, AIAA-2004-4960.*
5. Peterson, E.G., Nielson, G.C., Johnson, W.C., Cook, K.S., Barron, J.G. (1968). Generalized coordinate grain design and internal ballistics evaluation program. In *Proceeding of CPIA/AIAA 3rd Solid Propulsion Conference June 1968.*
6. Willcox, M.A., Brewster, M.Q., Tang, K.C., Stewart, D.S. (2005). Solid propellant grain design and burnback simulation using a minimum distance function. In *Proceeding of 41st AIAA/ASME/SAE/ASEE Joint Propulsion Conference & Exhibit, Tucson, Arizona, 10–13 July 2005, AIAA 2005-4350.*
7. Malosio, M., Pedrocchi, N., Tosatti, L. M. (2009). Algorithm to offset and smooth tessellated surfaces. In *Computer-aided design and applications* (pp. 351–363).
8. Qu, X., & Stucker, B. (2003). A 3D surface offset method for STL-format models. *Rapid Prototyping Journal, 9*(3), 133–141.
9. Stucker, B., & Qu, X. (2003). A finish machining strategy for rapid manufactured parts and tools. *Rapid Prototyping Journal, 9*(4), 194–200.
10. Hartfield, R., Jenkins, R., Burkhalter, J., Foster, W. (2003). A review of analytical methods for solid rocket motor grain analysis. In *Proceeding of 39th AIAA/ASME/SAE/ASEE Joint Propulsion Conference & Exhibit, Huntsville, Alabama, July 2003, AIAA 2003-4506.*
11. Pavić, D., & Kobbelt, L. (2008). High-resolution volumetric computation of offset surfaces with feature preservation. *Computer Graphics Forum, 27*(2), 165–174.
12. Koç, B., & Lee, Y. S. (2002). Non-uniform offsetting and hollowing objects by using biarcs fitting for rapid prototyping processes. *Computers in Industry, 47*, 1–23. 2002.

Part II
Advances in Aerodynamics

Chapter 6
A Prediction of Aerodynamics of Arbitrary Shape Non-fragmenting Space Debris During Decay Without Ablation

Sayavur I. Bakhtiyarov, Ramiz S. Gurbanov, Eldar T. Abdinov, and Nadir Yilmaz

Nomenclature

$\dot{\gamma}_w$	The wall shear rate
A	The reference area
C_D	The drag coefficient
C_L	The lift coefficient
Jo	The moment of inertia
l	Reference length
m	The mass of debris
Δp	Pressure drop
μ	Molecular viscosity
μ_T	Turbulent viscosity
ρ	The mass density of the fluid

S.I. Bakhtiyarov (✉)
Department of Mechanical Engineering, New Mexico Institute of Mining and Technology, Socorro, NM, 87801, USA
e-mail: sayavur.bakhtiyarov@nmt.edu

R.S. Gurbanov • E.T. Abdinov
Department of Theoretical Mechanics, Azerbaijan State Oil and Industry University, Baku, Azerbaijan

N. Yilmaz
Department of Mechanical Engineering, Howard University, Washington, DC, 20059, USA

T.H. Karakoç et al. (eds.), *Advances in Sustainable Aviation*,
DOI 10.1007/978-3-319-67134-5_6

6.1 Introduction

A safe operation of space assets requires knowledge of both the natural and man-made space environment and methods to predict the hazards to space systems in these environments. Notably, one environment of concern to users of space is the man-made orbital (space) debris environment. The growth and evolution of the space debris environment has been well documented. The danger to manned and unmanned space systems is obvious. Some specifics concerning that environment as they exist today:

- More than 22,000 objects larger than 10 cm are currently tracked by the US Space Surveillance Network. Only about 1000 (4.5%) of these represent operational spacecraft. The estimated population of particles between 1 and 10 cm in diameter is approximately 500,000. The number of particles smaller than 1 cm probably exceeds tens of millions.
- Most orbital debris resides within 2000 km of Earth's surface in low Earth orbit (LEO). Within this volume, the amount of debris varies significantly with altitude. The greatest concentrations of debris are found near 800–850 km.
- In LEO (below 2000 km), orbital debris circle the Earth at speeds of between 7 and 8 km/s. However, the average impact speed of orbital debris with another space object will be approximately 10 km/s. Consequently, collisions with even a small piece of debris involve considerable energy and have the potential to do considerable damage.
- The higher the altitude, the longer the orbital debris will typically remain in Earth's orbit. Debris left in orbits below 600 km normally fall back to Earth within several years. At altitudes of 800 km, the time for orbital decay is often measured in decades. Above 1000 km, orbital debris normally will continue circling Earth for a century or more.
- The Chinese antisatellite test (Fengyun, January 2007) and Iridium-Cosmos collision (February 2009, colliding at 11.7 km/s) nearly doubled the measurable environment.

By all measures, many now believe this environment has reached a critical point in certain LEO orbits, the so-named Kessler syndrome wherein the debris environment essentially becomes self-generating through debris-to-debris collisions within this belt (collision rate and debris creation higher than the natural debris atmospheric reentry removal rate).

NASA and the international space agencies have adopted guidelines and assessment procedures to reduce the number of nonoperational spacecraft orbiting the Earth. The most practiced method of post-mission disposal is to allow the reentry of the spacecraft from natural orbital decay or controlled entry. In order to hasten orbital decay, it is necessary to lower the perigee altitude so that atmospheric drag will cause the spacecraft to enter the earth's atmosphere more rapidly. However, in such cases, the surviving debris impact footprint cannot be guaranteed to avoid inhabited landmasses. Controlled entry normally is achieved by using more propellant with a larger propulsion system to cause the spacecraft to enter the atmosphere at a steeper flight path angle. The vehicle will then enter the atmosphere at a more precise latitude and longitude, and the debris footprint can be positioned over an uninhabited region, generally located in the ocean.

Numerous technological concepts have been developed and considered for active debris removal. For example, for small debris (smaller than 10 cm in diameter) in LEO, such concepts include the use of thin film or lasers (ground-based or space-based) to lower their orbit in order to ease their decay by reentry into the Earth's atmosphere. The most of the concepts for removal large debris are based on retrieval methods such as returning them to earth or recycling them in space. Therefore, a development of methods to predict an aerodynamics of arbitrary shape space debris is important aspect of the space debris removal process.

6.2 Method

A new engineering method is proposed to predict an aerodynamics of arbitrary shape non-fragmenting space debris without ablation. In this method, a complex variable method ("linearization of single-bonded area") was applied to derive a universal formula for velocity of arbitrary shape fragments. This technique allows describing the space fragments (debris) of various shapes, sizes, and masses. This method introduces a function

$$\psi = u + \frac{\Delta p}{4\mu l}\left(x^2 + y^2\right) \tag{6.1}$$

which can be transformed to Laplace equation with the boundary condition $\psi = \frac{\Delta p}{4\mu l}\left(x^2 + y^2\right)$ on the stream contours. A proposed technique has the following aerodynamics interpretation. The problem of determination of the debris velocities in the viscous fluid medium is reduced to the problem of determination of flow of the fluid around a prism with the same cross section area rotating with the angular velocity $\omega = \frac{\Delta p}{2\mu l}$. In order to find the function ψ (x, y) for arbitrary cross section area, an interior of this area in z plane ($z = x + iy$) is reflected into interior of the unique circle $\xi -$ plane ($\zeta = \xi + i\eta = \rho e^{i\theta}$) using the following power series:

$$z = a_0 + a_1\zeta + a_2\zeta^2 + \cdots + a_n\zeta^n + \ldots \tag{6.2}$$

Using a Schwarz's integral, the function $f(\zeta)$ is determined in a unit circle by the given real part on the contour [1–3]. Then taking into the account the boundary conditions, we are able to determine both a function $f(\zeta)$ and the flow function $\psi = \frac{1}{2}\omega z \cdot \bar{z}$:

$$\psi = \omega\left[\frac{1}{2}\sum\nolimits_{k=0}^{\infty} a_k^2 + \sum\nolimits_{k=1}^{\infty} \rho^k \cos\, k\theta \sum\nolimits_{s=0}^{\infty} a_s a_{s+k}\right] \tag{6.3}$$

Combining (6.1) and (6.3), we find the debris velocities in ζ-plane:

$$\psi = \frac{\Delta p}{2\mu l}\left[\frac{1}{2}\sum_{k=0}^{\infty} a_k^2\left(1-\rho^{2k}\right)+\sum_{k-1}^{\infty}\rho^k \cos\, k\theta \sum_{s=0}^{\infty} a_s a_{s+k}\left(1-\rho^{2s}\right)\right] \tag{6.4}$$

The distribution of velocities *u(x,y)* in *z* plane is constructed according to [4, 5]:

$$x=\sum_{k=0}^{\infty} a_k \rho^k, \qquad y=\sum_{k=1}^{\infty} a_k \rho^k \sin\, k\theta \tag{6.5}$$

The results of numerical calculations by Eq. 6.4 for the breakup fragments with various cross sections were validated by formula for geometrical shapes given in [6, 7]. An equivalent radius of the object of arbitrary shape was defined as [6, 7]:

$$r_e' = \frac{S^3}{4\pi^2 J_0 r_H} = \frac{2r_H}{\xi'} \tag{6.6}$$

where $\xi' = 8\pi^2 J_0/S\chi^2$ is the shape coefficient. For example, for the right *n*-angle shape:

$$\xi' = \frac{\pi^2}{3n^2}\left(1+3\cot^2\frac{180^\circ}{n}\right) \tag{6.7}$$

In the above-described method, the linear size characteristics of cross section of arbitrary shape objects can be determined by different ways, such as by the exact solution of Eq. 6.1, based on the Boussinesq's equation or using a *conform representation* method. In this research project, all *conform representation* method has been applied. The turbulent flow of air will be studied using a model of superposition of the molecular and turbulent viscosities [8–10]:

$$\overline{\tau}_w (1-y) = (\mu+\mu_t)\frac{du}{dy} \tag{6.8}$$

It will be assumed that the turbulent flow occurs by the two-layered Prandtl-Taylor structure model. The equivalent viscosity of fluid will be determined as: $\mu = \mu_{\text{equiv}} = \overline{\tau}_{\varpi}/\dot{\gamma}_w$. The flow resistance will be estimated by the Blasius formula [2, 3]. A comparison of the two-layered Prandtl-Taylor structure model with laminar Darcy and Blasius turbulent models is shown in Fig. 6.1.

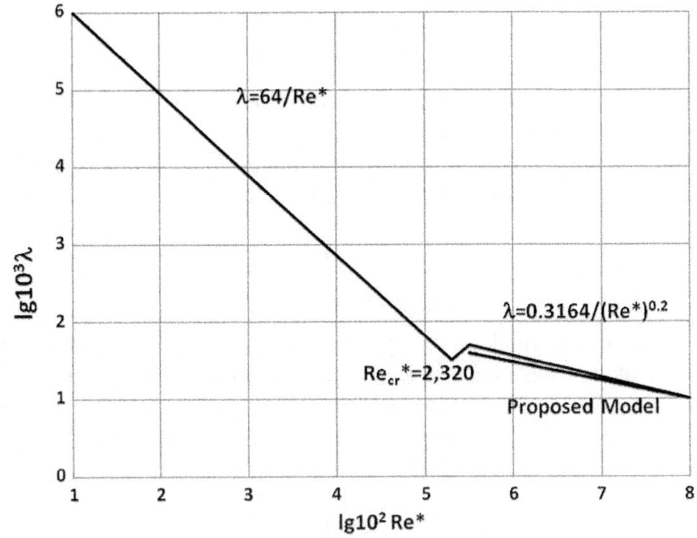

Fig. 6.1 A comparison of the proposed model with laminar Darcy and Blasius turbulent models [3]

6.3 Results and Discussion

The proposed technique was incorporated to the following debris motion equations in *2D* Cartesian coordinate system as:

$$m\frac{dv_x}{dt} = -F_{Dx} + F_{Lx} \tag{6.9}$$

$$m\frac{dv_y}{dt} = -G - F_{Dy} + F_{Ly} \tag{6.10}$$

where m is the mass of the projectile, $\vec{v}$ is the projectile velocity, $\vec{G}$ is the force of gravity (projectile weight), $\vec{F}_D$ is the drag force, $\vec{F}_L$ is the lift force, $\frac{d\vec{v}}{dt} = \vec{a}$ is an acceleration of the projectile, and t is a time. Drag and lift are two components of the aerodynamic force F_A. A drag force is the component of the surface force parallel to the flow direction. If we assume that the drag force is approximately proportional to the square of the projectile's speed relative to the air, then we can write:

$$F_{\mathrm{D}} = -\frac{C_{\mathrm{D}} A_{\mathrm{F}} \rho}{2} V^2 \tag{6.11}$$

where C_{D} is the drag coefficient, ρ is the mass density of the fluid, and A is the reference area (usually the face area A_{F}).

A lift force is the component of the aerodynamic force that is perpendicular to the approaching flow direction. It can be estimated using the following formula:

$$F_{\mathrm{L}} = \frac{C_{\mathrm{L}} A \rho}{2} V^2 \tag{6.12}$$

where A is the area normal to the lift force and C_{L} is the lift coefficient which is a function of the angle of attack, which is an angle between the fragment's reference line (chord) and the oncoming flow, Mach number and Reynolds number. Kuchemann [11] proposed an empirical relationship for predicting a lift force to drag force ratio:

$$\frac{F_{\mathrm{L}}}{F_{\mathrm{D}}} = \frac{4\,(M+3)}{M} \tag{6.13}$$

where M is the Mach number. The formula (6.13) was verified by wind tunnel tests, and it is considered quite accurate.

Equations 6.9 and 6.10 were solved numerically. A general algorithm of the simulations is presented in Fig. 6.2. Theresults of the numerical simulations are shown in Figs. 6.3, 6.4, and 6.5. A parameter $k = \sqrt{\frac{2mg}{C_D A \rho}}$ was used to present the simulation results.

6.4 Conclusions

Using a complex variable method ("linearization of single-bonded area"), a universal formula for velocity of arbitrary shape fragments was derived. This technique was incorporated with the projectile 2D motion equations and found to be a useful approach to predict trajectories of space fragments. A parameter $k = \sqrt{\frac{2mg}{C_D A \rho}}$ was introduced to present the simulation results. It is shown that the range of the landed debris significantly depends on the debris mass and drag coefficient, but it is less affected by the initial velocity.

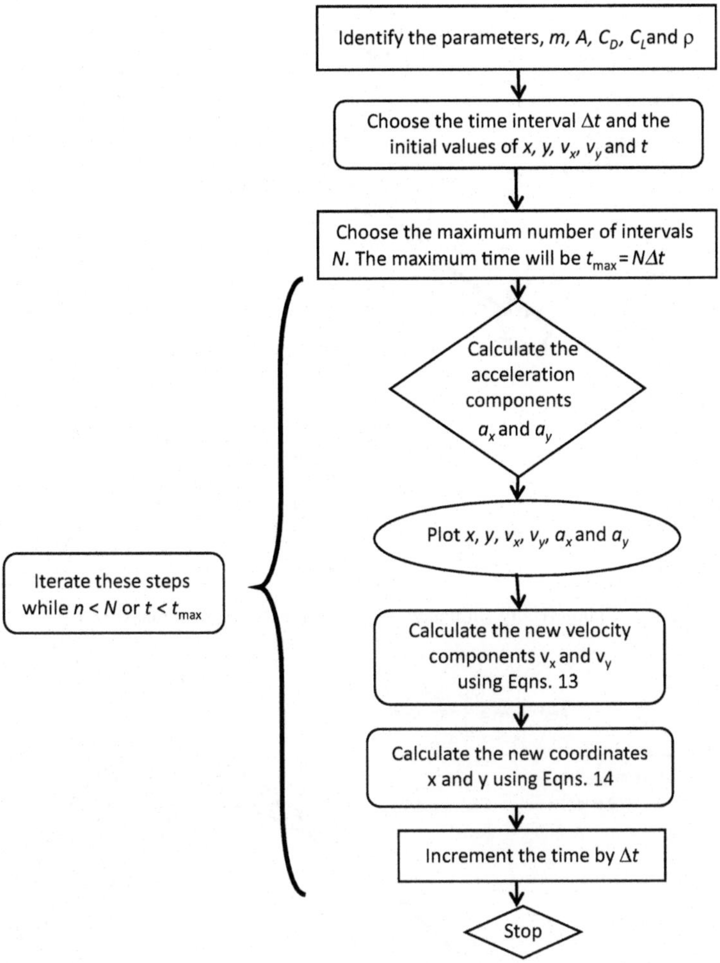

Fig. 6.2 General algorithm of numerical analysis

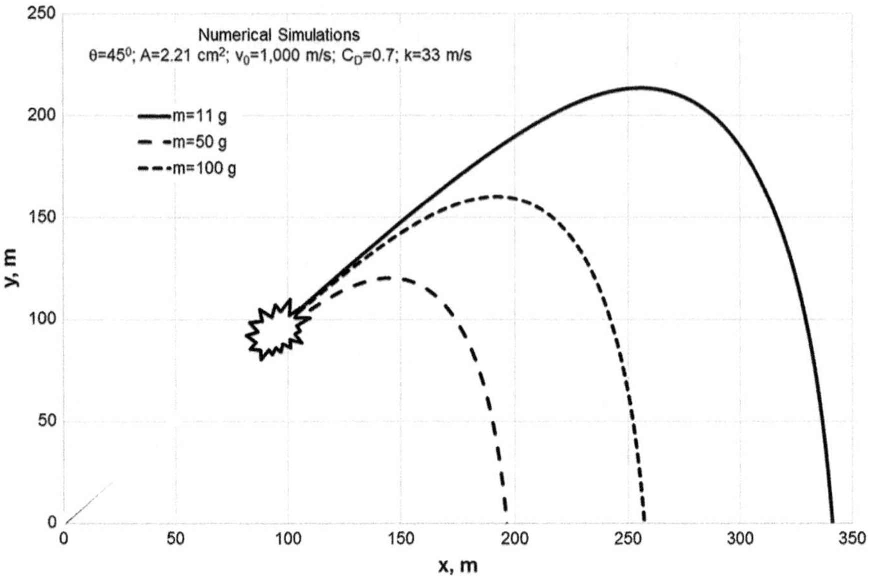

Fig. 6.3 Variation of numerically simulated trajectories with fragment mass

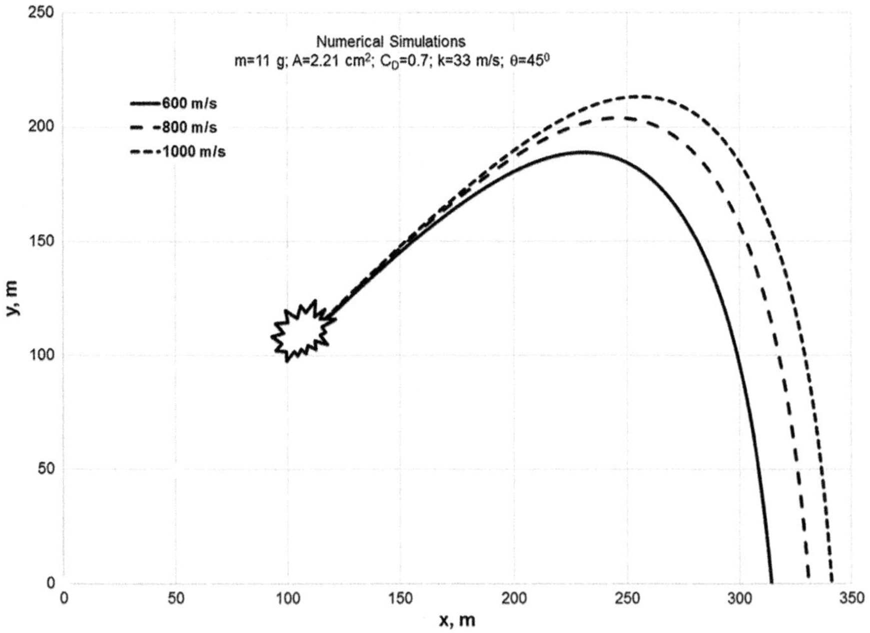

Fig. 6.4 Variation of numerically simulated trajectories with fragment velocities

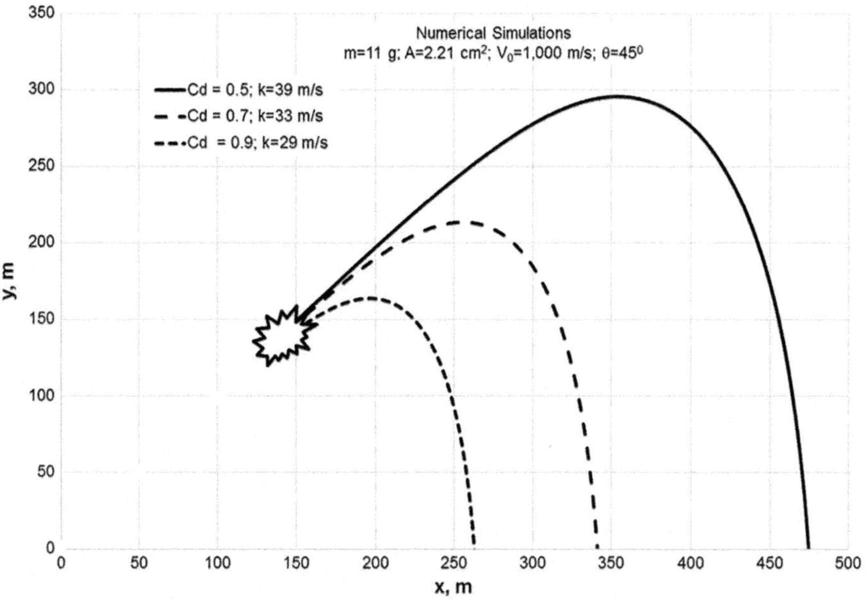

Fig. 6.5 Variation of numerically simulated trajectories with drag coefficient

References

1. Leibenzon, L. S. (1953). *Collected works* (Vol. 3, p. 680). Moscow: Academy of Sciences of USSR Publishers. (in Russian).
2. Leibenzon, L. S. (1947). *Flow of natural fluids and gases in porous medium* (p. 244). Moscow: Gostekhizdat. (in Russian).
3. Loitsyansky, L. G. (1973). *Mechanics of fluids and gases* (p. 848). Moscow: Nauka.
4. Mirzadjanzade, A. K. (1959). *Hydrodynamics of viscoplastic and viscous fluids in oil production applications* (p. 409). Baku: Azerneftneshr. (in Russian).
5. Lavtentyev, M. A., & Shabat, B. V. (1973). *Methods of complex variable functions theory* (p. 736). Moscow: Nauka. (in Russian).
6. Gurbanov, R. S., & Abdinov, E. T. (1987). A study of laminar flow in pipes and channels of arbitrary cross sections using a complex variable functions theory. *DAN Az SSR, 43*(3), 37–42.
7. Polya, G., & Szego, G. (1962). *Izoperimetric inequalities in mathematical physics* (p. 336). Moscow: Springer. (in Russian).
8. Dvoyeglazov, B. F., & Vachagin, K. D. (1974). Experimental study of dispersion systems flow in non-circular channels. *Proceedings of Kazan Institute of Chemical Engineering, 53*, 138–141.
9. Adler, Y. P., Markov, E. V., & Granovski, Y. V. (1971). *Design and optimization of experiments* (p. 284). Moscow: Nauka. (in Russian).
10. Schiller, L. (1936). *Fluids flow in pipes* (p. 230). Moscow: ONTI-NKTp. (in Russian).
11. Küchemann, D. (2012). *The aerodynamic design of aircraft, AIAA Education Series*. Reston: American Institute of Aeronautics & Astronautics.

Chapter 7
Performance of a Propeller-Driven Cargo Airplane

Kasım Biber

7.1 Introduction

An airplane has been designed to meet the needs of cost-effective and relatively short-range utility operations. It has a maximum takeoff weight of 8618 kg (19,000 lb) and payload of 3402 kg (7500 lb). The airplane has an un-pressurized cabin and single-pilot operation at a cruising altitude of 3050 m (10,000 ft) with a range over 250 nautical miles. Its propulsion system consists of a twin-pack engine driving a single propeller through a combining gearbox, resulting in a centerline thrust. Figure 7.1 shows the top and side views of the airplane. It is projected to be certified under FAR23 commuter category regulations as in [1].

For the preliminary design, wind tunnel tests were conducted on 1/12 scale model of this airplane. Stability and control characteristics were obtained for flaps up and deflected cases as presented in [2]. Based on the test results, improvements were made in airplane design.

For the present study, the airplane performance was predicted for takeoff, climb, cruise, and landing flight segments. The prediction was made in a custom-developed computer program. The program calls engine and propeller performance models and drag polar as input for flaps 0-, 15-, and 33-deg cases. The diameter ratio of fuselage to propeller is 0.65 for the cargo airplane, and, therefore, the drag polar was corrected in a subroutine to account for the effect of propeller slipstream as in [3]. Landing gear was extended down throughout the flight. From start to finish, the flight time, distance, and fuel flow were calculated in weight, altitude, and temperature loops. Within these programming loops, integration was performed to find the total flight time, distance, and fuel flow. The computer calculation was

K. Biber (✉)
Bartin University, Faculty of Engineering, Department of Mechanical Engineering,
Bartin 74100, Turkey
e-mail: kbiber@bartin.edu.t

T.H. Karakoç et al. (eds.), *Advances in Sustainable Aviation*,
DOI 10.1007/978-3-319-67134-5_7

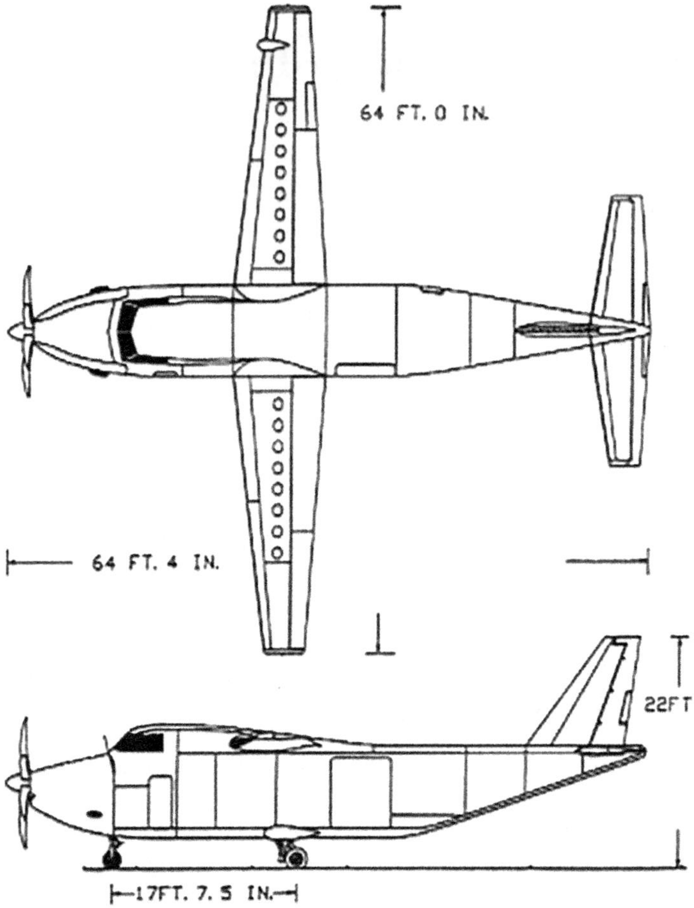

Fig. 7.1 Airplane top and side views

checked against single-point hand calculations. Among others, [4] provides the performance equations used for the calculations in the present analysis.

7.2 Takeoff

The dual-pack engine operation provides maximum shaft horsepower (SHP) of 2700 during takeoff. However, in case one engine fails, the other engine has the capability to provide enough power to complete the mission.

Takeoff has flight segments, described for flight test certification of airplanes, as in Federal Aviation Regulations in the United States presented in [1]. It is a combination of takeoff distance and takeoff flight path. Takeoff distance consists

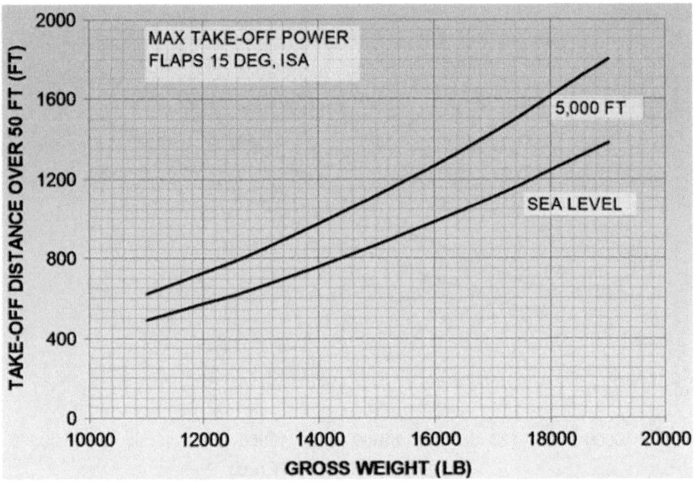

Fig. 7.2 Takeoff distance variation with gross weight at sea level and 5000 ft altitudes. Maximum takeoff power at ISA and flaps at 15°

of a ground run and a climb to 50-ft height. Takeoff flight path includes acceleration at 400-ft height and a final climb to 1500-ft height. Figure 7.2 shows takeoff distance variation with gross weight at sea level and 5000 ft altitudes. The data is presented for standard day temperature (ISA) and maximum takeoff engine power. The lift-off speed is $V_{LOF} = 1.1V_{S1}$ and climb-out speed $V_2 = 1.2V_{S1}$. The stall speed, V_S, is calculated with flaps 15°. For maximum takeoff weight of 19,000 lb, the takeoff distance is 1381-ft at sea level and 1804-ft at 5000-ft altitudes.

Balanced field length was also calculated. It is considered as a balance between the distance to continue the takeoff following the recognition of an engine failure at V_1 speed and the distance required to stop if the takeoff should be aborted. For the example airplane, the engine failure speed, V_{EF}, is also considered as a propeller control system failure speed, most critical to producing thrust, considering all single point failures. This speed is related to the stall speed, and it must not be less than $1.05V_{S1}$ or greater than $1.2V_{S1}$. This is under some special provisional conditions of various FAR 23 Aviation rules, due to the novel and unusual design feature of the airplane. Figure 7.3 shows the balanced field length variation with gross weight at sea level and 5000 ft altitudes. For One Engine Inoperative (OEI) condition, the engine has a special rating called the APR. Unlike the normal takeoff with both engines operational, the balanced field length was calculated to clear 35-ft screen height. It is 1928 ft at sea level and 2456-ft at 5000-ft altitude for maximum takeoff weight.

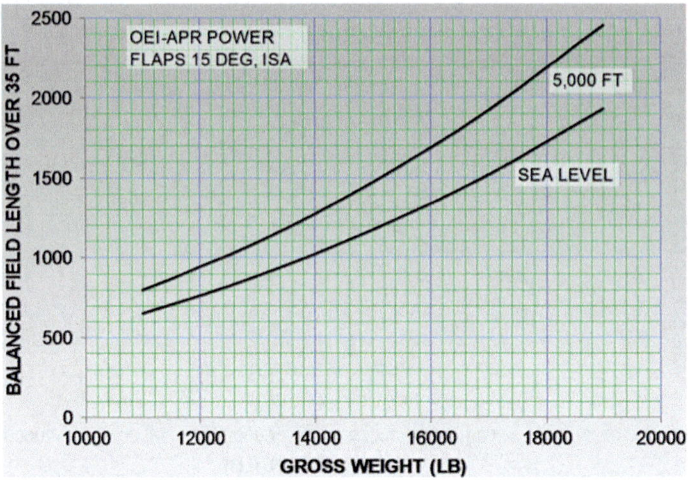

Fig. 7.3 Balance field length variation with gross weight at sea level and 5000 ft altitudes. Takeoff and APR power at ISA and flaps at 15°

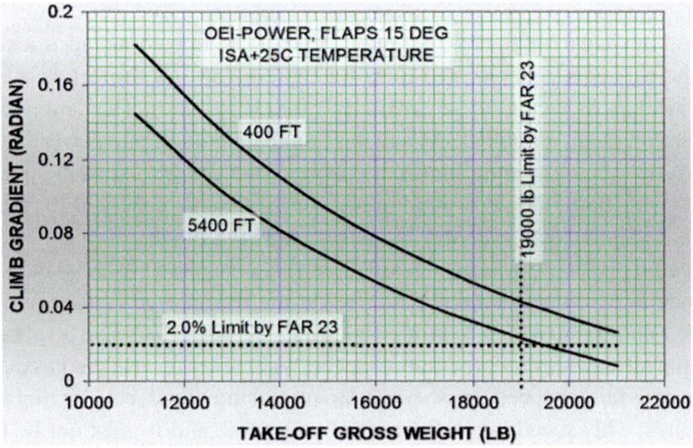

Fig. 7.4 Second segment climb gradient variation with gross weight for 400 and 5400 ft altitudes. OEI power at ISA+25 °C and flaps at 15 deg

7.3 Climb

The second segment climb gradient is considered critical for flying capability of the airplane. It is determined at 50 ft screen height when the speed $V_2 = 1.2V_{S1}$. Figure 7.4 shows the variation of second segment gradient with gross weight at 400 and 5400-ft altitudes with OEI at ISA+25C temperature. For its maximum weight, it is clearly shown that the airplane meets and exceeds the FAR 23 requirement of 2% gradient at 5400-ft altitude.

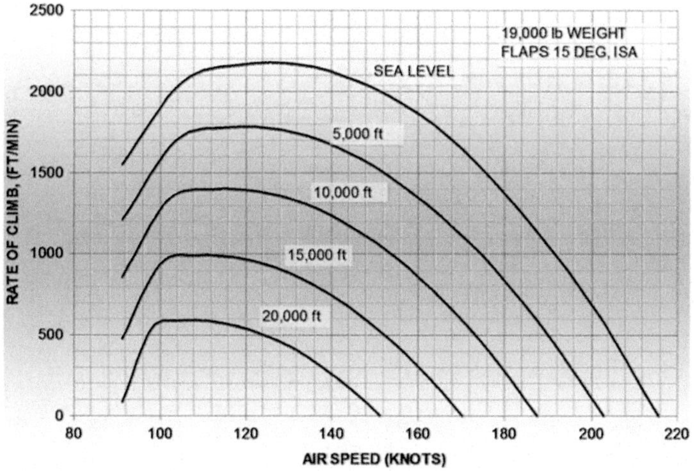

Fig. 7.5 Rate of climb variation with airspeed at altitudes from sea level to 20,000 ft. Climb power at ISA and flaps up

Climb performance was determined for altitudes from sea level to 20,000-ft altitude. Climb engine power was used. Figure 7.5 shows the rate of climb variation with airspeed at ISA temperature and maximum takeoff weight with flaps up. Maximum rate of climb 2163 ft/min occurs at sea level, when the airspeed at 130 knots. It decreases with increase of altitudes, as expected.

7.4 Cruise

Cruise performance was determined at an altitude of 10,000 ft with a typical cruise speed of 165 knots. The power was set to cruise rating. Fuel flow data was obtained from engine performance to calculate specific range data, as in Fig. 7.6 for various weights.

Figure 7.7 shows the payload-range chart for 600 gal of fuel. The airplane is initially considered to have its maximum payload of 7500 lb with no fuel on board. The fuel is added to so that the required range of 250 knots can be flown. At the end of the mission range, the airplane has reached its maximum allowable ramp or taxi weight. To increase the range beyond the mission value, it is necessary to start trading payload for fuel. When the range becomes 1034 nautical miles, the traded payload is equal to 2722 lb, corresponding to 600 gal of fuel. For further increase of range, 4778 lb of payload should be offloaded. This results in a marginal increase of range that makes the total range equal to 1082 nautical miles.

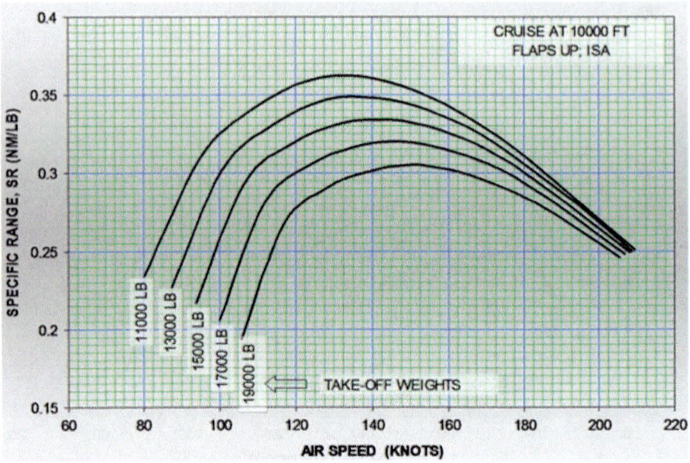

Fig. 7.6 Cruise-specific range variation with airspeed at airplane weights from 11,000 to 19,000 lb. flaps up and ISA conditions

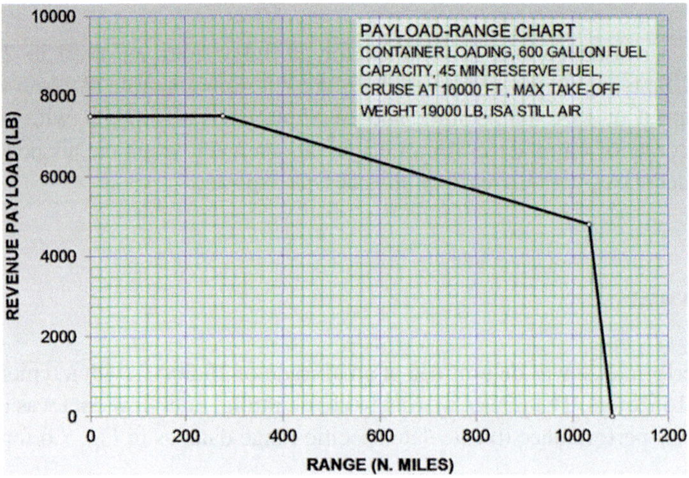

Fig. 7.7 Payload-range chart for 600 gal of fuel. Cruise at 10,000-ft altitude with 165 knots. Weight 19,000 lb with container loading and 45-min reserve. Cruise power at ISA and flaps up

7.5 Landing

Landing distance was calculated from 50-ft height to a full stop on the ground. It consists of air and ground distances. The approach speed at 50 ft height, $V_3 = 1.3V_{S1}$. The touchdown speed on the ground, $V_{TD} = 1.15V_{S1}$. The stall speed V_{S1} was calculated with flaps down 33°. Figure 7.8 shows the variation of landing distance with gross weight at sea level and 5000-ft altitudes. For the maximum weight, the distance is 2128-ft at sea level and 2393-ft at 5000-ft altitude.

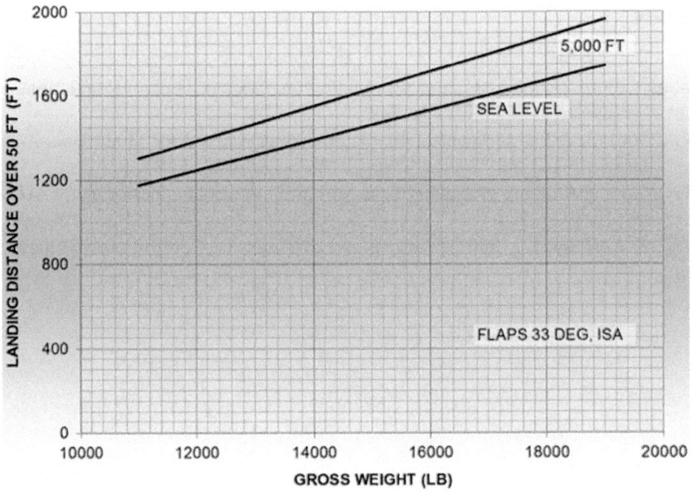

Fig. 7.8 Landing distance variation with gross weight at sea level and 5000 ft altitudes. Idle power at ISA and flaps 33°

Table 7.1 Summary of airplane performance results for dual-pack engine power

Performance parameter	English	SI
Maximum takeoff weight	19,000 lb	8618 kg
Maximum landing weight	19,000 lb	8618 kg
Empty weight (containerized freight)	9000 lb	4082 kg
Maximum cruise speed at 10,000 ft (3048 m), ISA	205 kts	380 km/h
Normal cruise speed at 10,000 ft (3048 m), ISA	180 kts	334 km/h
Maximum range cruise at 10,000 ft (3048 m), ISA	153 kts	283 km/h
Takeoff distance over 50 ft (12 m), 19,000 lb, ISA	1525 ft	465 m
Landing distance over 50 ft (12 m), 19,000 lb, ISA	1745 ft	532 m
Maximum rate of climb at sea level, 19,000 lb, ISA	1861 ft/min	567 m/min
Rate of climb at 5000 ft, 19,000 lb, ISA	1721 ft/min	525 m/min
Rate of climb at 5000 ft, 19,000 lb, ISA+20 °C	1661 ft/min	506 m/min

7.6 Conclusion

Flight performance of a FAA FAR23 commuter airplane equipped with a dual-pack engine driving a single propeller through a unique combining gearbox is presented as summarized in Table 7.1. The analysis includes the use of a custom-developed computer program that includes engine and propeller performance models with airplane drag polar. With maximum takeoff weight of 19,000 lb, it is shown in the analysis that the airplane meets and exceeds 2% requirement of second segment climb gradient at APR engine rating and ISA+25 °C temperature. High-performance flap system allows short takeoff and landing distance.

References

1. Anonymous. (2000). Federal Aviation Administration, Code of Federal Regulations 14, Parts 1 to 59, Part 23, Revised as of January 1, 2000.
2. Biber, K. (2006). Stability and control characteristics of a new FAR23 airplane. *AIAA Journal of Aircraft, 43*(5), 1361–1368.
3. Biber, K. (2011). Estimating propeller slipstream drag on airplane performance. *AIAA Journal of Aircraft, 48*(6), 2172–2174.
4. Lan, C. E., & Roskam, J. (1980). *Airplane aerodynamics and performance*. Ottawa: Roskam Aviation.

Chapter 8
Numerical Investigation of an S-Duct Diffuser at Different Inlet Boundary Conditions

Samet Aslan and D. Funda Kurtulus

Nomenclature

AIP	Aerodynamic interface plane (duct engine face)
DC (θ)	Steady circumferential distortion parameter
H	Height [m]
K	Thrust variation to PR ratio
L	Length [m]
M	Mach number
$\dot{m}$	Mass flow rate [kg/s]
P	Total pressure [Pa]
p	Static pressure [Pa]
PR	Pressure recovery
q	Dynamic pressure [Pa]
s	Centerline length [m]
SFC	Specific fuel consumption [kg/s*N]
T	Temperature [K]
X	Engine thrust [N]

S. Aslan
TEI TUSAS Engine Industries, 26003, Eskisehir, Turkey

Aerospace Engineering Department, METU, 06800, Ankara, Turkey

D.F. Kurtulus (✉)
Aerospace Engineering Department, METU, 06800, Ankara, Turkey
e-mail: kurtulus@metu.edu.tr

T.H. Karakoç et al. (eds.), *Advances in Sustainable Aviation*,
DOI 10.1007/978-3-319-67134-5_8

Subscripts

∞	Settling chamber
AIP	Engine face (AIP) conditions
amb	Ambient, outlet conditions

Greek

Θ	The lowest average sector angle
$\eta_{\sigma i}$	The incompressible flow ratio

8.1 Introduction

Air is required by the air-breathing engines and it is supplied by an inlet or intake. The outer flow goes over the inlet and its presence influences the outer flow structure. Furthermore, the internal flow goes through the inlet, and it influences directly the engine performance and stability [1].

The duty of an engine inlet is described in General Electric Company *Installation Handbook for Turbojet Engines* [2], as "This process must be accomplished with the least possible loss in total-pressure or head, with the best attainable flow distribution, and with the least amount of drag."

An inlet is supposed to recover most of the available freestream total pressure at the compressor face in order to achieve higher thrust levels. Moreover, to prevent surge and stability issues, at the aerodynamic interface plane, the flow conditions are expected to be as uniform as possible. Variation of the flow conditions from uniformity is described as distortion [1, 3, 4]. The inlet pressure recovery has direct impact on engine thrust as

$$\frac{\Delta X}{X} = K\frac{\Delta P}{P_{\infty}} \tag{8.1}$$

where X is engine thrust and K is a coefficient that depends upon the particular engine.

Some of the inlet total pressure is lost due to wall friction, turbulent mixing, and shock waves inside the duct. Adverse pressure gradient causes flow separation and turbulent mixing [1, 3].

Circumferential distortion has critical impact on compressor stability. The operating point is found to move toward the surge line as total pressure distortion increases [4–6].

Main causes of total pressure distortion are flow separations, which may occur at the lip or sidewalls, and wakes. Due to high diffusion rates, boundary layer may thicken and finally separate at the inner side of the bends. Adverse streamwise pressure gradient appears because of increasing cross-sectional area, and this situation may end up with flow separation. At the separation regions, local flow conditions change dramatically. Moreover, during the supersonic flights, shock and boundary layer interactions create distortion at the inlet [5–7].

Turning of the flow through the bends results in centrifugal forces, and the centrifugal forces form secondary flows. The magnitude of the pressure gradient at the first bend occurs as a result of the centrifugal force proportional to $\rho u^2/R$. Thus, pressure at the outer region of the bend becomes higher than the pressure at the inner region. As a result, this pressure gradient creates secondary flows toward inside turn. The low-momentum boundary layer fluid is transferred from outside toward inside turn by these resulting secondary flows. Hence, at the inner side of the first bend, a low total pressure region appears. Although the same phenomena are expected to reverse that distortion through the second turn, at the end of the first turn, much of the low-momentum fluid is located at the outside of the second bend. Therefore, the strong pressure gradient that expected to reverse the situation does not appear. As a result, secondary flows are formed because of the bends, and they lead to large flow distortion at AIP [5, 6] (Fig. 8.1).

In an experimental study, Wellborn et al. [8] have investigated compressible flow through a serpentine inlet at Mach = 0.6 and Re = 2.6×10^6. The symmetrical flow structure has been observed about the geometrical symmetry plane. At the end of the bends, streamwise flow separation regions have been caught. Secondary flows appeared as a result of the duct curvature, and a pair of counterrotating vortices due to those secondary flows has been observed at the AIP. The low-momentum fluid of the boundary layer has been found to be conveyed by these vortices toward the center of the duct.

In a numerical study, Harloff et al. [9] investigated compressible flow within a diffusing S-duct and a circular-to-rectangular transition duct. PARC3D CFD code was used to solve 3D RANS equations, and Baldwin-Lomax and k-ε turbulence models were used for closure. Although the computed total and surface static pres-

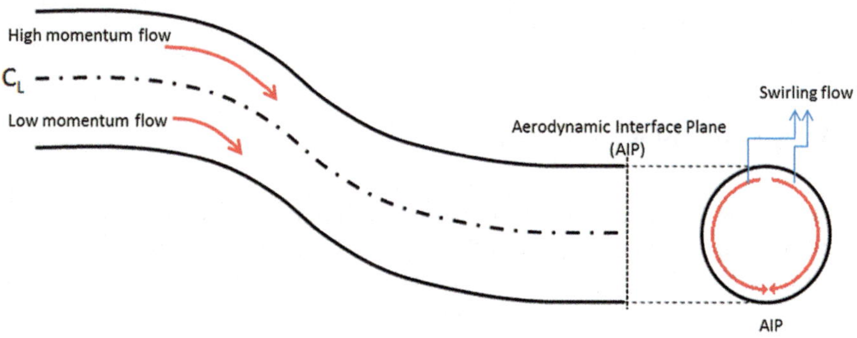

Fig. 8.1 Effect of centrifugal forces

sure field were accurate for the transition duct, in order to predict strong cross flow, separated boundary layer turbulence modeling is needed as an enhancement. Results of the S-duct computations were similar to experimental results; however, the length and angular extend of the boundary layer separation were underpredicted with both of the turbulence models. Furthermore, in order to compute strong cross flow and separated boundary layer, more advanced turbulence models were also needed.

With the same turbulence models, Harloff et al. [10] have shown that transverse velocities were predicted to be higher than the experimental transverse velocities in the region of the counterrotating vortices and less than the experimental transverse velocities in the core flow region.

Wellborn et al. [11, 12] demonstrated a large 3D region of separated flow within a serpentine inlet. It is seen that the separated region was blocking the cross section of the flow; thus, increased velocities were observed at the rest of the cross-sectional area. Larger shear stress levels and larger total pressure losses occurred because of the increased velocity. Furthermore, strong pressure-driven secondary flows are induced because of the duct curvature, and they evolved into two counterrotating vortices. Low-momentum boundary layer flow is convected to the center by the use of these vortices. PARC3D code was used to solve 3D RANS equations. Turbulent structure was modeled using Baldwin-Lomax and k-ε turbulence models. Calculations showed cross flow total pressure nonuniformity and separation regions. The reattached location found in computations was at the further position than the experimentally obtained. Furthermore, analyses predicted the amount of low-momentum flow that is transferred by the cross flow is less than the experimental value. It is stated that in order to overcome this issue, more advanced turbulence models are needed.

8.2 Geometry

Numerical simulations are compared with the experimental study that is conducted at TEI, which is investigation of an S-duct diffuser. The diffuser had been designed for a micro turbojet engine-powered aircraft. The experiments related to the present study were conducted in an open-circuit, atmospheric pressure, and low-speed subsonic blowout wind tunnel [13].

8.3 S-Duct Inlet

Design speed of the S-duct inlet is around Mach 0.6. The geometry was manufactured using rapid prototyping method. Figure 8.2 shows the schematic view of the S-duct inlet, and the geometrical parameters of the S-duct diffuser are listed in Table 8.1. At the designed flight conditions, the selected engine is expected to suck air with a mass flow rate of about 0.7 kg/s.

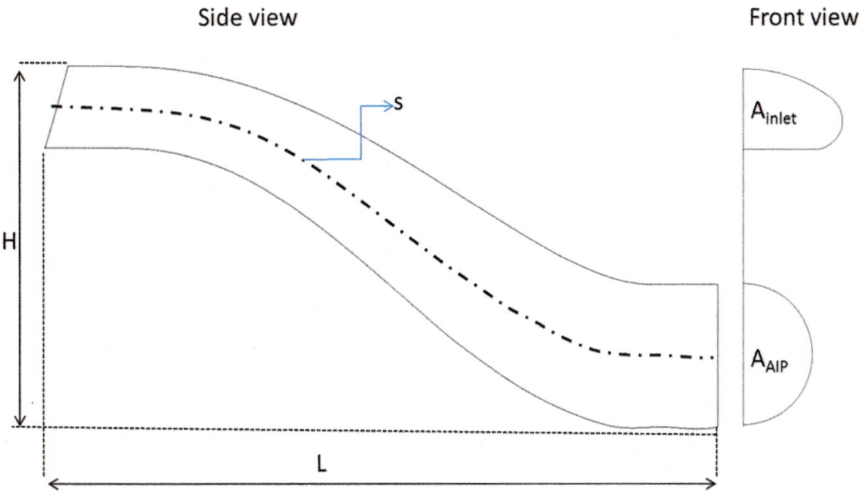

Fig. 8.2 Schematic view of the S-duct inlet

Table 8.1 Geometrical parameters of the S-duct model

Quantity	Value	Description
A_{inlet}/A_{AIP}	0.86	S-duct diffusion ratio
H/L	0.53	S-duct height to length ratio
s/L	1.08	S-duct centerline to length ratio

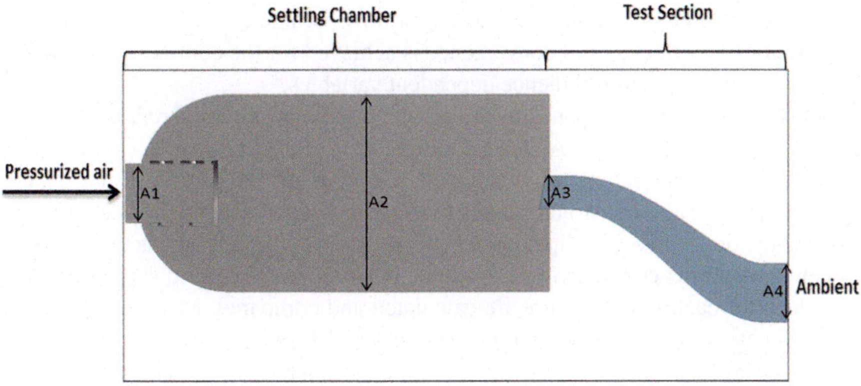

Fig. 8.3 Schematic view of the test setup [13]

8.4 Test Section

While the outlet section is left free to the atmosphere, inlet side is mounted impermeable by a flange to the wind tunnel settling chamber. Figure 8.3 shows the schematic view of the test section.

Table 8.2 Area ratios for the test section

Quantity	Value
A1/A2	0.088
A2/A3	13.40
A3/A4 (=$A_{\mathrm{inlet}}/A_{\mathrm{AIP}}$)	0.86

Table 8.3 Mesh size and y^+ relation

Case	Turbulence model	Mesh size	y^+ (mean)
1	Realizable k-ε	656736	18
2	Realizable k-ε	1449600	4.28
3	Realizable k-ε	3172820	1.45
4	Realizable k-ε	4682400	0.35

The cross-sectional areas of the settling chamber and the S-duct diffuser are called as A1, A2, A3, and A4. The ratios of those cross-sectional areas are listed in Table 8.2.

8.5 Numerical Approach

Numerical simulation of serpentine inlet involves the application of conservation laws of the mass (continuity), momentum, and energy for the discretized control volume domain. Each conservation equation is solved at every finite volume iteratively until the desirable convergence in flow parameters is achieved. Steady-state solution is pursued until convergence is achieved for the continuity, x, y, and z momentum, energy, and turbulence dependent variables.

3D Navier-Stokes equations are solved with the help of commercially available general-purpose CFD code ANSYS Fluent. Since the geometry is symmetrical, finite volume computations are conducted at only one half portion of the intake. Computational grid is constructed using the unstructured tetrahedral cells with prism layers at the duct walls. No slip boundary condition is applied at the duct walls. Grid refinement study is conducted using coarse, medium, and fine grid structures. The y^+ values for each grid structure are calculated and compared. Mean y^+ values of all computations along the walls are shown in Table 8.3.

It is found that the calculated aerodynamic parameters do not change dramatically for the given mesh sizes. Nonetheless, the fine mesh structure, which is about 4.7 M cells, is used for the rest of the computations. This mesh size is chosen because of the smallest $y+$ values are achieved for this condition, so that it gives more reliability for near wall solutions (Fig. 8.4).

Pressure-based solver is used for turbulent flow field survey. Four different RANS turbulence models, which are one-equation model Spalart-Allmaras, two-equation models Realizable k-ε and k-ω SST, and seven-equation model Reynolds stress model, are used to overcome closure issue. SIMPLE algorithm is chosen for the pressure-velocity coupling. For spatial discretization, all the variables except pressure, second-order upwind scheme is used, and for the pressure, PRESTO! scheme is preferred.

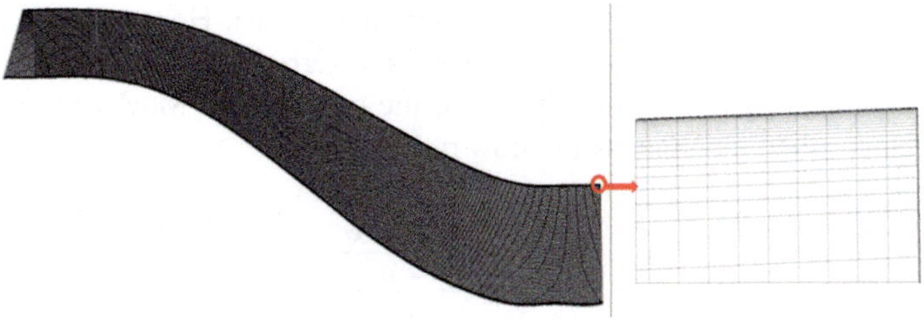

Fig. 8.4 Mesh structure

At the final step of this study, the influence of the inlet boundary condition on flow structure is investigated. From the previous analyses, it is known that lip separation occurs at the entrance of S-duct. In order to see how this lip separation influences the AIP profile, an additional hypothetical case is conducted using uniform profile as inlet boundary condition, which attains the specified mass flow rate. By this way, the relation between the lip separation and the engine interface plane profile is surveyed.

8.6 Data Reduction

The data in the current study are given in non-dimensional form. Mean values of aerodynamic results are presented. Pressure coefficient is given by Eq. 8.2:

$$C_{\mathrm{p}} = \frac{P - P_{\mathrm{amb}}}{P_{\infty} - P_{\mathrm{amb}}} \tag{8.2}$$

Here, P_{∞} is the total pressure at the plenum chamber, P is the pressure at the measurement point, and P_{amb} is the ambient (atmospheric) pressure.

8.7 Pressure Recovery

While the flight Mach number is above 0.5, the inlet operates as a diffuser such as decelerating the flow and increasing static pressure. Considering this process, the pressure recovery term is used to mention the efficiency of the intake. However, this definition is irrelevant at other Mach values, and therefore total pressure ratio is used instead of static pressure ratio to define the pressure recovery. Due to boundary layer shear stress, separation and shock waves; total pressure decreases across the intake.

In the light of the above definition, the pressure recovery (PR) is formulated as:

$$\mathrm{PR} = \frac{P_{\mathrm{AIP}}}{P_{\infty}} \tag{8.3}$$

where P_{AIP} is the mean total pressure at the engine face. However, this equation becomes irrelevant since as Mach number goes to zero that definition goes to unity. Therefore, another parameter is defined for the cases where Mach number is near zero, which is called incompressible flow ratio:

$$\eta_{\sigma i} = \frac{P_{\mathrm{AIP}} - p_{\infty}}{P_{\infty} - p_{\infty}} \tag{8.4}$$

where p_{∞} is the freestream static pressure [3].

8.8 Compatibility Features

Although pressure recovery is an important parameter to define the efficiency of the inlet, there are other aspects of the flow influenced by the inlet and important for the engine operation such as distortion. Different aircraft engine manufacturers found their own distortion descriptors for that purpose. Rolls-Royce developed one of the most widely used descriptor which is used to describe the steady circumferential distortion [3]. Steady circumferential distortion parameter DC(θ) is defined as given in Eq. 8.5, and θ sector definition for distortion coefficient is presented in Fig. 8.5:

$$\mathrm{DC}\,(\theta) = \frac{P_{\mathrm{AIP}} - P_{\theta}}{q_{\mathrm{AIP}}} \tag{8.5}$$

This parameter is obtained by measuring the difference between the engine face or AIP average total pressure (P_{AIP}) and the lowest average total pressure in any sector defined by a critical angle of θ (P_{θ}), and then by dividing this difference by the average dynamic pressure at the engine (q_{AIP}) face (Fig. 8.5).

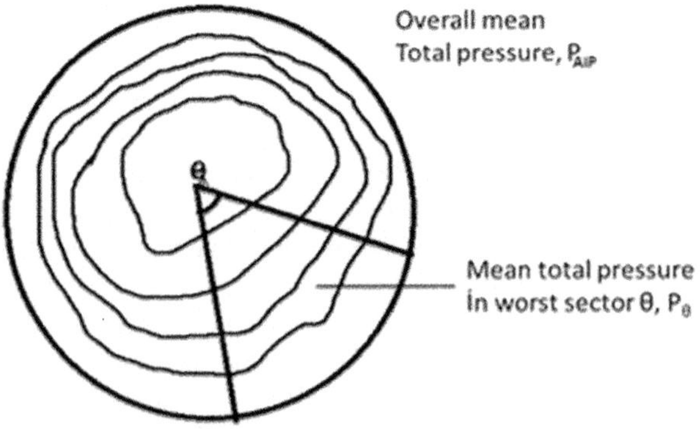

Fig. 8.5 Illustration of total pressure contours and θ sector for definition of distortion coefficient

The lowest pressure region is found after tracing all of the AIP area with sectors of θ as shown in Fig. 8.5. Afterward, the lowest average sector values found are put into the into the Eq. 8.5 [4]. It is stated that the sector angle value must be significant enough that the compressor blades feel the impact of distortion, thus flow reaches steady state [3]. A commonly used coefficient is DC(60), and in the current study, the results with $\theta = 60°$ will be given. Hereafter, DC(60) will be the total pressure steady-state distortion descriptor [6].

8.9 Results

Realizable k-ε turbulence model is used for mesh dependency calculations because of its certain mathematical background and ability to give satisfactory results for many engineering fluid mechanics problems. After deciding the mesh structure, different turbulence models are used for computations and their capabilities are compared. Furthermore, numerical studies are also compared with the hypothetical analysis with uniform inlet mass flow boundary condition in comparison with the numerically obtained inlet velocity profile.

8.9.1 Grid Refinement Study

Results of grid resolution study and corresponding y^+ values are listed in Table 8.4. Figures 8.6 and 8.7 show the variation of top and bottom wall static pressure across the S-duct length for different grid structures. Figures 8.6 and 8.7 show that the coarsest grid results vary from the others. Especially, the top wall static pressure distribution at the medium RKE and fine RKE solutions almost coincide with each other. It can be stated that the computations are not influenced by the mesh structure after 3 M mesh size. However, the rest of the analyses are conducted using 4.5 M mesh size to achieve the lowest y^+ values and a better near wall solutions.

Figure 8.8 demonstrates symmetry plane streamlines and outlet pressure recovery contours for different mesh sizes.

The general flow structure is seen similar for all four cases. Reverse flow due to lip separation and also separation regions after the bends is seen at the symmetry plane streamlines. Outlet PR contours show high values at the bottom side and low values near the top of the wall. Top low-pressure region is expected due to the

Table 8.4 Grid refinement study analyses properties

Cases	Turbulence model	Mesh size	y+ (mean)
1	Realizable k-ε	656736	18
2	Realizable k-ε	1449600	4.28
3	Realizable k-ε	3172820	1.45
4	Realizable k-ε	4682400	0.35

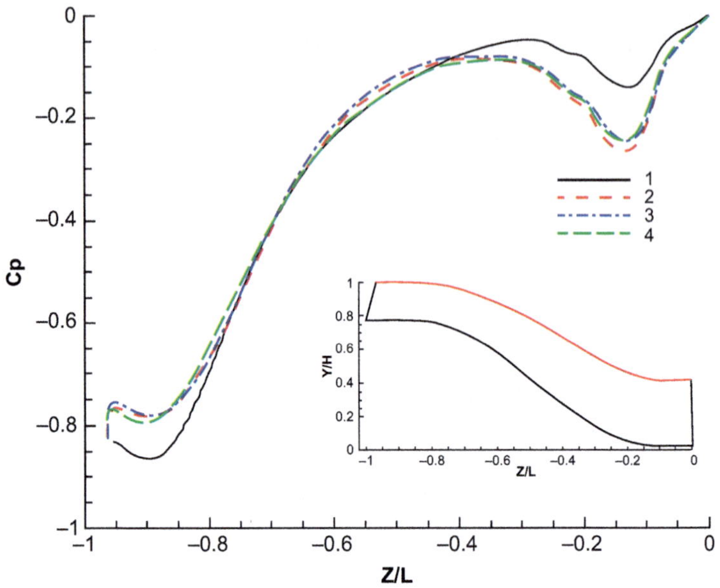

Fig. 8.6 Top wall static pressure through the duct for different mesh sizes (Cases 1–4)

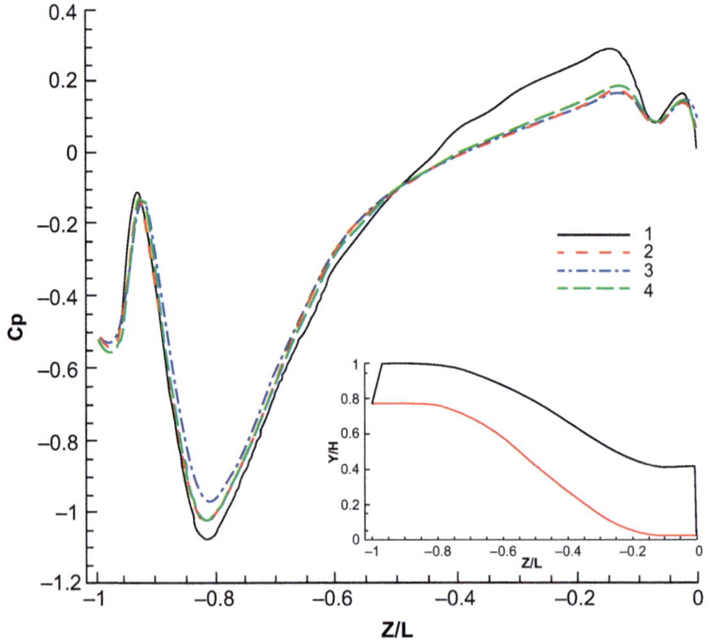

Fig. 8.7 Bottom wall static pressure through the duct for different mesh sizes (Cases 1–4)

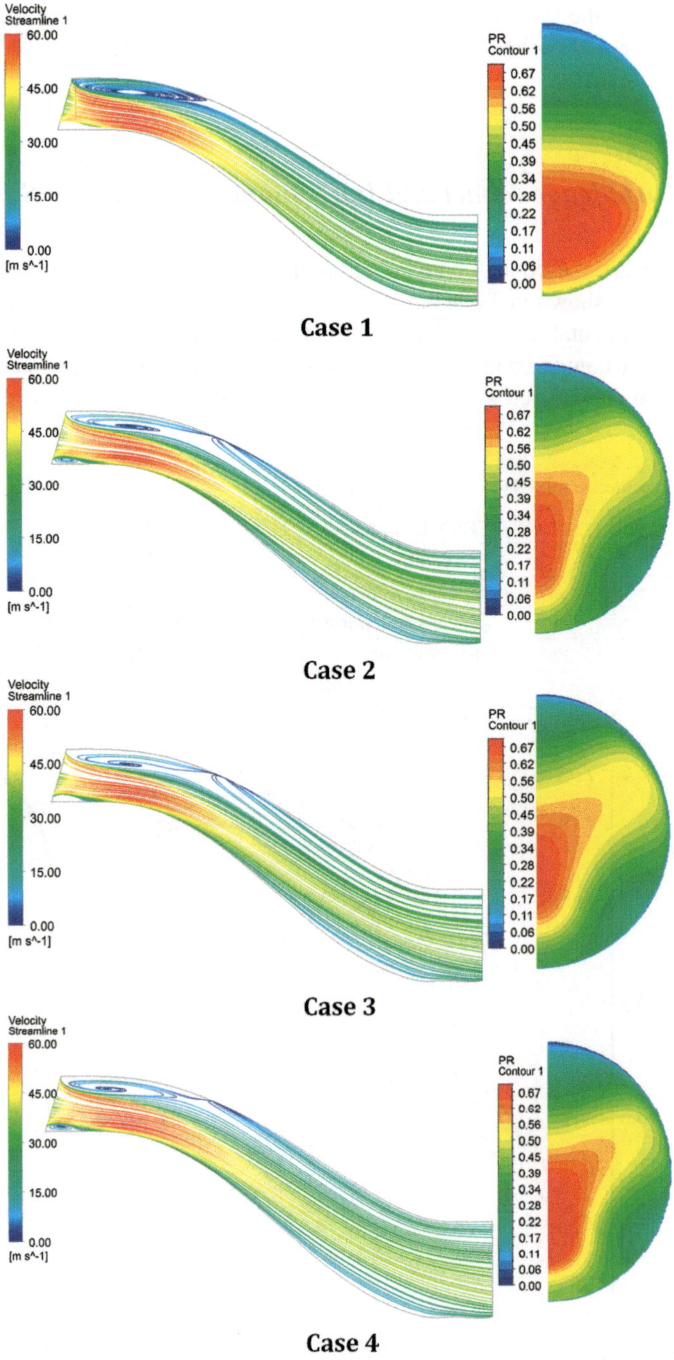

Fig. 8.8 Symmetry plane streamlines and outlet PR contours for different mesh sizes (Cases 1–4)

separation after the second turn. Another low-pressure region appears at the right bottom section and pushes high-pressure zone upward. This zone appears because of the swirling flow.

8.9.2 *Turbulence Model and Inlet Boundary Condition Study*

The fine mesh is used to compare different turbulence models, and the cases investigated are shown in Table 8.5.

The experimental top wall C_p distribution and the computed top wall C_p distribution are compared in Fig. 8.9. Theshown numerical results are obtained from studies that are conducted with different turbulence models and uniform inlet profile

Table 8.5 Turbulence models

Case	Turbulence model	Mesh size
4	Realizable k-ε	4682400
5	Spalart-Allmaras	4682400
6	Shear stress transport	4682400
7	Reynolds stress model	4682400
8	Realizable k-ε (uniform mass flow inlet BC)	4682400

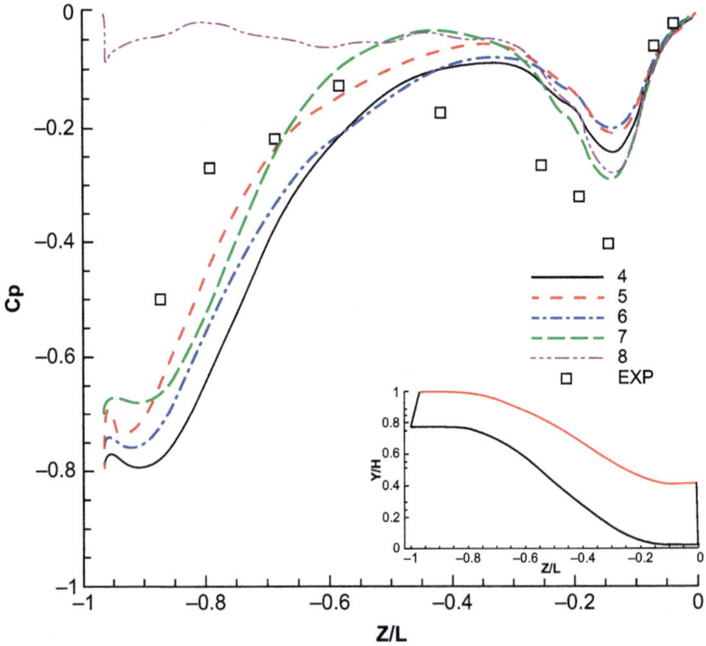

Fig. 8.9 Top wall static pressure through the duct for different turbulence models and BCs

boundary condition. It is seen that all of the computations, except Case 8 with uniform boundary condition, show similar C_p variation with experiment through the duct length. Lip separation causes lower pressure at the entrance region, and it increases gradually through the reattachment zone. Afterward, due to the separation through the second bend, C_p decreases again. However, computations overpredict the impact of the lip separation and underpredict the impact of the second turn separations. Although, local minimum pressure position is captured correctly, local maximum C_p position is shown at more upstream position than the experiment. The closest local maximum C_p location is captured by Case 7. The result of uniform inlet profile case (Case 8) is obviously different from the other computations and also from the experiment. This proves that the lip separation at the entrance exists, and it causes a pressure drop which does not appear for the hypothetical uniform entrance boundary condition (Fig. 8.9). However, it is noted that the pressure drop through the second turn is seen also for this case. Numerical results for the bottom symmetry plane show similar results. Pressure drops due to lip separation, first bend separation, are remarkable (Fig. 8.10).

Figure 8.11 demonstrated the symmetry plane streamlines and outlet pressure recovery contours for different turbulence models.

Figure 8.11 shows that for the uniform inlet profile condition, neither the symmetry plane separations and nor the AIP PR variation are strong. On the

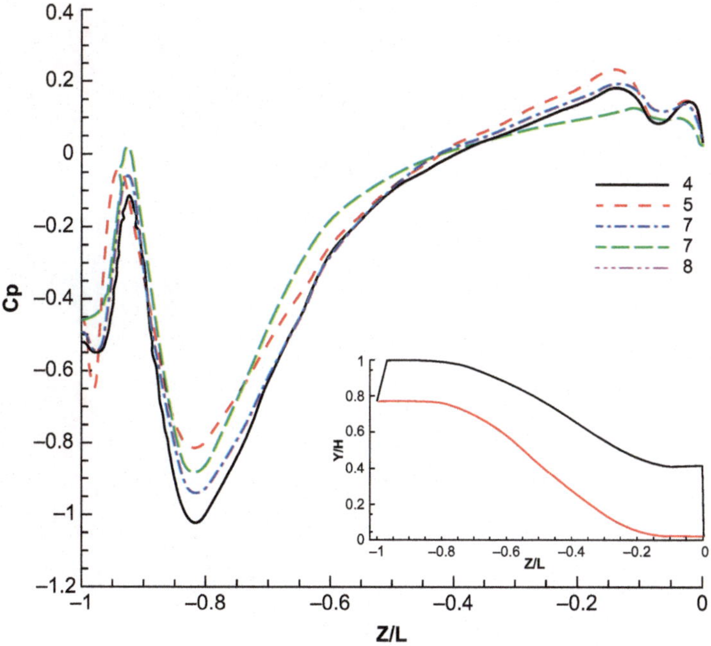

Fig. 8.10 Bottom wall static pressure through the duct for different turbulence models and BCs

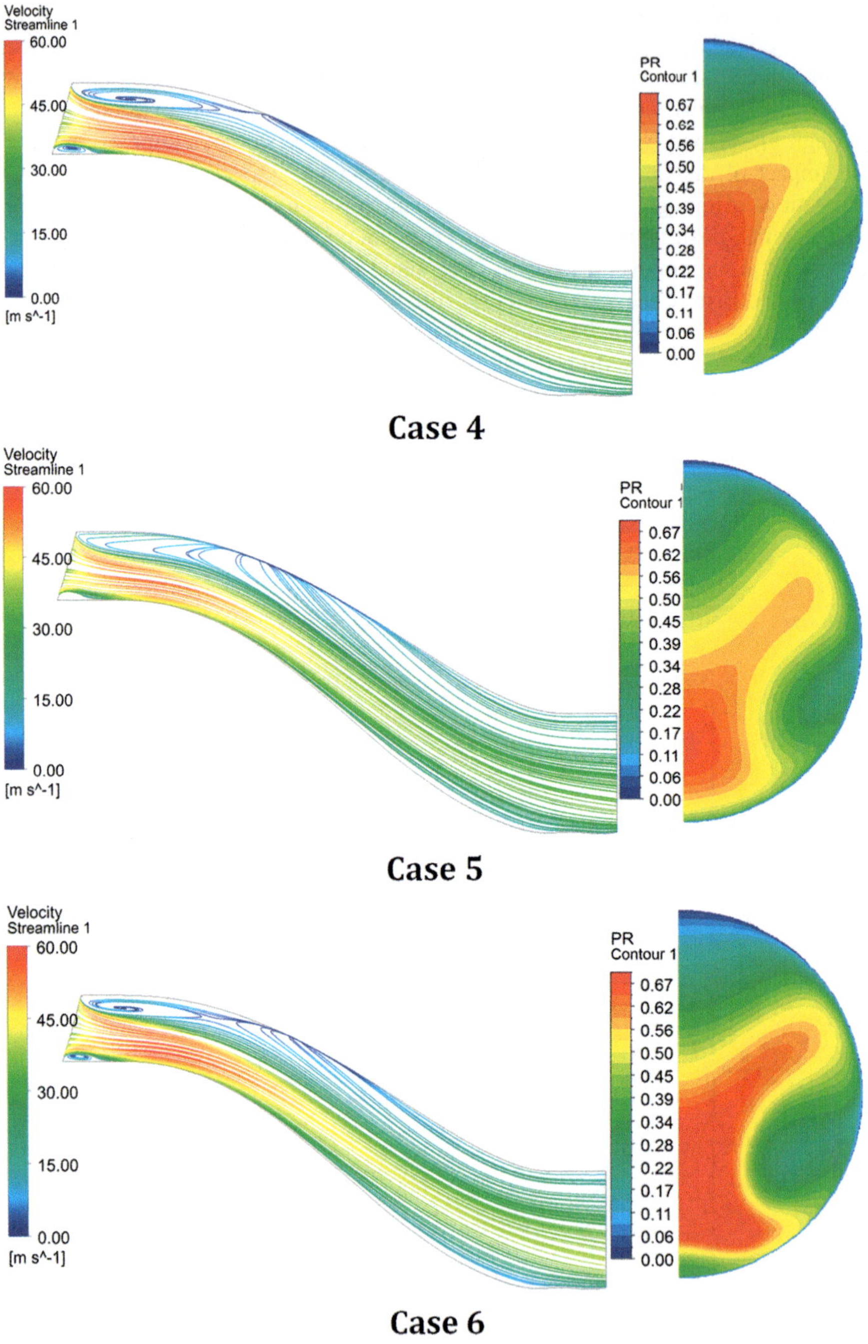

Fig. 8.11 Symmetry plane streamlines and outlet PR contours for different turbulence models and BCs (see Table 8.5 for the case numbers 4–8)

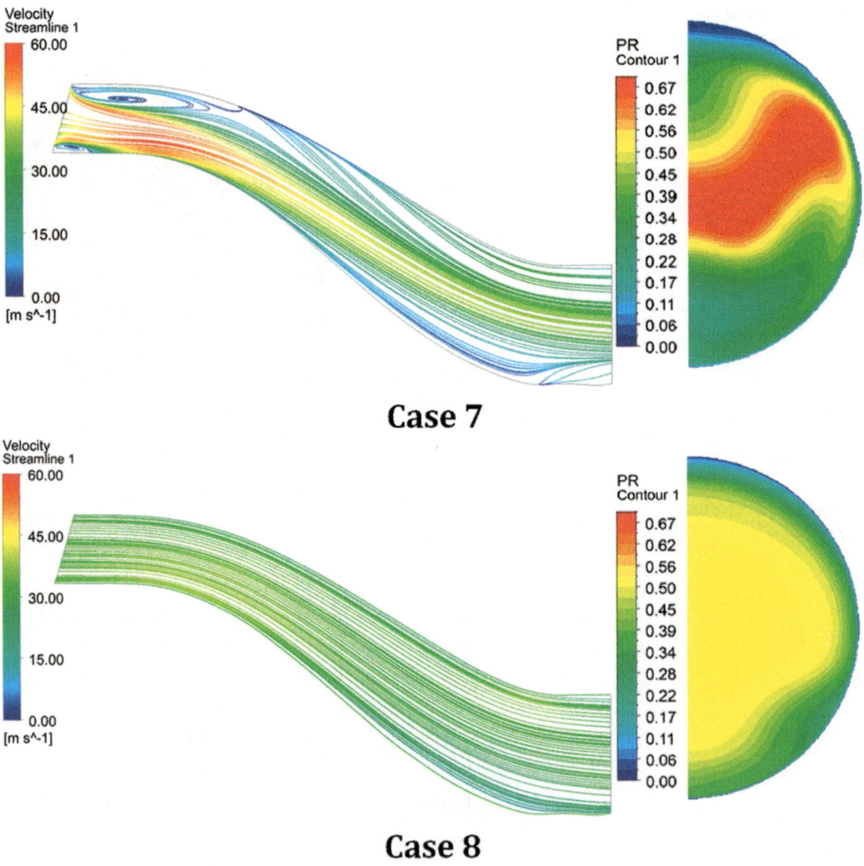

Fig. 8.11 (continued)

contrary, for the other cases where the inlet condition is lip separated flow structure, outlet pressure distortion is obvious. Hence, it can be concluded that unless the inlet flow is separated, adverse pressure gradient through the bends is not that much effective on flow structure. Also, the influence of the vortices related to secondary flow on the AIP flow structure is weak. Even though all of the turbulence models are successful to capture the lip separation and reattachment regions, only the Reynolds stress model indicated accurately the detachment at the bottom side due to swirl.

Figure 8.12 illustrates pressure recovery (PR) at the outlet plane obtained from the experiment.

The experimental pressure recovery contour reveals similar distribution with the RSM computations Case 7. However, RSM model underpredict the PR values at the lowest PR locations and overpredict the PR values at the highest PR locations. It can be stated that RSM model capture the flow structure, but it overpredicts the effect of the separations.

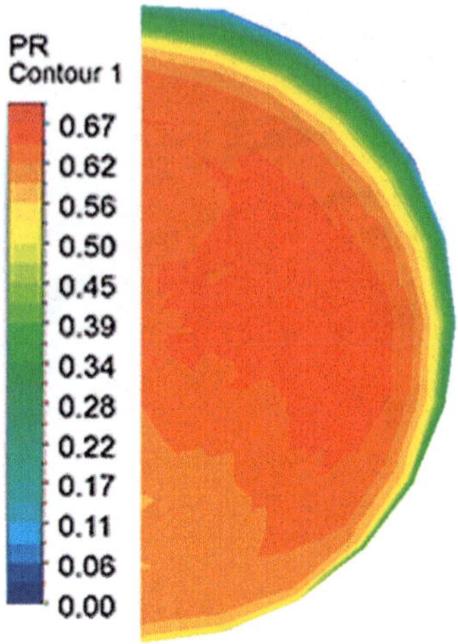

Fig. 8.12 Experimental pressure recovery contour at the outlet plane

Table 8.6 Performance parameters

Case	$\dot{m}$ (kg/s)	PR	$\eta_{\sigma,i}$	DC(60)
1	0.121	0.9944	0.459	0.235
2	0.119	0.9943	0.442	0.110
3	0.119	0.9943	0.441	0.104
4	0.119	0.9943	0.449	0.150
5	0.119	0.9943	0.447	0.170
6	0.119	0.9945	0.462	0.219
7	0.119	0.9944	0.455	0.146
8	0.121	0.9944	0.455	0.034
Experiment	0.121	0.9954	0.583	0.043

Table 8.6 showsthe calculated values of the area averaged parameters over AIP from different numerical simulations and from the experiment.

It is seen that numerical analyses give sufficient results for wall statics pressure values. On the other hand, computed performance parameters are away from the experimental data. It was stated that the outlet PR profile computed using RSM turbulence model was close to the experimental one: however, gradients were different. It is also seen that the experimental DC60 calculation is much smaller than the numerical values.

8.10 Conclusion

In this study, flow structure in a diffusing S-duct is numerically investigated. Analyses are conducted to simulate a previously experimental study [13]. Ansys Fluent 14 commercial CFD code is used to solve 3D Navier-Stokes equations. Influence of the turbulence models on computations has been surveyed. Moreover, the entrance flow structure impact on the internal flow has been investigated.

Lip separation is seen at the entrance of the duct. Also, streamwise flow separation regions were caught through the bends of the diffuser. Flow separation is one of the main contributors of pressure loss with secondary flows. Secondary flow structures are observed as a result of spatial pressure variations. Pressure loss and distortion are seen at the aerodynamic interface plane. It is found that the lip separation strengthens the effect of the wall separations inside the duct and increases the AIP distortion drastically.

Numerical results approach the experimental wall pressure values considerably well. Also, the effect of wall separation and secondary flows at the AIP is caught by the analyses. Reynold stress model gives the closest results to the experimental measurement obtained. However, numerical analyses overpredict the pressure loss and pressure gradient across the AIP.

Acknowledgments The authors would like to thank TEI – TUSAS Engine Industries for the help for conducting the experiments. They also would like to thank Ender Hepkaya and Sefa Yilmazturk for their support.

References

1. Seddon, J., & Goldsmith, E. (1985). *Intake aerodynamics: An account of the mechanics of flow in and around the air intakes of turbine-engined and ramjet aircraft and missiles*. London: Collins.
2. General Electric Company Flight Propulsion Division. (1951). *Installation handbook for turbojet engines*. Evendale: Aircraft Gas Turbine Division, General Electric.
3. Goldsmith, E. (1993). *Practical intake aerodynamic design*. Washington, DC: American Institute of Aeronautics and Astronautics.
4. Bissinger, N. C. and Breuer, T. (2010). Basic principles – gas turbine compatibility – intake aerodynamic aspects. Encyclopedia of Aerospace Engineering. John Wiley & Sons, Ltd. Chichester, West Sussex, U.K.
5. Harrison, N., Anderson, J., Fleming, J., & Ng, W. (2013). Active flow control of a boundary layer-ingesting serpentine inlet diffuser. *Journal of Aircraft, 50*(1), 262–271.
6. Anabtawi, A. J. (1999). Experimental investigation of boundary layer ingestion into diffusing inlets (Doctoral dissertation). Retrieved from ProQuest Dissertations and Theses.
7. Reid, C. The response of axial flow compressors to intake flow distortion. ASME Paper No. 69-GT-29, March 1969.
8. Wellborn, S. R., Reichert, B. A., Okiishi, T. H. An experimental investigation of the flow in a diffusing S-duct. 28th Joint Propulsion Conference and Exhibit. 6–8 July 1992, Nashville, Tennessee, AIAA-92-3622.
9. Harloff, G. J., Reichert, B. A., Wellborn, S. R. Navier-stokes analysis and experimental data comparison of compressible flow in a diffusing S-duct. AIAA Paper 92-2699, 1992.
10. Harloff, G. J., Reichert, B. A., Wellborn, S. R. Navier-stokes analysis and experimental data comparison of compressible flow in a diffusing S-duct. 13th International Conference on Numerical Methods in Fluid Dynamics. 6–10 July 1992, Rome, Italy, AIAA-92-3622.
11. Wellborn, S. R., Reichert, B. A., Okiishi, T. H. A study of the compressible flow through a diffusing S-duct. Washington, DC: National Aeronautics and Space Administrations, NASA TM-10641, December 1993.

12. Wellborn, S. R., Reichert, B. A., & Okiishi, T. H. (1994). Study of the compressible flow in a diffusing S-duct. *Journal of Propulsion and Power, 10*(5), 668–675.
13. Aslan, S., Yilmazturk, S., Kurtulus, D. F. Aerodynamic investigation of a serpentine inlet designed for a micro-turbojet engine powered aircraft. 1st International Symposium of Sustainable Aviation. 31 May – 3 June 2015, Istanbul, Turkey, ISSA-2015-129.

Chapter 9
Numerical Investigation of a Serpentine Inlet Validated with Experimental Results for Different Turbulence Models

Samet Aslan, D. Funda Kurtulus, Ender Hepkaya, and Sefa Yilmazturk

Nomenclature

AIP	Aerodynamic interface plane (duct engine face)
H	Offset [m]
L	Length [m]
M	Mach number
$\dot{m}$	Mass flow rate [kg/s]
p	Static pressure [Pa]
P	Total pressure [Pa]
PR	Pressure recovery
q	Dynamic pressure [Pa]
s	Centerline length [m]
T	Temperature [K]

9.1 Introduction

Serpentine inlets, reducing the aircraft observability and mass and increasing storage of the fuel tanks, are mostly used by defense industry for missile and military

S. Aslan (✉)
TEI TUSAS Engine Industries, 26003, Eskisehir, Turkey

METU Aerospace Engineering Department, 06800, Ankara, Turkey
e-mail: samet.aslan@tei.com.tr

D.F. Kurtulus
METU Aerospace Engineering Department, 06800, Ankara, Turkey
email: kurtulus@metu.edu.tr

T.H. Karakoç et al. (eds.), *Advances in Sustainable Aviation*,
DOI 10.1007/978-3-319-67134-5_9

purposes. Although the constraints for their unique geometry, the popularity is still increasing in aerospace industry.

Its importance begins with the preparation of the engine working conditions, so it has a vital role for the engine performance throughout flight.

The performance of the engine depends on the flow entering through the inlet. The flow has to be smooth, and the design has to provide high-pressure recovery so that it can produce high trust regarding minimum losses.

Moreover distortion is another subject affecting the performance of the engine which is the variation of the expected uniform flow from real case. The parameter of the change in this flow movement and angle is defined as swirl.

Boundary layer separation and secondary flows are the main flow characteristics investigated in a serpentine inlet. Both experimental and numerical studies show that these two phenomena affect the inlet performance at the compressor face.

An experimental study (Wellborn et al. [1]) conducted at Mach $= 0.6$ and $\mathrm{Re} = 2.6 \times 10^6$ shows that after the bend of the serpentine inlet, flow separations occur. Due to the curvature of the duct, secondary flow changes into counter-rotating eddies at the AIP which convects low momentum fluid of the boundary layer toward the center of the duct.

A numerical study on compressible flow inside a serpentine duct with cross-sectional transition, circular to rectangular, is investigated by Harloff et al. [2]. PARC3D CFD code is used to solve 3D RANS equations with two different turbulence models, namely Baldwin-Lomax and k-ε. According to the results, total and surface static pressure has a good agreement with the experiments. However, turbulence models were not enough to predict strong cross flow and separated boundary layer. It is concluded that enhancements are needed to predict the length and angle of the boundary layer separation. Another similar work done by Harloff et al. [3] shows that turbulence models predict the flow in the region of counter-rotating eddies in contrast to the flow in the core.

Wellborn et al. [4, 5] investigated the separated flow in the 3D region of a serpentine inlet both numerically and experimentally. Due to the flow separation, a blockage is formed in the S-duct which increases the core velocities when passing over the separated flow region. This core velocity increase result in larger shear stress levels and larger total pressure losses. Due to duct curvature, secondary flows turn into counter-rotating eddies. The numerical results show that the reattachment for the flow is obtained further than experimental results. This proved that development in turbulence models is needed to solve this complex flow.

An experimental study made by Aslan et al. [6] on a S-duct inlet designed for a micro-turbojet engine powered aircraft shows that flow separation causes total pressure loss because of the characteristics of the flow.

The objective of the current study is to find out flow characteristics such as separation, swirl, and secondary flows of a serpentine inlet using different turbulence models and compare them with experimental results. Distortion and pressure loss are the parameters to define the performance of the models at aerodynamic interface plane.

9.2 Case Definition

The RAE M2129 experimental diffuser that was designed by Royal Aircraft Establishment was used in numerical validation studies. The experimental data and the model were provided by QinetiQ. This data set was generated after a large number of experiments, and in many numerical investigations, the results of these experiments have been used to validate CFD codes. The experiments were carried out in RAE Bedford (UK) wind tunnel. This tunnel has a cross section of 13 × 9 ft and serves as a closed-circuit, low-speed facility at atmospheric test conditions. In order to obtain a flow by pressure difference between the inlet-outlet, an ejector is placed at the rear of inlet. The experiments were performed at the free stream conditions of 0.21 Mach. The data includes mean quantities on the duct walls at the engine face [7, 8].

9.2.1 M2129 Diffuser Geometry

Geometry of the S-duct diffuser is shown in Fig. 9.1. Here, AIP to throat section area ratio is defined as $A_{AIP}/A_{throat} = 1.40$, where the throat diameter is equal to 128 mm. Centerline curvature causes the offset (H), and for that study, H is defined as 1.065 D_{throat}. The length of the duct (L) is 4.668 D_{throat}.For the M2129 S-duct configuration, the centerline and the diameter are defined by

$$y = 0.0 \tag{9.1}$$

$$z = -0.15L\left(1.0 - \cos\left(\frac{\pi x}{L}\right)\right) \tag{9.2}$$

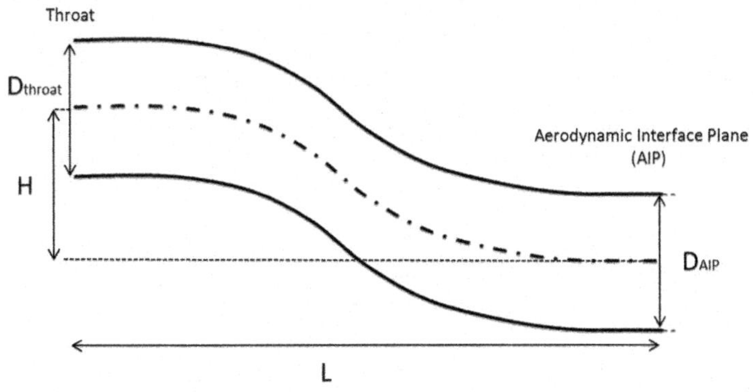

Fig. 9.1 M2129 diffuser geometry

Table 9.1 Geometrical parameters defining the RAE M2129 S-diffuser model [8]

Parameters	Value [m]	Description
D_{throat}	0.1288	Throat diameter
D_c	0.1440	Highlight (entry) diameter, capture area
D_{AIP}	0.1524	Diameter at the engine face
X_{AIP}	0.4839	Engine face position
L	0.4572	Duct length

$$D = D_{\text{throat}} + (D_{\text{AIP}} - D_{\text{throat}})\left(3.0\left(1.0 - \frac{x}{L}\right)4 - 4.0\left(1.0 - \frac{x}{L}\right)3 + 1.0\right) \quad (9.3)$$

where x is between 0 and L and where L is the duct length. The throat diameter is shown as D_{throat}, whereas engine face diameter is named as D_{AIP}. The location of the aerodynamic interface plane is called x_{AIP}. Curvature of the diffuser gives an offset in the z-direction of 1.065 D_{throat}. Both upstream and downstream parts of the S-duct are extended in parallel direction to the freestream. All geometrical parameters are given in Table 9.1.

9.3 Numerical Approach

In three-dimensional simulations of serpentine inlet, conservation of mass, momentum, and energy is solved numerically. The conservation equations iterated for each finite volume during the solution until the flow field parameters reached an acceptable level at desired monitoring points. A steady solution of the diffuser problem is achieved by convergence of mass; x-, y-, and z-direction momentum; energy; and turbulent variables. The continuity, momentum and energy equations governing the fluid system are given below [9]. Continuity equation:

$$\vec{\nabla} \cdot \left(\rho \vec{V}\right) = 0 \quad (9.4)$$

where $\vec{V}$ is velocity and ρ is density. Momentum equations:

$$x\text{-component}: \vec{\nabla} \cdot \left(\rho u \vec{V}\right) = -\frac{\partial p}{\partial x} + \frac{\partial \tau_{xx}}{\partial x} + \frac{\partial \tau_{yx}}{\partial y} + \frac{\partial \tau_{zx}}{\partial z} + \rho f_x \quad (9.5)$$

$$y\text{-component}: \vec{\nabla} \cdot \left(\rho v \vec{V}\right) = -\frac{\partial p}{\partial y} + \frac{\partial \tau_{xy}}{\partial x} + \frac{\partial \tau_{yy}}{\partial y} + \frac{\partial \tau_{zy}}{\partial z} + \rho f_y \quad (9.6)$$

$$z\text{-component}: \vec{\nabla} \cdot \left(\rho w \vec{V}\right) = -\frac{\partial p}{\partial z} + \frac{\partial \tau_{xz}}{\partial x} + \frac{\partial \tau_{yz}}{\partial y} + \frac{\partial \tau_{zz}}{\partial z} + \rho f_z \quad (9.7)$$

where u, v, and w are the x, y, and z components of the velocity, respectively, p is pressure, τ is shear stress, and $\vec{f}$ is body force per unit mass. Energy equation:

Table 9.2 Mesh size and y$^+$ relation

Mesh size	y$^+$
3.1 M	6–7
7.5 M	5
15.5 M	1–2

$$\vec{\nabla}\cdot\left[\rho\left(e+\frac{V^2}{2}\right)\vec{V}\right]=\rho\dot{q}+\frac{\partial}{\partial x}\left(k\frac{\partial T}{\partial x}\right)+\frac{\partial}{\partial y}\left(k\frac{\partial T}{\partial y}\right)+\frac{\partial}{\partial z}\left(k\frac{\partial T}{\partial z}\right)-$$
$$\frac{\partial(\mathrm{up})}{\partial x}-\frac{\partial(\mathrm{vp})}{\partial y}-\frac{\partial(\mathrm{wp})}{\partial z}+\frac{\partial(u\tau_{xx})}{\partial x}+\frac{\partial(u\tau_{yx})}{\partial y}+\frac{\partial(u\tau_{zx})}{\partial z}+\frac{\partial(v\tau_{xy})}{\partial x}+ \tag{9.8}$$
$$\frac{\partial(u\tau_{yy})}{\partial y}+\frac{\partial(u\tau_{zy})}{\partial z}+\frac{\partial(w\tau_{xz})}{\partial x}+\frac{\partial(u\tau_{yz})}{\partial y}+\frac{\partial(u\tau_{zz})}{\partial z}+\rho\vec{f}\cdot\vec{V}$$

where e is internal energy per unit mass, $\dot{q}$ is the rate of volumetric heat addition per unit mass, T is temperature, and k is the thermal conductivity.

ANSYS Fluent, that is, general-purpose computational fluid dynamics software, was used for all the 3D numerical simulations. As a simplification for reducing calculation time, only one symmetrical part of the duct is considered, and symmetry boundary condition was applied on the sliced plane. This assumption could be used by the help of geometrical and conditional symmetry of the duct. Due to complexity of the duct geometry (variable cross sections at each position, many splines, etc.), unstructured tetrahedral grids were preferred. This type of grid helps to capture the geometry and to obtain adaptive cells where solution requires. The mesh dependency was examined by using three different resolutions, namely coarse, medium and fine levels. Average values of y$^+$ with respect to mesh size are controlled and shown in Table 9.2.

The results of simulations showed that different number of meshes had similar results; however, for a reliable near wall solution, the smallest y$^+$ values were desired, and the simulations are continued with the fine resolution, 15.5 M cells (Fig. 9.2).

Pressure-based solver and SIMPLE algorithm for pressure-velocity coupling have been used for analyses. The turbulent flow field is modeled with three different well-known RANS models, Spalart-Allmaras (1-transport equation), Realizable k-ε (2-transport equations), and k-ω SST (2-transport equations), and the results are compared. As the spatial discretization, second-order upwind scheme is used for all variables, and as an exception for the pressure, PRESTO! scheme is applied. Air is selected for the flowing fluid and considered as ideal gas to take compressibility effects in to account.

In order to create freestream conditions at the inlet of the duct, a large domain that was modeled as a hemisphere is added to the fluid domain, and pressure far-field boundary condition is assigned to the outer surface of this imaginary domain and is set to 0.207 Mach and 100,018.8 Pa static pressure value. The temperature is given 280.2 K for this condition. The flat surface of imaginary domain is set as pressure outlet and considering the freestream conditions remains same static pressure and temperature conditions with the inlet boundary. The exit of the duct was also selected as pressure outlet but the required static pressure value is obtained

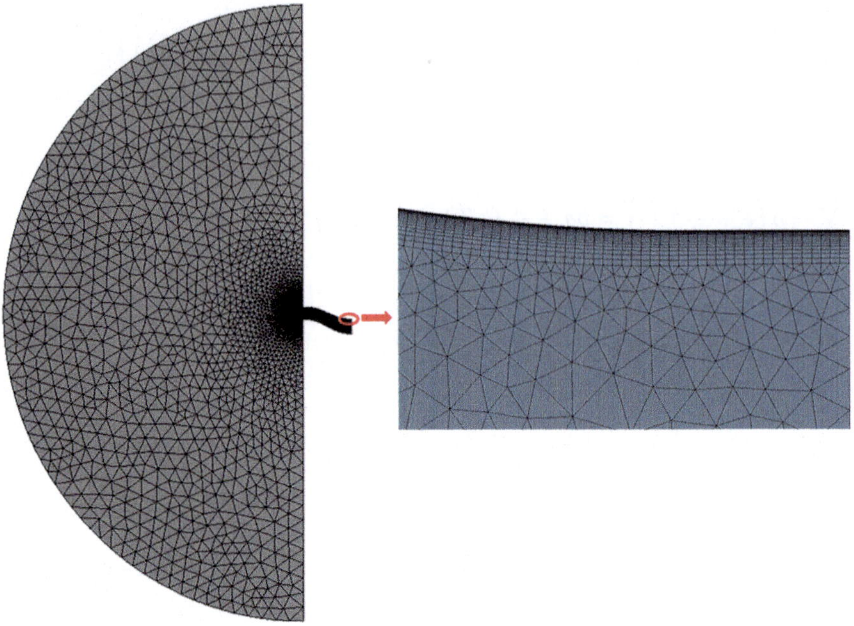

Fig. 9.2 Mesh structure

Table 9.3 Analyses properties

Name	Turbulence model	Mesh size	y^+
1	Realizable k-ε	3.1M	6–7
2	Realizable k-ε	7.5M	5
3	Realizable k-ε	15.5M	1–2
4	k-ω SST	15.5M	1–2
5	Spalart-Allmaras	15.5M	1–2

iteratively during the simulations. These iterative processes are continued until mass flow rates of simulations and the experiments reached the same value. No slip condition is applied to the walls, and symmetry boundary condition is used at the sliced face of the domain.

9.4 Results

In the numerical investigations of the duct, three different meshes are analyzed with Realizable k-ε turbulence model. This model is selected according to its mathematical background that gives the model name "Realizable" and its ability to provide acceptable results for wide range of fluid mechanics problem. On the other hand, selected turbulence models mentioned above are examined with the qualified grid. Results of grid resolution study with respect to turbulence model and y^+ values are listed at Table 9.3.

Figure. 9.3 demonstrates the symmetry plane Mach contours and outlet pressure recovery contours.

Results were found to be similar for all of the simulations. Maximum and minimum Mach number locations at symmetry plane coincide at each simulation. However, the extension of the maximum and minimum Mach zones varies from one case to the other. Likewise, at the outlet plane, PR contours show nearly the same results.

Experimental pressure recovery (PR) contours at the outlet plane is shown in Fig. 9.4.

Experimental and numerical pressure recovery contours agreed well with each other. However, PR gradients show smoother variation for experimental case, while computations show more drastic variations between regions.

Both computational and numerical results are listed in Table 9.4. PR and Mach number are calculated as the area averages at the outlet plane. It is found that Spalart-Allmaras turbulence model (Analysis 5) gives closer results to the experiment values.

Iteratively obtained pressure value matches the mass flow rates of simulations and experiments. Even though the flow rates match, Mach numbers calculated at the exit take values higher than 0.41. The deviation from the experimental value is about 3–4%, and this deviation is still in acceptable range. Moreover, the pressure recovery parameter calculated from the simulations is still in the range of experimental value. Another important result is that the pressure gradients at the outlet plane obtained from the numerical analysis have drastic differences with respect to experimental data; however, the area-weighted mean values are found to be very close to the experimental results.

9.5 Conclusion

Flow structure in a short, diffusing S-duct inlet had been investigated experimentally at QinetiQ facilities, previously. In this study, QinetiQ's work simulated using commercial CFD code ANSYS Fluent. For that study freestream Mach number was fixed at 0.207, and Reynolds number based on the nondimensional engine faces diameter of 777,000.

It is seen that computational results were agreed well with the experiments. While various turbulence models were used for turbulence closure, Spalart-Allmaras turbulence model was found to provide the most accurate results in terms of mass flow rate, Mach number, and pressure recovery for that case of.

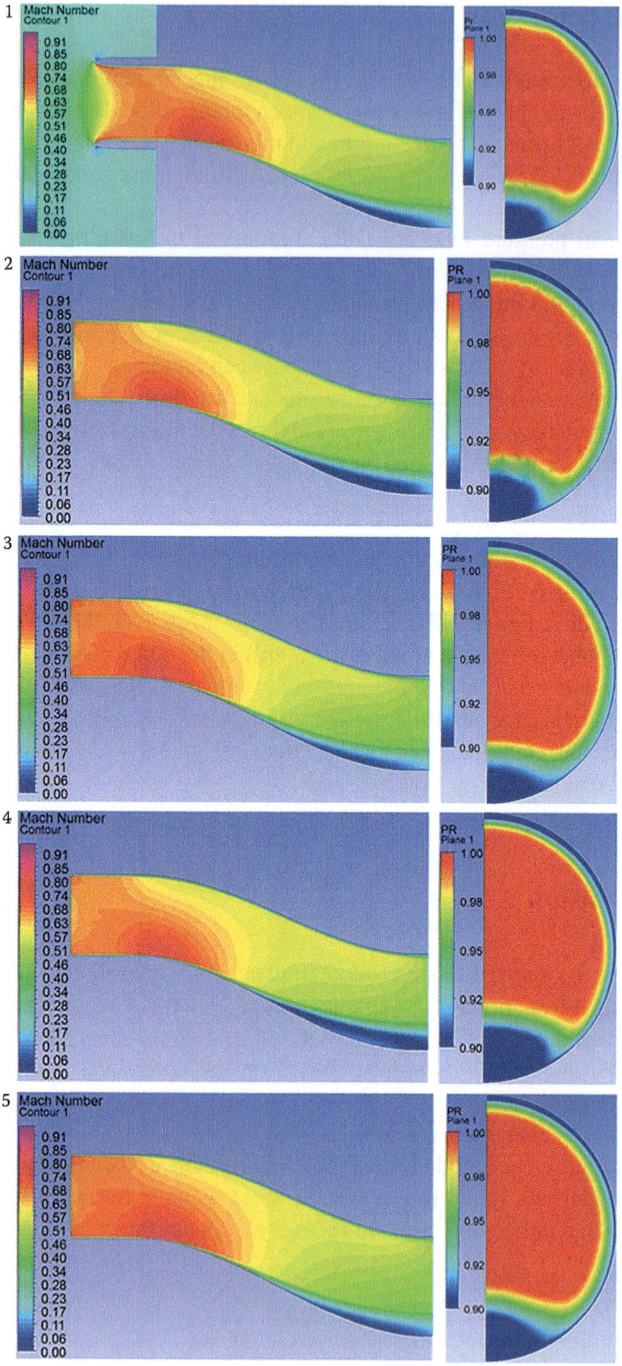

Fig. 9.3 Mach contours (*left column*) and PR contours (*right column*) for different analysis performed. (see Table 9.3 for analysis numbers 1–5)

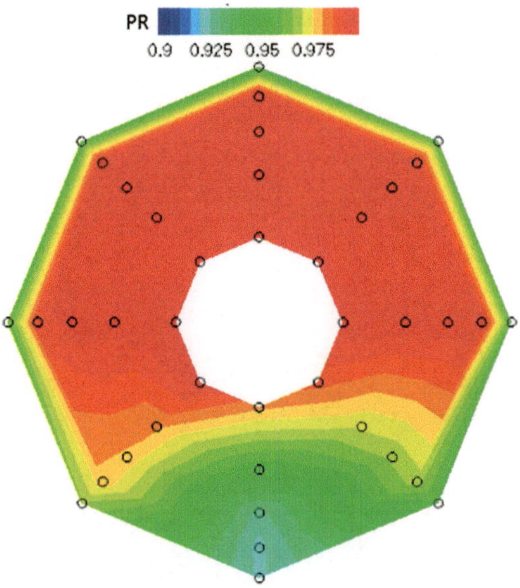

Fig. 9.4 Experimental pressure recovery contour at the outlet plane

Table 9.4 Comparison of numerical and experimental data

Name	PR	Mach	Mass flow
1	0.9763	0.414	2.846
2	0.9762	0.411	2.826
3	0.9772	0.420	2.887
4	0.9733	0.417	2.847
5	0.9754	0.419	2.871
Test	0.9798	0.395	2.873

References

1. Wellborn, S.R., Reichert, B.A., Okiishi, T.H. (1992). An experimental investigation of the flow in a diffusing S-duct. 28th Joint Propulsion Conference and Exhibit. 6–8 July, Nashville, AIAA-92-3622.
2. Harloff, G.J., Reichert, B.A., Wellborn, S.R. (1992). Navier-Stokes analysis and experimental data comparison of compressible flow in a diffusing s-duct. AIAA Paper 92–2699.
3. Harloff, G.J., Reichert, B.A., Wellborn, S.R. (1992). Navier-Stokes analysis and experimental data comparison of compressible flow in a diffusing s-duct. 13th International Conference on Numerical Methods in Fluid Dynamics. 6–10 July, Rome, AIAA-92-3622.
4. Wellborn, S. R., Reichert, B. A., & Okiishi, T. H. (1993). *A study of the compressible flow through a diffusing s-duct*. Washington, DC: National Aeronautics and Space Administrations., NASA TM-10641.
5. Wellborn, S. R., Reichert, B. A., & Okiishi, T. H. (1994). Study of the compressible flow in a diffusing S-duct. *Journal of Propulsion and Power, 10*(5), 668–675.
6. Aslan S., Yilmazturk S., Kurtulus D.F. (2015). Aerodynamic investigation of a serpentine inlet design for a micro-turbojet engine powered aircraft, International Symposium on Sustainable Aviation, Istanbul.
7. Air intakes for high speed vehicles, AGARD-AR-270, September 1991.
8. Application of CFD to high offset intake diffusers, GARTEUR AD/AG-43, October 2012.
9. Wendt, J. F., & Anderson, J. D. (2009). *Computational fluid dynamics: An introduction* (3rd ed.pp. 15–51). Berlin: Springer.

Chapter 10
Aerothermal Investigation of a Turbine Cooling Channel with U-Bend

Isa Kavas, Tolga Yasa, and D. Funda Kurtulus

Nomenclature

A	Heated surface area [m^2]
c	Specific heat
D	Diameter [m]
e	Rib height
H	Height of channel
h	Convective heat transfer coefficient [W/m^2K]
I	Current
k	Thermal conductivity
Nu	Nusselt number
P	Perimeter of channel cross section
Pr	Prandtl number
q	Heat flux [W/m^2]
Re	Reynolds number
Rke	Realizable k-epsilon model
RSM	Reynolds stress model
SST	Shear stress transport model
T	Temperature
V	Voltage

I. Kavas • D.F. Kurtulus
METU Aerospace Engineering Department, 06800 Ankara, Turkey

T. Yasa (✉)
Anadolu University, 26555 Eskisehir, Turkey
e-mail: tyasa@anadolu.edu.tr

T.H. Karakoç et al. (eds.), *Advances in Sustainable Aviation*,
DOI 10.1007/978-3-319-67134-5_10

v2f K-epsilon v2f model
W Width of channel
YS Low Reynolds k-epsilon Yang-Shih model

Greek

μ Dynamic viscosity [kg/(m·s)]

Subscripts

h Hydraulic
loss Heat loss
p Pressure
pl Plexiglass
sm Smooth wall configuration
tc Thermocouple
w Wall

10.1 Introduction

Turbine cooling is necessary to design compact and efficient gas turbine components since it lowers the material temperature to allowable limits. Cooling configurations are extensively investigated experimentally for the last decades. Effect of channel geometric parameters such as channel shape, rib shape, rib orientation, rib angle, number of heated walls, and channel Reynolds number is well characterized in the literature [1–7]. Experimental strategy for the cooling channel investigation commonly includes heating the walls of the channels and determination of internal surface temperature. For external and internal heat transfer in gas turbine, extensive information is provided in the literature [8].

Ribs or turbulators enhance heat transfer by affecting the flow parameters of the cooling channel. Ribs act like an obstacle for the flow and break the boundary layer creating secondary flows similar to forward and backward facing step. Periodic ribs break the laminar sublayer and create local wall turbulence because of separation and reattachment of flow between the ribs, which enhances the heat transfer on the surface [9]. Main incident that enhances the heat transfer is that flow reattaches the surface to recover the boundary layer creating a thin boundary layer. Detailed flow and heat transfer around the rib which is perpendicular to the flow is investigated experimentally [10]. The heat transfer distribution on the ribbed channel surface is determined with liquid crystal thermography, and the flow field is investigated with PIV and oil visualization techniques. The rib-induced secondary flows and pressure fluctuations at the rib vicinity are reported. Similar geometry with both 90° and 45°

rib angles is investigated conducting LES analysis [11]. Secondary flow structures are tried to be detected and related to heat transfer using vortex detection methods.

Depending on the design of cooling channels, cooling configurations often include serpentine passages. Characterization of flow inside smooth curved channels with square and circular cross section is studied experimentally in the literature [12–14]. Dean-type secondary flow structures develop due to the bend section influencing the flow field and heat transfer characteristics of downstream channel. Depending on the geometry of the bend, flow separates and reattaches the inner side wall at downstream channel. Wang et al. [15] conducted numerical analysis to determine the effect of U-turn geometry. The heat transfer results agreed with the experimental case of Chyu [16] for the sharp turn case. They stated that a sharp turn rather than circular one creates stronger swirl after the turn. Saha et al. [17] conducted RANS simulations to investigate the effect of turn shape on flow and heat transfer inside a two-pass channel. Addition of turning vane and periodic dimples and geometry variations on inner and outer walls at the turn region was the effecting parameters. Nine different U-bend geometries investigated for minimizing the pressure loss and maximizing the heat transfer. The results show that dimples at the U-bend section don't affect the secondary flows at the U-bend significantly and enhance the heat transfer moderately. Implementation of turning vane decreases both pressure loss and heat transfer, and symmetrical bulb-shaped thin inner wall performs better in terms of heat transfer and pressure loss. Hirota et al. [18] experimentally investigated the effect of inner wall clearance on the heat transfer inside a two-pass channel with 180° sharp turn using naphthalene sublimation technique. They concluded that there is a steep change in the detailed maps of heat transfer rates of the second pass outer wall depending on the clearance. Serpentine ribbed channels are widely investigated in order to improve the performance of cooling passages. Chen et al. [19] investigated heat transfer and pressure drop in two-pass channel with different turning vanes. Experiments using liquid crystal thermography and numerical analyses using RANS models (ORS, SST, k-ε) are compared. It is stated that Omega Reynolds Stress (ORS) model performs better, and configuration with turning vane creates less pressure loss and heat transfer compared to baseline. Combined effect of ribs and the turn geometry is addressed only in a few studies. Secondary flows due to rib angle affect the flow development through the turn region in multipass channels. Mochizuki et al. [20] experimentally investigated a two-pass ribbed channel with 180° sharp turn. The rib height to hydraulic diameter ratio was 0.09, and rib angles were 30°, 45°, 60°, 75°, and 90°. The results show that different rib arrangements create considerable differences in cooling performances of the entire channel. In the study, channel is equipped with ribs on both passages or one of them is remained smooth; it was concluded that applying the same rib configuration changes the pressure drop depending on the pass it is applied. The numerical simulation of the experimental geometry under stationary and rotating condition were performed with smooth and ribbed cases [21, 22]. Large eddy simulations showed that the heat transfer at the turn region and second pass increased due to the flow impingement and the turn-induced secondary flows at the stationary case, and it was stated that the pressure drop is much more sensitive to rib arrangements and rotation than the heat transfer. Heat transfer at the outer surface of the turn region is of interest in terms of tip cooling performance of the turbine blade. Wang et al. [23] conducted experiments using liquid crystal thermography with one smooth (reference) and seven different rib-roughened surfaces on outer wall. The Reynolds numbers were 20,000 and 26,000 based on hydraulic diameter. According to the results, it is indicated that the local heat transfer enhancement was dominated by the impingement of the flow for the smooth case, and ribbed configurations augment heat transfer rates with a penalty of nonuniform heat transfer distribution. Ribs at outer wall configurations cause a much larger pressure drop at the turn section, and 45° V-shaped ribs performs the best in terms of thermal performance.

Two-pass channel with square cross section will be used for internal cooling performance tests at Reynolds number of 20,000. Kavas [24] performed the experimental work with higher Reynolds numbers with detailed explanations and the numerical comparisons for Re = 20,000. Reynolds number is calculated based on the channel hydraulic diameter. The numerical simulations are compared with experimental data. Flow investigation to understand the flow-heat transfer relation downstream of the turn section is performed. Yasa et al. [25] numerically analyzed and showed the flow structures at the inter-rib and turning regions inside a square ribbed channel model at Re = 20,000. It is advantageous to perform detailed numerical analysis with these geometrical parameters because there are plenty of experimental and numerical studies regarding these geometrical parameters.

10.2 Experimental Apparatus and Procedure

10.2.1 Test Facility

The experiments were conducted at the turbine internal cooling test facility located at Tusaş Engine Industries (TEI). Figure 10.1 shows the experimental setup which is an open circuit suction-type wind tunnel. The facility consists of a blower, a settling chamber, an inlet section, and a test section. Air is sucked from the entrance of the inlet section, and a fully developed flow condition is achieved at the end of inlet section. Then, the flow passes through the test section before entering the settling

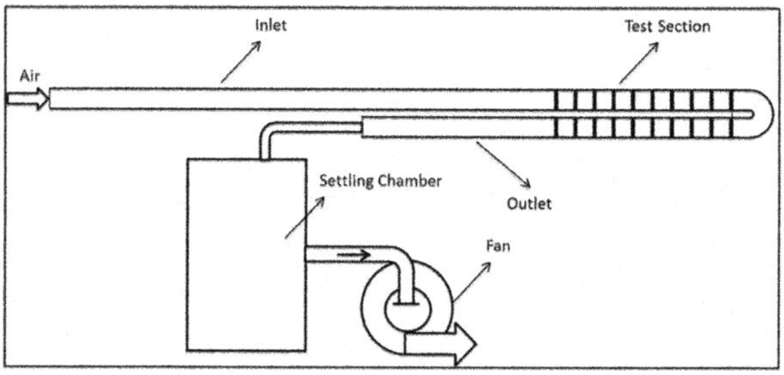

Fig. 10.1 Turbine internal cooling test facility

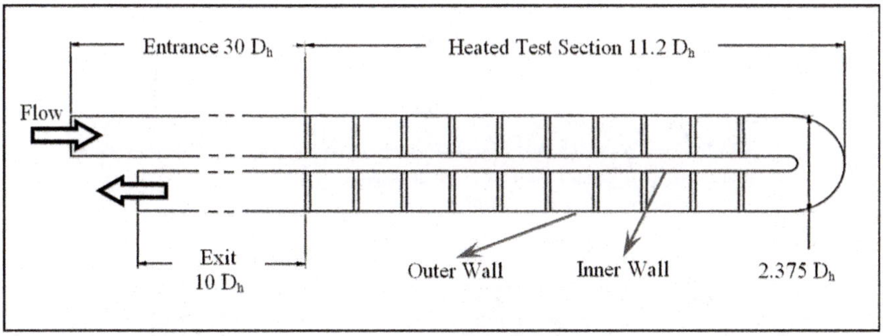

Fig. 10.2 Test section

Fig. 10.3 Tunnel heating

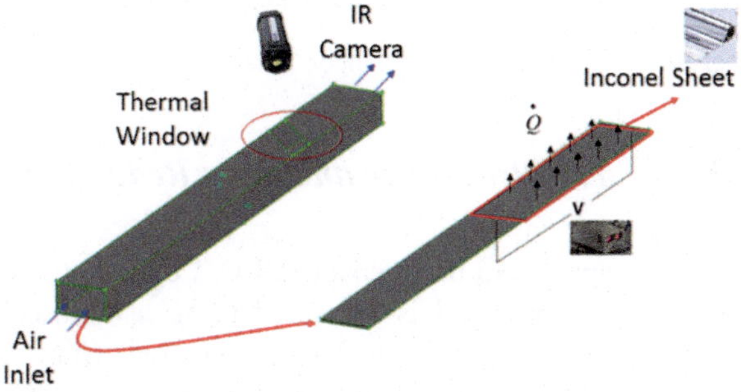

chamber and fan. The settling chamber is designed large enough (with a volume of 1 m^3) to dump down any flow oscillation generated by blower. The test section is a combination of two straight channels and a U-turn section. The channel cross section is square; therefore, the aspect ratio of the channel is unity.

The dimensions of the test section are presented in Fig. 10.2. The straight parts of test section contain turbulent generators called as *rib* on the bottom wall only. Totally ten ribs are located at each straight parts of the test section. The ribs are attached to the bottom wall with zero clearance. They also have square cross section where the rib height is 10% of the channel height. The ribs are installed with an orientation of 90° to the side wall of the test section as well as the flow direction. The periodic distance between the two ribs is set ten times of the rib height.

10.2.2 *Tunnel Heating Strategy*

Foil heater technique is applied to obtain the heating condition. The heated test section part is heated from the bottom wall. Voltage is applied from two ends of the Inconel sheet, and constant heat flux boundary condition is provided from the bottom wall (Figs. 10.3 and 10.4).

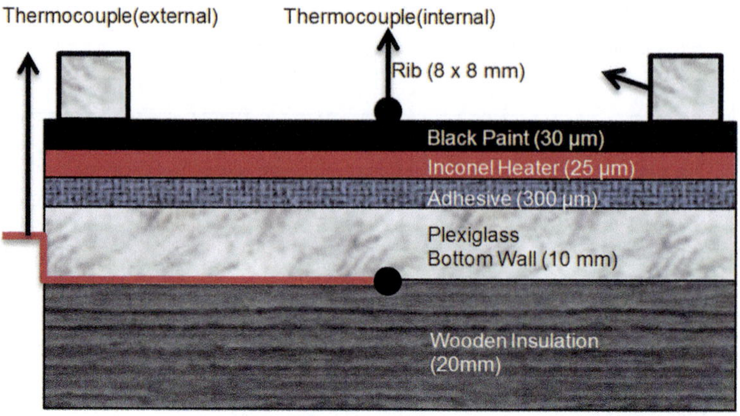

Fig. 10.4 Inconel foil heating composite (not to scale)

10.2.3 Test Procedure and Data Reduction Method

Heat flux applied to the Inconel foil was determined using Eq. 10.1. The voltage across the Inconel foil is measured using a multimeter. Voltage (*V*) across the Inconel does not change in lateral direction. Current (*I*) across the Inconel is determined simply by DC power supply. Voltage (*V*) across the Inconel foil is obtained applying the multimeter at both ends of the heated section as shown in Fig. 10.3. k_{pl} is the thermal conductivity of plexiglass material which is taken constant as 0.19 W/mK [26, 27]. T_{w} is the local temperature of the heated surface which is obtained using the IR camera data. T_{tc} is the temperature measured by loss thermocouples, and A is the heated area for the Inconel foil. T_{g} is the bulk temperature for air. D_{h} is the hydraulic diameter of the channel which is calculated by Eq. 10.5, and k_{air} is the thermal conductivity of air which is determined using the properties of air at atmospheric pressure:

$$q = V \cdot I \tag{10.1}$$

Losses are determined with Eq. 10.2 using 1D conduction assumption across plexiglass. Local temperature just above the thermocouple location is used to determine the losses, and loss heat flux q_{loss} is assumed to be the same for the whole heated area:

$$q_{\mathrm{loss}} = k_{\mathrm{pl}} \cdot (T_w - T_{\mathrm{tc}}) \tag{10.2}$$

Local heat transfer coefficient is calculated between two rib turbulator using the thermal information:

$$h = (q - q_{\mathrm{loss}}) \,/\, \left[A \cdot \left(T_w - T_g\right)\right] \tag{10.3}$$

The local Nusselt number is calculated using the hydraulic diameter of channel and air thermal conductivity corresponding to bulk temperature (Eq. 10.4). Nusselt number is normalized using the smooth wall measurement results:

$$\mathrm{Nu} = \frac{hD_\mathrm{h}}{k_\mathrm{air}} \tag{10.4}$$

$$D_\mathrm{h} = \frac{4WH}{P} \tag{10.5}$$

Reynolds number (*Re*) is determined using the bulk velocity which is obtained by the differential pressure measurements. Velocity is obtained from the difference between total and static pressure as shown in Eq. 10.6:

$$u = \sqrt{\frac{2\Delta p}{\rho_\mathrm{air}}} \tag{10.6}$$

Prandtl number (*Pr*) is taken as 0.69 using Eq. 10.7:

$$\mathrm{Pr} = \frac{c_p\mu}{k} \tag{10.7}$$

10.3 Numerical Setup

Rib turbulators are implemented to the gas turbine blade cooling channels to increase the convective heat transfer. The ribs create flow separation, impingement, and secondary flows through the channels; hence it is very important to predict the flow and heat transfer numerically.

10.3.1 Model Geometry and Boundary Conditions

A simplified version of experimental model is built to reduce computational cost. As the heat transfer and flow become periodical after fourth rib inside a straight ribbed channel, less number of ribs are implemented to the numerical model [10]. The numerical model is a square channel with five and six ribs on one wall in the first and second pass, respectively. Numerical model has the same cross section (80×80 mm) and other geometrical parameters as the experimental setup. As in the experimental setup, the description of the numerical geometry is depicted in Fig. 10.5.

Inlet velocity measurement of the experimental campaign and constant temperature of 300 K is taken as inlet boundary conditions. At the outlet, pressure outlet boundary condition is applied. Figure 10.6 shows the spanwise averaged, symmetry

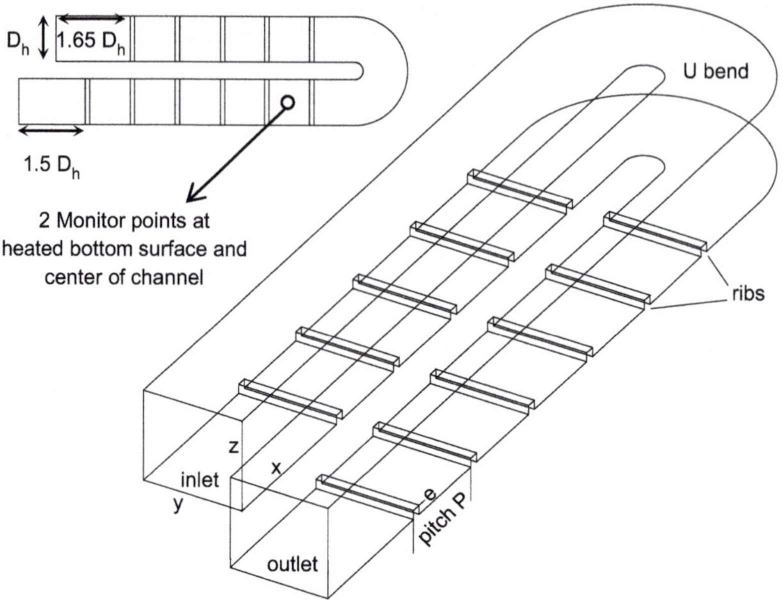

Fig. 10.5 Top view and 3D view of numerical setup

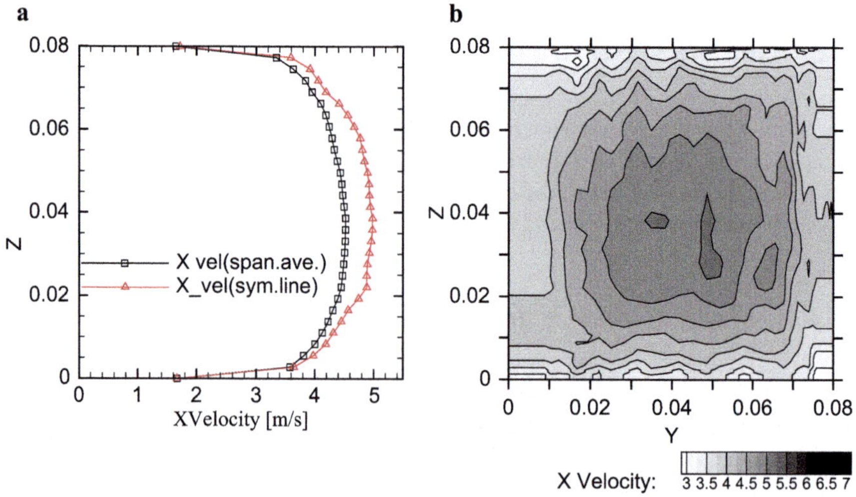

Fig. 10.6 (**a**) Spanwise averaged and (**b**) 2D contour of streamwise velocity

line and 2D contour of measured inlet velocity profile. Since there is no turbulence intensity measurement, it is given as 5% with a hydraulic diameter of 80 mm [28]. Neumann boundary condition is applied at the inter-rib surfaces and U-turn region of bottom wall considering constant heat flux. The rest of the surfaces are accounted for no-slip adiabatic wall.

10.3.2 Numerical Approach and Mesh Generation

Numerical analyses are conducted with ANSYS FLUENT 14.0 software. Steady-state solutions were obtained with pressure-based solver algorithm. SIMPLE (Semi-Implicit Method for Pressure-Linked Equations) algorithm is selected for pressure velocity coupling. PRESTO (Pressure Staggering Option) is selected for spatial discretization. Second-order discretization is used for all equations. Enhanced wall treatment is activated for the realizable k-ε model and the Reynolds stress model. Realizable k-ε model is worth using since the initial studies claim that it performs better for separated flows and flows with complex secondary features [29, 30]. Likewise, RSM model has additional two equations to solve for dissipation rate, and it has a larger potential for accurate predictions for complex flows [31, 32]. In the present case, flow over the ribs and the flow through the U-bend are expected to create separated flow, shear layers, secondary flows, and relatively high dissipation rates. Therefore, one can expect better performance in terms of flow solution which is connected to the heat transfer distribution on the surface. Enhanced wall treatment is checked when it is applicable since grids resolving the boundary layers are expected to perform better. For all numerical analysis, computational domain consists of fully structured multiblock mesh structure with refinement near the walls and rib vicinities. In order to capture the boundary layer flow, y^+ is set below unity for all surfaces with an exception of the surface of the first rib at the second pass, which has y^+ value of approximately 1.7. Contours of y^+ values for ribbed case are shown in Fig. 10.7a.

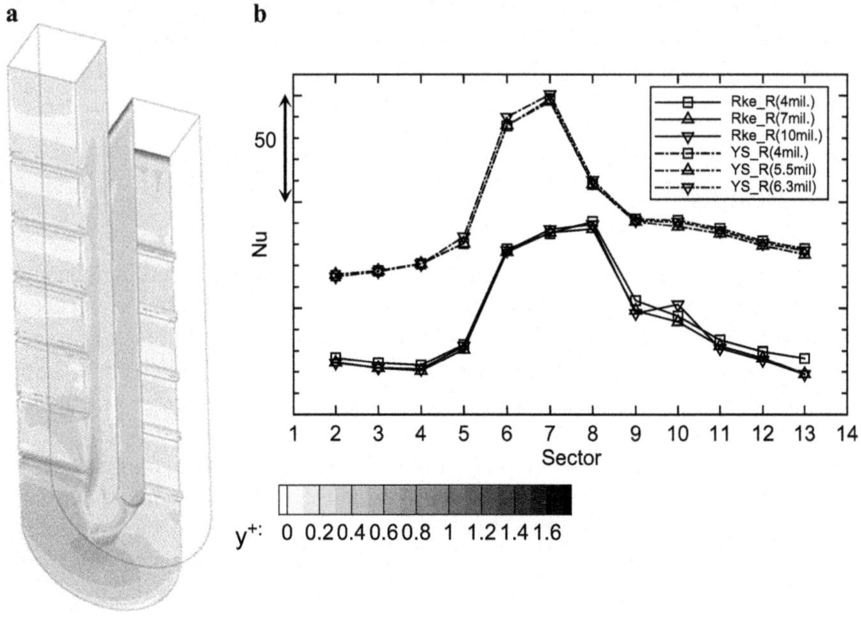

Fig. 10.7 (**a**) y^+ contours and (**b**) mesh independency of Rke and YS models of ribbed channel model

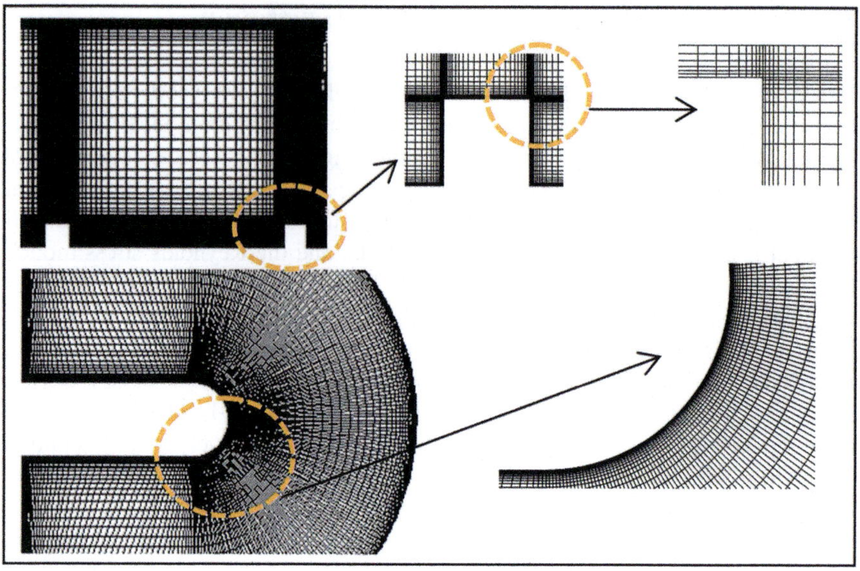

Fig. 10.8 Mesh structure at inter-rib and turn regions

Mesh independency is performed for k-ε and low Re YS turbulence models for ribbed channel model. The grids of smooth cases are generated with a similar logic that of the ribbed models. Inter-rib area averaged Nu results of two models are presented in Fig. 10.7b.

First cell thickness is set as 0.25% of the rib height, and grid is generated with a growth rate of 1.2. Figure 10.8 shows the mesh structure for inter-rib and turn regions. Figure 10.9 shows the selected mesh structures for the analysis. There are 4.2 and 2.5 millions of elements for ribbed and smooth cases, respectively.

10.4 Results

In order to assess the heat transfer enhancement of rib-roughened cooling channel, Nusselt number distribution of smooth (unribbed) channel is obtained experimentally as a reference. The results of Nusselt number distribution and area averaged Nusselt number measurements are depicted in Fig. 10.10. Although there is no rib presented in the smooth channel, the average Nusselt number is calculated respecting the regions presented for the rib-roughened case.

In the first rib section, redeveloping of flow results in high Nu values up to fifth to sixth inter-rib space. Then, flow becomes periodical and present almost constant heat transfer coefficient up to the U-bend section. The results also show that near the outer wall region, the Nu appears to be higher compared to the near inner wall region. This can be explained by two-dimensional heat transfer of inner wall region.

Fig. 10.9 Mesh structure for smooth (*top*) and ribbed (*bottom*) cases

The heater is installed as a single piece which results in a heat flux generation below the mid-wall. The wall is made of plexiglass; hence the conduction through the mid-wall is not well established. Therefore, some part of the generated heat flux goes in spanwise direction so more heat flux directed to the near inner wall region.

10.4.1 Comparison with Numerical Predictions

Performances of five turbulence models are compared with the experimental Nusselt number results of the ribbed case for a single inter-rib space. Low Re YS, v2f, and realizable k-ε models are selected later for smooth channel simulations. The numerical results of ribbed channel are normalized by using the results of smooth channel. The flow field downstream of a single rib solved by low Re YS and v2f models is compared with the literature. Finally, Nu, Nu_{sm}, and heat transfer enhancement factor (Nu/Nu_{sm}) results are analyzed.

In Fig. 10.11, spanwise averaged Nusselt numbers in flow direction over fourth inter-rib space are plotted for all turbulence models and experiments. Additionally, the Nu distribution of the channel midline (symmetry plane) is plotted for some cases in order to characterize the difference between both approaches. v2f model results in the best approximation to the real values. It predicts well the flow redevelopment region but underpredicts when there is a recirculation. SST model also provides the same trend than the results of v2f model but underpredicts the results. Low Re YS model overpredicts the Nu numbers, whereas all others underpredict the heat transfer coefficient. Low Re YS model is the only model where there is clear difference between midline and averaged results. This difference

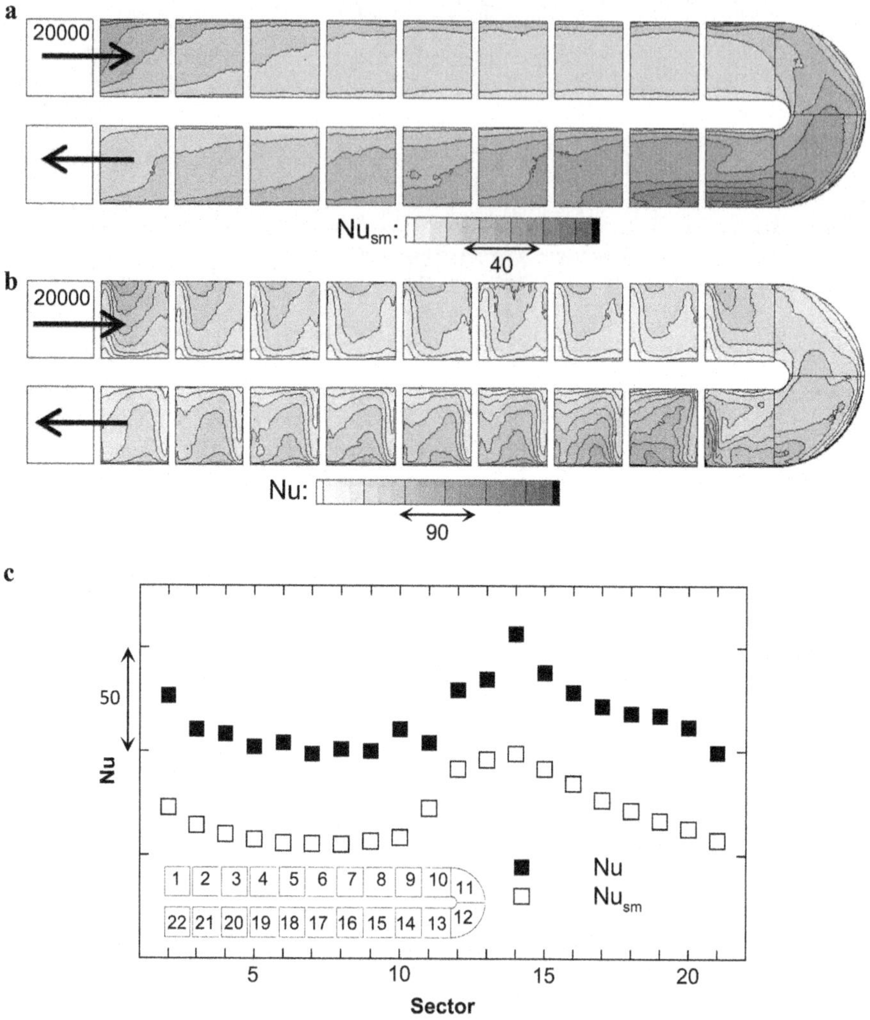

Fig. 10.10 (**a**, **b**) Distribution and (**c**) inter-rib averaged Nu results (experimental)

means that the secondary flows generated at the edges of bottom wall and lateral walls are stronger than the other models. The location of the maximum heat transfer coefficient also corresponds to the flow reattachment and varies depending on the turbulence model. Reattachment distance from the leading rib wall is about 3e for the experiments and low Re YS model and 2.5e for Rke model. The reattachment location is determined at 5e from the downstream of leading rib for v2f model.

Streamtraces along fourth inter-rib space are presented for v2f and low Re YS models in Fig. 10.12. The size of the circulation region is directly related with the flow reattachment. Inter-rib region velocity measurements of Rau et al. [33]

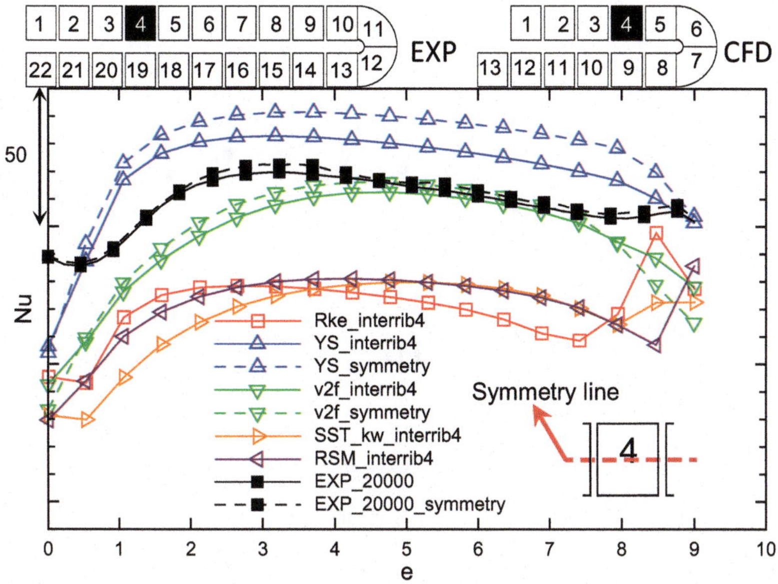

Fig. 10.11 Spanwise averaged Nu along fourth inter-rib

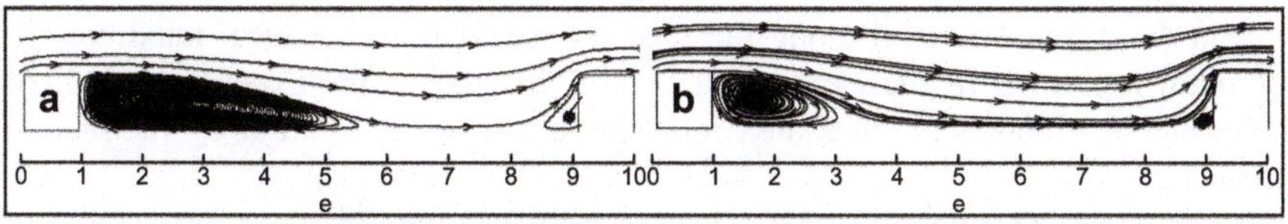

Fig. 10.12 Spanwise fourth inter-rib streamtraces for (**a**) v2f and (**b**) low Re YS

at a height of 0.1e from the heated wall reports that the flow impingement occurs at about 3.7e from the rib rear wall for P/e = 9 and Re = 30,000. However, in the simulation, it is observed that the v2f model predicts much wider recirculation region downstream of the rib (4.5e from the rib rear wall), whereas the low Re YS model prediction of the recirculation region shows that the reattachment distance is about 2.2e from the rib rear wall.

The overall area averaged Nusselt number distribution is compared in Fig. 10.13. Low Re YS and v2f models are in relatively better agreement at the straight parts of the channel when compared to other turbulence models. On the other hand, they overpredict the Nusselt number at turn region. Low Re YS and v2f models cannot capture the decreasing heat transfer behavior through the first pass; however, Rke, k-ω SST, and RSM models show a decreasing pattern. Consequently, Rke, k-ω SST, and RSM models underpredict the heat transfer inside a straight channel with 90° ribs. In other words, separated flow-heat transfer can be estimated with low Re YS and v2f models when these five models are considered. The heat transfer increases because of rib implementation, and U-turn geometry at the downstream of the bend

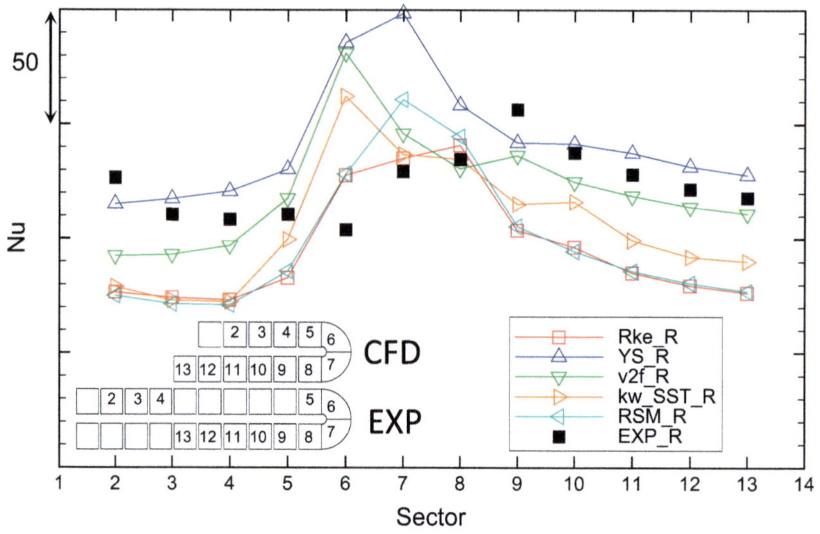

Fig. 10.13 Inter-rib area averaged Nu results

is not captured by any of these models. Only v2f model shows just a small increase, and the heat transfer trend along the channel model makes a local maximum at the ninth region as in the experiment results. Similar to the experiments, all of the turbulence models predicts the decreasing heat transfer trend along the second channel.

Three of five turbulence models are selected for smooth wall computations, and the results are compared with one of the experiments. Area averaged Nu results along the smooth channel configuration are illustrated in Fig. 10.14. Rke and low Re YS models are in a good agreement with the experimental results at the first passage. However, when flow enters the U-turn and in the second channel, all models fail to predict the correct heat transfer coefficients. Rke is the most successful to capture the trend, but it overpredicts the Nu value. Additionally, the Nu number decays more rapidly compared to the measurements. The v2f model shows totally different physics than the real case in whole channel.

Nu, Nu_{sm} of the Rke, v2f, and low Re YS model are demonstrated in Fig. 10.15. In smooth wall configurations, heat transfer distribution is almost constant at the first passage. Heat transfer increase at the surfaces near the outer wall at the turn region shows up because of the higher momentum fluid that impinges at the outer regions [15]. In all smooth channel simulations, sharp heat transfer increase occurs at the outer region of just downstream of the U-turn. The flow coming from the turn region impinges at the outer wall and creates the highest heat transfer region for the whole smooth channel. U-turn effect on heat transfer diminishes relatively earlier at the v2f model results.

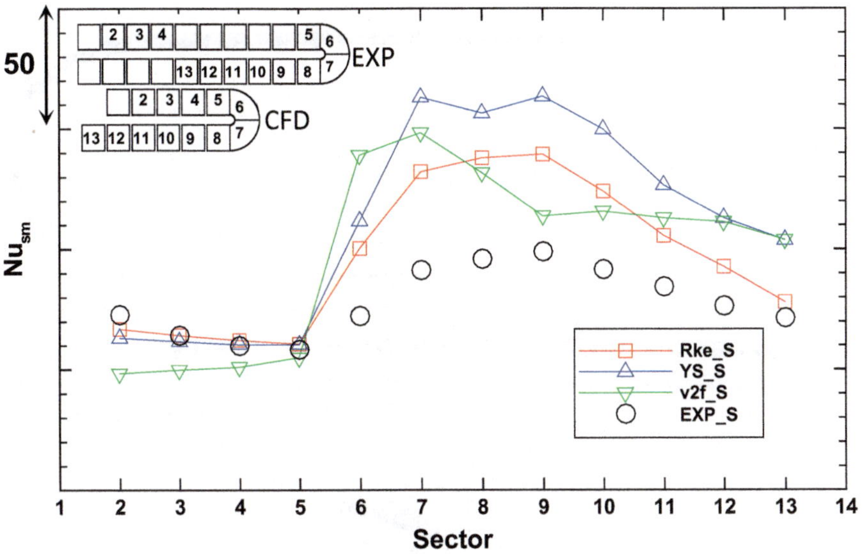

Fig. 10.14 Inter-rib area averaged Nu_{sm} results

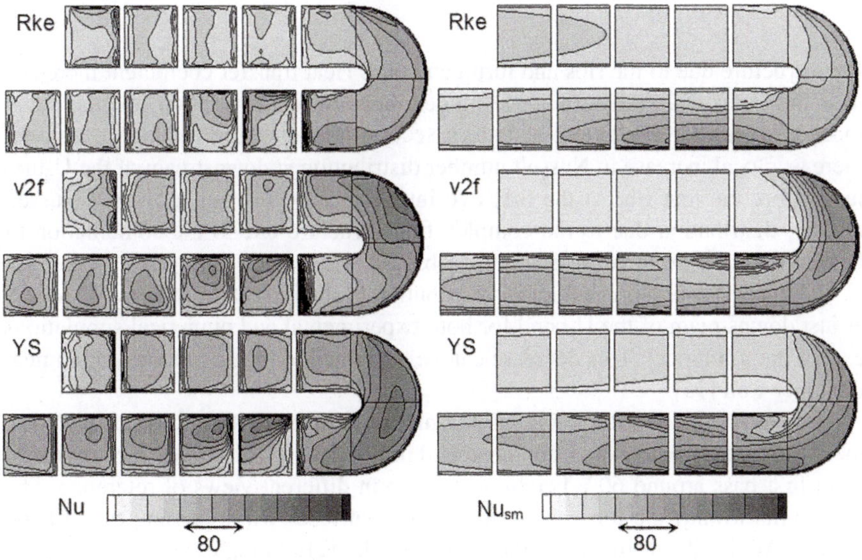

Fig. 10.15 Nu and Nu_{sm} distributions for three turbulence models

At the first passage, all models show almost periodic distribution of heat transfer coefficient after second rib. At the turn region, turning geometry has a significant effect, and Dean vortices are generated due to the bend. That local heat transfer increase is because of the impingement of flow to the heated surface and the complex

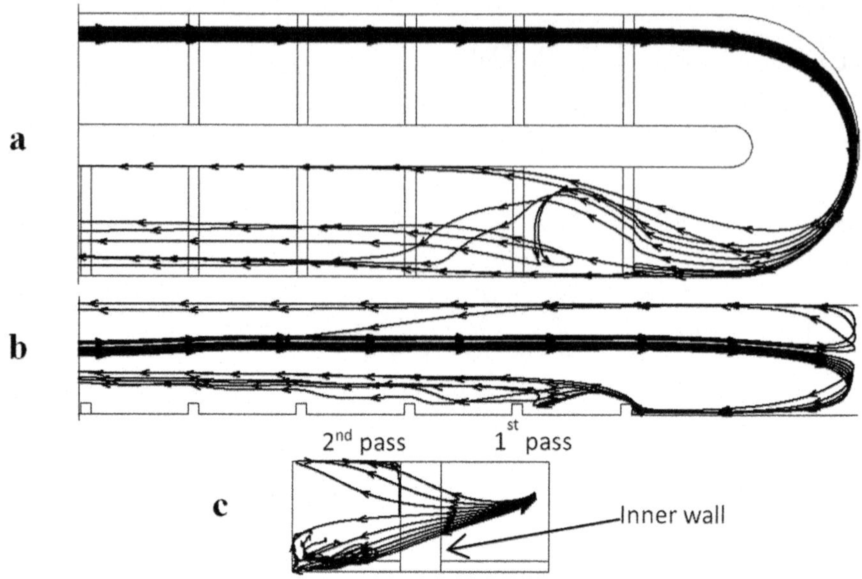

Fig. 10.16 Streamtraces at the bend region (**a**) *top* view, (**b**) *front* view, (**c**) *right* view

flow structure due to the ribs and turn geometry. Heat transfer coefficient increases near the outer wall. All turbulence models have different location for maximum local heat transfer coefficient at U-turn section. For all three turbulence models, there is a local increase in Nusselt number distribution at downstream of the U-turn just before the first rib. At the first two inter-rib areas, there is a distinct Nusselt number distribution due to the complex flow structure due to the combination of U-bend-induced Dean vortices and 90° ribs.

Relatively lower Nusselt number distribution is obtained at the inner wall vicinity at just downstream of the U-bend for both experimental and numerical simulations except the v2f model. This decrease can be attributed to the circulating region near the inner wall [15].

Unique Nu distribution at eighth and ninth inter-rib spaces can be attributed to the different flow behavior as demonstrated in Fig. 10.16. The streamtraces released from first pass around 60% height are shown in different views of reference. The flow splits toward the top and heated bottom surfaces while passing the U-bend region. At the downstream of the bend, a relatively complex flow structure is observed at eighth and ninth inter-rib spaces. Flow goes toward the inner wall and circulates around the eighth inter-rib area. After the third rib, heat transfer pattern gets relatively similar to the one of first passage. However, difference of flow field due to U-turn can still be effective through the whole second passage by creating an asymmetrical heat transfer distribution. The Nu distribution close to the outer side wall is observed to be higher than the average values for all numerical results.

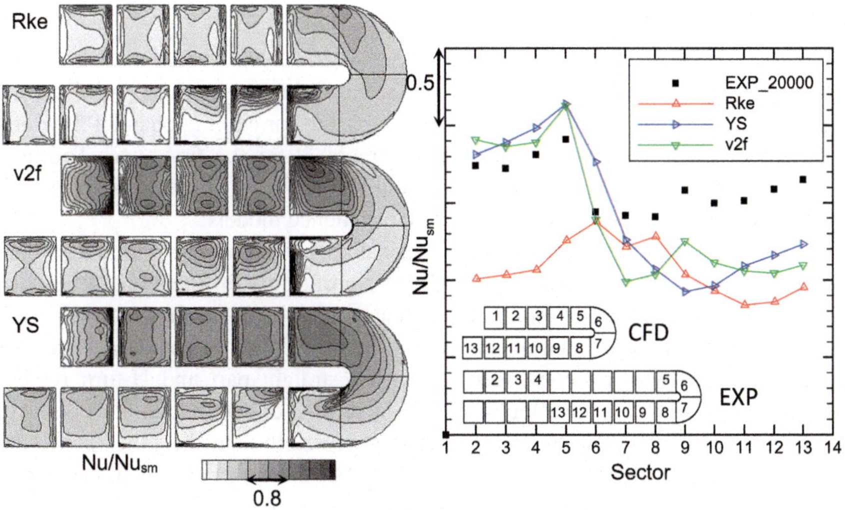

Fig. 10.17 Nu/Nu_{sm} distributions and inter-rib averaged results

Normalized heat transfer results of ribbed channel results show the augmentation of heat transfer due to rib implementation to the cooling duct. Figure 10.17 presents the normalized heat transfer distribution and inter-rib area averaged results over the channel model. The outputs are also compared with the experimental data.

The effect of ribs at the first pass still reaches the turning part. That sharp decrease at the bend shows that unribbed turn area shows the similar heat transfer behavior for smooth and ribbed cases. In other words, the ribs at the upstream channel have less significant effect on U-turn section when compared to the first pass. Nu/Nu_{sm} distribution is relatively lower at the second passage. Nu/Nu_{sm} increases constantly due to the redevelopment of flow to the rib-roughened channel. The heat transfer enhancement level of first passage is almost obtained at 12th to 13th sectors for the experiments.

Although the Nu distribution is well predicted by the Rke and low Re YS model compared to the v2f, the cooling effectiveness results show a different conclusion. In the first channel, the cooling effectiveness results of low Re YS and v2f models get close to the experimental results. The Rke model fails to predict the effectiveness value correctly because it estimates high level of Nu for the smooth channel case. On the other hand, Rke model is quite successful when the cooling effectiveness is predicted at the U-turn region. In the second pass, all three models underpredict the experimental results.

The overall trend of the cooling effectiveness of the second channel is only captured by low Re YS turbulence model. The local maximum at the second inter-rib region at the second pass (ninth inter-rib area) can be captured by only v2f model.

10.5 Conclusions

The present work refers to the heat transfer inside a two-pass one wall ribbed channel with circular U-bend. An experimental setup for cooling channel flow and heat transfer investigations is designed and operated for the first time. The heat transfer and heat transfer enhancement results are obtained by experiments, and numerical analysis is experienced using five turbulence models.

The results of experimental and numerical investigation show that the heat transfer behavior of a two-pass channel shows significantly different characteristics in first and second pass. Solutions of the first inter-rib areas at first pass, U-bend, and downstream of the U-bend of second pass are compared. In general, turbulence models come up with different results for first straight part and U-turn region. Solutions for second pass are affected mostly by turning effects, and models perform in different manner when compared to first pass results. On the other hand, the smooth channel results show similar trend for all three turbulence models with different deviations from experimental results.

For smooth case, heat transfer decreases along the first pass until the thermal boundary layer is developed. Turning effect gradually increases the Nusselt number level, and its effect remains through the whole second pass. The turning effect gradually increases the heat transfer for both smooth and ribbed cases through the U-bend. That effect causes the abrupt decrease in the heat transfer enhancement factor starting with the bend at 11th sector (for experimental geometry).

All v2f, low Re YS, and Rke models give higher Nusselt number results for the smooth wall case. Among the three, Rke has the closest results as it captures the heat transfer perfectly at the first pass and gives higher Nusselt number results at the U-bend section and second pass. v2f and low Re YS models show relatively good agreement with the experimental results for ribbed case. The other turbulence models underpredict the 90° ribbed channel heat transfer. The heat transfer in the U-bend region is overpredicted by v2f and low Re YS models. The sudden decrease of Nu/Nu_{sm} at the U-bend section is predicted larger than the experimental results. The paper points out the necessity of the ribs at the U-bend vicinity regarding the heat transfer enhancement with the ribs with respect to the case without ribs. Regarding the turbine cooling channel, rib implementation has to be with good care because cooling channel designers have to consider the necessity of ribs and augmentation of heat transfer due to ribs. In terms of turbulence modeling, authors want to emphasize that the performance of RANS modeling needs to be improved for heat transfer prediction at cases having ribs which represent forward and backward facing step. The combination of ribs on the wall and the turning of the channel is another challenge from which heat transfer prediction needs to be revealed.

Acknowledgments The authors would like to thank TEI (Tusas Engine Industries) for the help of conducting the experiments and acknowledge the financial support of the European Commission to the project RBF4AERO.

References

1. Han, J. C., Glicksman, L. R., & Rohsenow, W. M. (1978). An investigation of heat transfer and friction for rib roughened surfaces. *International Journal of Heat and Mass Transfer, 21*(8), 1143–1156.
2. Han, J. C. (1984). Heat transfer and friction in channels with two opposite rib-roughened walls. *ASME Journal of Heat Transfer, 106*(4), 774–781.
3. Han, J. C., Park, J. S., & Lei, C. K. (1985). Heat transfer enhancement in channels with turbulence promoters. *Journal of Engineering for Gas Turbines and Power, 107*(3), 629–635.
4. Han, J. C. (1988). Heat transfer and friction characteristics in rectangular channels with rib turbulators. *ASME Journal of Heat Transfer, 110*(2), 321–328.
5. Han, J. C., & Park, J. S. (1988). Developing heat transfer in rectangular channels with rib turbulators. *International Journal of Heat and Mass Transfer, 31*(1), 183–195.
6. Han, J. C., Ou, Q., Park, J. S., & Lei, C. K. (1989). Augmented heat transfer in rectangular channels of narrow aspect ratios with rib turbulators. *International Journal of Heat and Mass Transfer, 32*(9), 1619–1630.
7. Park, J. S., Han, J. C., Huang, Y., & Ou, S. (1992). Heat transfer performance comparisons of five different rectangular channels with parallel angled ribs. *International Journal of Heat and Mass Transfer, 35*(11), 2891–2903.
8. Han, J. C., Dutta, S., & Ekkad, S. V. (2012). *Gas turbine heat transfer and cooling technology* (2nd ed.). Boca Raton: CRC Press.
9. Ekkad, S. V., & Han, J. C. (1997). Detailed heat transfer distributions in two-pass square channels with rib turbulators. *International Journal of Heat and Mass Transfer, 40*(11), 2525–2537.
10. Çakan, M. (2000). Aero-thermal investigation of fixed rib-roughened internal cooling passages. Ph. D. Thesis, Universite Catholique de Louvain.
11. Vass, P. (2005). Large Eddy Simulation of a ribbed duct flow with FLUENT: Effect of rib inclination. Ph. D. Thesis.
12. Rowe, M. (1970). Measurements and computations of flow in pipe bends. *Journal of Fluid Mechanics, 43*, 771–783.
13. Azzola, J., Humphrey, J. A. C., Iacovides, H., & Launder, B. E. (1986). Developing turbulent flow in a U-bend of circular cross-section: Measurement and computation. *Journal of Fluids Engineering, 108*(2), 214–221.
14. Schabacker, J., Bölcs, A., Johnson, B.V. (1998). PIV investigation of the flow characteristics in an internal coolant passage with two ducts connected by a sharp 180° bend. Int Gas Turbine & Aeroengine Congress & Exhibition.
15. Wang, T. S., & Chyu, M. K. (1994). Heat convection in a 180-deg turning duct with different turn configurations. *Journal of Thermophysics and Heat Transfer, 8*(3), 595–601.
16. Chyu, M.K. (1989). Regional heat transfer and pressure drop in two-pass and three-pass flow passages with 180-degree sharp turns. Gas Turbine and Aeroengine Congress and Exposition.
17. Saha, K., & Acharya, S. (2013). Effect of bend geometry on heat transfer and pressure drop in a two-pass coolant square channel for a turbine. *Journal of Turbomachinery, 135*(2), 1–12.
18. Hirota, M., Fujita, H., Syuhada, A., Araki, S., Yoshida, T., & Tanaka, T. (1999). Heat/mass transfer characteristics in two-pass smooth channels with a sharp 180-deg turn. *International Journal of Heat and Mass Transfer, 42*(20), 3757–3770.
19. Chen, W., Ren, J., & Jiang, H. (2011). Effect of turning vane configurations on heat transfer and pressure drop in a ribbed internal cooling system. *Journal of Turbomachinery, 133*(4), 1–11.
20. Mochizuki, S., Murata, A., & Fukunaga, M. (1997). Effects of rib arrangements on pressure drop and heat transfer in a rib-roughened channel with a sharp 180 deg turn. *Journal of Turbomachinery, 119*(3), 610–616.
21. Murata, A., & Mochizuki, S. (2004). Large Eddy simulation of turbulent heat transfer in a rotating two-pass Smooth Square channel with sharp 180° turns. *International Journal of Heat and Mass Transfer, 47*(4), 683–698.

22. Murata, A., & Mochizuki, S. (2004). Effect of rib orientation and channel rotation on turbulent heat transfer in a two-pass square channel with sharp 180° turns investigated by using large eddy simulation. *International Journal of Heat and Mass Transfer, 47*(12–13), 2599–2618.
23. Wang, C., Wang, L., & Sundén, B. (2015). Heat transfer and pressure drop in a smooth and ribbed turn region of a two-pass channel. *Journal of Applied Thermal Engineering, 85*, 225–233.
24. Kavas, İ. (2015). Cooling performance investigation of a two-pass rib-roughened channel. M. Sc. Thesis, METU.
25. Yasa, T., Kavas, İ., Yılmaztürk, S., & Kurtuluş, D. F. (2016). *Flow field investigation of rib-roughened serpentine channel. Sustainable aviation energy and environmental issues chapter 29* (pp. 351–361). Switzerland: Springer International Publishing.
26. "No Title." [Online]. Available: http://www.plexiglas.com/export/sites/plexiglas/.content/medias/downloads/sheet-docs/plexiglas-mc.pdf
27. "No Title." [Online]. Available: http://www.plexiglas-shop.com/pdfs/en/211-1_PLEXIGLAS_GS_XT_en.pdf
28. Xie, G., Li, S., Zhang, W., & Sunden, B. (2013). Computational fluid dynamics modeling flow field and side-wall heat transfer in rectangular rib-roughened passages. *Journal of Energy Resources Technology, 135*(4), 042001.
29. Reynolds, W.C. (1987). Fundamentals of turbulence for turbulence modeling and simulation. Lecture Notes for Von Karman Institute Agard Report No. 755.
30. Shih, T. H., Liou, W. W., Shabbir, A., Yang, Z., & Zhu, J. (1995). A new k-ε Eddy-viscosity model for high Reynolds number turbulent flows – model development and validation. *Journal of Computers and Fluids, 24*(3), 227–238.
31. Launder, B. E. (1989). Second-moment closure: Present... and future? *International Journal of Heat and Fluid Flow, 10*(4), 282–300.
32. Launder, B. E., Reece, G. J., & Rodi, W. (1975). Progress in the development of a Reynolds-stress turbulence closure. *Journal of Fluid Mechanics, 68*(3), 537–566.
33. Rau, G., Çakan, M., Moeller, D., & Arts, T. (1998). The effect of periodic ribs on the local aerodynamic and heat transfer performance of a straight cooling channel. *Journal of Turbomachinery, 120*(2), 368.

Part III
Energy and Environmental Issues

Chapter 11
Environment-Friendly Technology of Airport's Sewerage

Sergii Shamanskyi and Sergii Boichenko

11.1 Introduction

Several types of sewage waters are generated during airport functioning. They can be divided into industrial, domestic, and rain waters. It is known that 1 m^3 of discharged sewages, which have been not sufficiently purified, can result in a situation when from 10 to 50 m^3 of natural clean water becomes unsuitable for many types of usage.

In [17], treatment technologies, employed at treatment plants, depend on sewage water types and their compositions and can differ substantially. Quality of treated waters can also differ depending on technology. Considerable amount of harmful sediments are created during all types of treatment processes. They also require to be utilized. In [2], at the same time, treatment plants are objects, where it is possible to get additional alternative energy sources and other valuable materials.

Therefore, searching for environmentally friendly and energy-efficient ways to organize airport sewage systems is an actual task.

11.2 Research Methods

There was a system approach applied while doing this research. The approach was used to analyze different methods of anaerobic digestion intensification and to form new technological requirements to digestion organization. The approach was also used to analyze highs and lows of conventional methods of microalgae cultivation and to form propositions to organize the cultivation more effectively. There were

S. Shamanskyi (✉) • S. Boichenko
National Aviation University, Kyiv, Ukraine
e-mail: shamanskiy_s_i@ukr.net; chemmotology@ukr.net

T.H. Karakoç et al. (eds.), *Advances in Sustainable Aviation*,
DOI 10.1007/978-3-319-67134-5_11

applied deductive methods for studying four different stages of anaerobic digestion and optimum characteristics of the mediums to accomplish them more efficiently. The same methods were applied for researching different techniques of using photobioreactors for microalgae cultivation. There were applied inductive methods to develop new components of the technological process.

11.2.1 Analysis of the Research and Publications

While developing sewerage of an airport and other airline enterprises, it is necessary to take into account several factors, such as amount and composition of sewage waters, which are discharged, possibility and reasonability of treating them on local treatment plants, possibilities of different treatment technologies application, possibility and reasonability of valuable materials extraction out of the sewage water subsequent using of these materials, possibility of obtaining and utilizing additional energy resources at treatment plant, etc.

Quality of treated sewage water determines its subsequent influence on environment. In accordance with the concept of development of the state enterprise "Boryspil International Airport," when the airport achieves its full handling capacity of 66.5 million passengers per year, quantity of just sanitary sewage water will make up to 18,500 m^3 per day. When discharging this treated water insufficiently, considerable amount of pollution will get into the environment. At such amounts of sewage water treating, about 180 m^3 of sludges will be produced during treatment process. These sludges will need utilization. In the conditions of insufficient supply with freshwater, the question arises about possibility of sewage water recycling and using for technical purposes. Water recycling dictates the necessity of increasing treated water quality. In accordance with the conception, prospective electrical power demand of the airport will go up to more than 88 MW; its thermal power demand on heating and hot water supply will exceed 280 MW. In the conditions of shortage in power resources arises a question about partial power supplying, using own alternative energy sources.

Conventional sewage water treatment facilities usually apply such methods as mechanical treatment, biological treatment, and biochemical treatment. Using these methods, treatment plants are not always able to provide sufficient purification of the water, especially when pollution content in received sewage water substantially varies. It is advisable to apply additional purification facilities for aftertreatment of sewage waters. Methods of nature purification are often used for these purposes. In [8, 12], some of the most popular installations are used: special open oxidation ponds, where different aquatic organisms are cultivated; special agricultural tanks filled with hydroponics, where different aquatic plants are grown; special soil filter systems, where different soil plants can grow; and others [10].

One of the main drawbacks of these methods is that all of them require large areas both for their placement and for sanitary protective zones, which have to be established around them.

One more drawback, which can be named, is that these methods can give little benefits apart from doing final sewage water treatment. On the other hand, sewage water can be directly used as medium for aquatic organism cultivation. Different algae can be used as cultivated organisms. It has many advantages.

Among the algae cultures, which can be cultivated, the most perspective can be microalgae [21]. Some of their types are very rich on lipids. The next culture can be recomended for cultivating, which have percentage of lipids in them:

- *Botryococcus braunii* – 25 ... 85%;
- *Neochloris oleoabundans* – 35 ... 54%;
- *Stichococcus* sp. – 40 ... 59%;
- *Nannochloropsis* sp. – 31 ... 68%;
- *Dunaliella tertiolecta* – 36 ... 42%;
- *Dunaliella salina* – 16 ... 44%;
- *Haematococcus pluvialis* – 25 ... 45%;
- *Scenedesmus dimorphus* – 16 ... 44%;
- *Prymnesium parvum* – 22 ... 38%;
- *Tetraselmis suecica* – 20 ... 30%;
- *Chlorella* sp. – 28 ... 32%;
- *Chlorella vulgaris* – 14 ... 22%;
- *Isochrysis galbana* – 22 ... 38%;
- *Euglena gracilis* – 14 ... 20%.

In [24] due to the high lipid content, these microalgae can be the sources of motor fuel of the third generation.

Conventional facilities for microalgae cultivation, so-called bioreactors, can be divided into two types: the open type and the closed type. The main drawbacks of open-type bioreactors are vulnerability to weather conditions, substantial losses of CO2 during cultivation, and complicacy in providing continuous cultivation and for closed-type bioreactors – complicacy in feeding CO2 inside the reactor and limitedness in productivity because of low CO2 solubility in water.

Integral part of any water treatment process is producing some amount of sludge. At sanitary sewage water treatment plants, the sludge can be of different types. But primarily, it is row sludge from primary settlers and activated sludge from secondary settlers. Altogether, the total volume of them can reach 1% of the volume of treated water. Compositions of these sludges are different. The first one contains from 65% to 75% of organic substances and the second one from 70% to 75%.

In [11], organic part of both sludges contains primarily hydrocarbons, proteins, and fats. Altogether they make up to 85% of all organic part with respect to the mass. The rest is a group of different lignin-humus compounds.

In [3], utilization of these sludges is an actual task. Among the known methods of sludge utilization, it is possible to name these methods:

- Combustion
- Using as material for construction purposes
- Using as food additives for some animals
- Using in agriculture as organic fertilizer and others

Combustion method has become widespread. Burning of sludges is carried out mainly in multiple-bedded furnaces or drum-type furnaces with fluidized bed. This is a quick method to get rid of the sludges. Moreover, ash, which is formed during combustion, has characteristics, which allow using it for different purposes. For example, adding this ash to sludge before dehydration allows accelerating the process with using fewer amounts of other chemicals.

However, it has shortcomings, an economic factor in particular, because efficiency of the process is questionable. It depends on two main characteristics of the sludge: its specific lower heat value and its humidity. Heat value, in turn, depends on percentage and composition of organic part of the sludge. Amount of thermal energy, which can be obtained while burning 1 kg of the sludge, can be calculated by the formula

$$Q_{\mathrm{burn}} = q_{\mathrm{org}} \cdot m_{\mathrm{org}}, \, [\mathrm{MJ/kg}]\,, \tag{11.1}$$

where q_{org} is the law heat value of the sludge's dry organic part in MJ/kg and m_{org} is the mass of the dry organic part in 1 kg of the wet sludge (kg/kg). Mass of the dry organic part, which is included in 1 kg of the sludge, can be defined by the formula

$$m_{\mathrm{org}} = \left(1 - \frac{w}{100}\right) \cdot \frac{c}{100}, \, [\mathrm{kg/kg}]\,, \tag{11.2}$$

where w is the humidity of the sludge (percent of the water in the sludge) and c is the content of organic matter in dry part of the sludge with respect to the mass, in %. Amount of thermal energy, which will be consumed on water evaporation of 1 kg of the sludge, can be calculated by the formula

$$Q_{\mathrm{evapor}} = q_{\mathrm{evapor}} \cdot \frac{w}{100}, \, [\mathrm{MJ/kg}]\,, \tag{11.3}$$

where q_{evapor} is the total consumption of thermal energy on water heating and evaporation, in MJ/kg.

Sanitary sludges usually have specific heat value q_{org} from 16.8 to 27.8 MJ per 1 kg of their dry organic matter. Content of organic matter c makes up from 65% to 75% of the sludge's dry part.

Let us calculate energy balance of raw sludges, which are just out of settlers. Humidity of raw sludges w can vary from 96% to 98%. Mass of its dry organic part, according to the formula 11.2, will be in the range of

$$m_{\mathrm{org}}^{\mathrm{raw}} = \left(1 - \frac{(96 \div 98)}{100}\right) \cdot \frac{(65 \div 75)}{100} = 0,013 \div 0,03, \, (\mathrm{kg/kg})\,.$$

The specific amount of thermal energy, obtained from burning such sludge, according to the formula 11.1, will be in the range

$$Q_{\mathrm{burn}}^{\mathrm{raw}} = (16.8 \div 27.8) \cdot (0.013 \div 0.03) = 0.218 \div 0.822, \, (\mathrm{MJ/kg})\,.$$

In [11], total consumption of thermal energy on heating and evaporating of 1 kg of water, which is in the sludge, is $q_{\text{evapor}} = 4.2$ MJ/kg. So the amount of thermal energy on evaporating, while burning 1 kg of raw sludge, according to the formula 11.3, will be in the range

$$Q_{\text{evapor}}^{\text{raw}} = 4.2 \cdot \frac{(96 \div 98)}{100} = 4.032 \div 4.116, (\text{MJ/kg})\,.$$

As one can see, in the case of burning raw sludge, even in theory, the amount of energy, which is needed for evaporating, is much bigger than the amount of energy obtained from burning. It means that raw sludge cannot be burned independently. It requires a lot of additional fuel to fill the gap in energy balance. After mechanical dewatering, the sludge can have humidity from 75% to 85%. Let us calculate the same energy balance for this sludge. Mass of its dry organic part, according to the formula 11.2, will be in the range of

$$m_{\text{org}}^{\text{dewater}} = \left(1 - \frac{(75 \div 85)}{100}\right) \cdot \frac{(65 \div 75)}{100} = 0,0975 \div 0,1875, (\text{kg/kg})\,.$$

The specific amount of thermal energy, obtained from burning in this case, according to the formula 11.1, will be in the range

$$Q_{\text{burn}}^{\text{dewater}} = (16.8 \div 27.8) \cdot (0.0975 \div 0.1875) = 1.638 \div 5.1375, (\text{MJ/kg})\,.$$

The amount of thermal energy on evaporating in this case, according to the formula 11.3, will be in the range

$$Q_{\text{evapor}}^{\text{dewater}} = 4.2 \cdot \frac{(75 \div 85)}{100} = 3.15 \div 3.57, (\text{MJ/kg})\,.$$

One can see from these calculations that positive energy balance, even in theory, can be achieved only while incinerating dewatered sludge with high specific heat value of its dry organic matter.

Environmental impact of combustion method can also be far from favorable. Burning products are discharged into the atmosphere. If the sludge contains salts of heavy metals, they go into the atmosphere too. To prevent atmospheric pollution, it is necessary to apply costly equipment for burning products purification.

While utilizing sludges for construction purposes, high energy-consuming pre-treatment is also needed, in particular drying.

Activated sludge is very rich with proteins and contains a lot of different vitamins, especially vitamin B12. It makes possible to use it as a food additive to some animals' diets.

One of the most promising ways of sludge utilization can be using them as organic fertilizers. The sludges contain also a large amount of nutrients, which are valuable for agricultural fields.

At the same time, the sludges can contain a lot of pathogenic microorganisms and also salts of heavy metals. In addition, raw sludges are inclined to rotting processes, emitting methane into the atmosphere. All of these facts restrain using sludges as fertilizers.

One of the best ways to exterminate pathogens and, at the same time, to stabilize sludges, eliminating possibility of rotting, is anaerobic fermentation of sludges in methane tanks. If the process takes place at thermophilic conditions, it allows exterminating over 98% of pathogens.

In [1, 4], it is understandable that salts of heavy metals have to be removed from the sludges before fermentation and using them as fertilizers [20]. It is because they inhibit the process of fermentation and have very negative effect on agriculture. The presence of heavy metals is unwanted for almost all methods of sludge utilization. For example, after incineration, they can be discharged into the atmosphere along with combustion materials. In [14], there are methods of extracting heavy metals from sludges, but this is not the issue of this research.

Due to the presence of large amounts of organic substances in sludges, it allows obtaining energy-valuable biogas during fermentation. In [5], this biogas can be used for own purposes of a sewage treatment plant or can be sold to other gas consumers.

It may be possible to say that the main shortcoming of anaerobic fermentation is that conventional technologies of fermentation and, consequently, constructions of methane tanks are not perfect.

At conventional technologies, all the fermentation process is organized in the way that raw sludge, which is going to be fermented, goes into the methane tank and is mixed with the sludge, which is being fermented already. Mixing of these two sludges disturbs microflora, which is responsible for fermentation. It causes inhibitory effect on it. It results in slowing down of the fermentation and in decreasing completeness of organic matter decomposition. These, in turn, results in decreasing of biogas yield. In many cases, such technologies are considered to be not efficient.

So, it is a very actual task to develop new technologies of fermentation, which are based on results of latest scientific researches of processes, occurring in a methane tank during fermentation [26].

In [25], there is a large variety of methane tank constructions, which are classified according to different characteristics. There are also very many methods of intensification of fermentation processes in methane tanks. The methods can be combined in the next two groups [13].

The first group can be called biological methods. It includes cultivating and using new more efficient microorganisms, responsible for fermentation, like "Methanobacterium kadomensis St.23," obtained during laboratory research [26]; adding different catalysts to the fermented sludge, which accelerate the process [6]; and providing immobilization of microorganisms, when they are attached to different additional materials, such as fiberglass, pebbles, and so forth, which the inside capacity of a methane tank is filled with. Immobilization prevents efflux of microorganisms along with fermented sludge during unload process [15]; applying common fermentation of different types of wastes, so that they help each other to be fermented easier and others.

The second group can be called technological methods. It includes agitation of the sludge, so that microorganisms were distributed more evenly over the inner space of the methane tank and forming of dead spaces was prevented, where fermentation does not take place due to stagnation; application of different types of sludge pretreatment, such as mechanical destruction of clods in it, acid hydrolysis, alkaline hydrolysis, thermal hydrolysis, ultrasound irradiation, and so forth; and fermentation organized as several-stage process, so that it will be possible to prevent inhibitory effect of different groups of microorganisms on each other.

Some authors consider the last method as very difficult to realize. It is because they believe that designs of methane tanks have to be very complicated and it would be necessary to use additional also complicated devices [25].

In order to form propositions relative to new technology, let us consider technological methods.

The major resulting products of anaerobic fermentation are methane and carbon dioxide. They are formed as a result of decomposition of organic matter, which is present in the sludge. Over 90% of it can be decomposed. Researches, which have been made recently, show that rate of decomposition and percentage of decomposed organic matter at the end of the process depends on initial composition of the sludge and conditions, under which the process of fermentation takes place. The research allows applying new approaches to intensify the process, to make it more efficient and to propose new technology.

In conventional fermentation technologies, the process is usually organized in the next ways [9].

The first one – a methane tank consists of just one capacity, and all fermentation takes place in it. The process is continuous. It means that after equal intervals, portions of the sludge, which is being fermented, are unloaded out of the methane tank, and, at the same time, new portions of raw sludge are loaded inside. After that, raw sludge is mixed with fermented one to be involved in the fermentation process. The obvious drawback of this organization is that raw sludge is constantly mixed with fermented one, and, after regular unloading of the methane tank, in the unloaded portion of fermented sludge, there is always a certain concentration of unfermented (raw) sludge. It reduces biogas yield, because part of organic matter remains not decomposed and allows some pathogenic microorganisms to be carried out from the methane tank and to end up in the final fermented sludge. It raises environmental risks while using this sludge as fertilizer.

The second one – a methane tank consists of two capacities. In this case, intensive fermentation takes place in the first capacity with intensive biogas releasing. The second capacity serves for waning the process and for compacting the sludge, when water is partly separated from it. Such organization allows obtaining more condensed sludge with lower humidity and therefore allows decreasing volumes of sludge concentration tanks, for subsequent sludge dehydration. But the same drawbacks as in the previous way of fermentation are present here. It is because in spite of having two capacities, the main part of the process almost entirely occurs in the first one.

The third one – a methane tank consists of more than two capacities. The process is very similar to the second way of organizing fermentation. Additional capacities are added to the methane tank for better sludge concentration. It allows decreasing volume of fermented sludge due to lower humidity, but it also leads to necessity to create significantly more complicated design of the methane tank. Even so, characteristics of fermentation itself are not substantially improved.

Composition of microorganisms, which are responsible for fermentation, their activity, rate of reproduction, and, as a result, final products of their metabolism not only depend on raw sludge composition but also depend on conditions in the methane tank. Latest researches show that vital functions of microorganisms in anaerobic fermentation differ from the same functions during different aerobic processes substantially. Relationships between aerobic microorganisms are usually established in a way that some of them are predators and some are victims. So they compete with each other. In anaerobic processes, waste products of one type of organisms become nutrients for others. So there is no such competition [9]. All fermentation can be considered as several consecutive, relatively independent metabolic processes, when each of it involves different kinds of microorganisms [27].

It is possible to distinguish the next four groups of anaerobic microorganisms.

The first group provides hydrolysis of organic matter, when complicated compounds are decomposed to more simple organic substances, which are accessible for fermentation. Proteins hydrolyze into soluble peptones; hydrocarbons hydrolyze first into polysaccharides, then to oligosaccharides, and then to monosaccharides. Fats hydrolyze into glycerides, phospholipids, and others.

The second group provides acidogenesis, during which it produces volatile fatty acids, long-chain fatty acids, and alcohols. For this production, they use waste products, which were formed by the first group.

The third group provides acetogenesis. By using volatile fatty acids, long-chain fatty acids, and alcohols, the group produces acetate (acetic acid).

The final fourth group provides methanogenesis. Using acetate, it produces methane, which is the basic component of final biogas.

Among the characteristics of the medium that define activity and production rate of each group of microorganisms are medium temperature, its pH, concentration of oxygen, and others. For some groups of microorganisms, favorable values of these characteristics are different, sometimes substantially [16]. Especially different conditions require the second and the fourth groups of microorganisms, which are responsible for acidogenesis and methanogenesis. At conventional fermentation technologies, when all four parts of the process occur in a single capacity, it is impossible to create favorable conditions for all groups. So they usually try to benefit methanogenic organisms, because they are more susceptible to medium characteristics and grow more slowly. In this case, because of having different favorable conditions, activity of acidogenic microorganisms decelerates, and the amount of created volatile fatty acids, long-chain fatty acids, and alcohols decreases. Having less amount of nutrition, acetogenic organisms produce less amount of acetate. Because acetate serves as part of the nutrition for methanogenic organisms, final biogas yield goes down.

It is possible to create more favorable conditions for acidogenic microorganisms in the capacity. But it also makes no good. In these conditions, vital activity of the second group speeds up. It leads to producing more acids and alcohols that, in turn, causes substantial dropping of medium pH. Low value of medium pH inhibits vital activity of methanogenic microorganisms and also leads to producing less of biogas. It is obvious that the solution should be in searching such conditions, where there is equilibrium between vital activities of acidogens and methanogens.

But there also can be other solutions. According to results of latest research, if acidogenic and methanogenic anaerobic microorganisms are separated in space vital activity, both of them can be increased considerably. It is possible to conclude about favorable parameters of the medium not only for both of these groups of organisms but for the all four of them [18].

So, to intensify the process more effectively, it is necessary to divide all the process into four consecutive stages: hydrolysis, acetogenesis, acidogenesis, and methanogenesis. After dividing, it is necessary to use an individual capacity for each stage. In each capacity, it is necessary to create favorable conditions for microorganisms, responsible for accomplishing the stage.

At the stage of hydrolysis, medium conditions should stimulate decomposition of complicated organic compounds. At this stage, different pretreatments of sludge can be applied, such as mechanical grinding of the sludge, adding acids or alkalis to it, heating and keeping for some time at high temperature, and ultrasound irradiation of the sludge [7].

It is obvious that mechanical grinding can do no good for organic decomposition intensification. It is because sewage water sludges initially consist of very small particles.

Alkalis, while being added to sludges, stimulate decomposition better than acids.

Keeping the sludge for some time at high temperature gives good results; decomposition goes much faster, but it requires a lot of energy on heating process.

Irradiation with ultrasound causes quicker hydrolysis than other methods [7]. Ultrasonic cavitation arises under the influence of ultrasound. The cavitation causes mechanical destruction of organic matter. Intensity of destruction depends on cavitation intensity and, therefore, on intensity of irradiation. Increase of irradiation intensity leads to increase of energy expenditure. Adding barbotage of liquid medium with gas to irradiation allows increasing intensity of the cavitation [19]. It can be explained this way. Gaseous bubbles, which enter the medium, become cavitation nucleus, intensifying the process considerably with less additional expenditure of energy. The rate constant of organic matter destruction increases up to 2.6 times and biological matter up to 4.3 times.

Increase of pressure in the medium on the one hand intensifies bubble collapse and, thereby, increases local pressure rising, which in turn stimulates intensification of hydrolysis; on the other hand, it raises cavitational threshold, hampering appearance of cavitational bubbles themselves. Optimum value of excessive pressure, at which value rate constant of organic matter destruction becomes maximal, is 0.5 · 105 Pa. In comparison with conducting the same process at atmospheric pressure, the constant organic constituent increases up to 1.2 times and the constant of biological constituent up to 1.5 times [19].

Barbotage ought to be carried out with inert gas, which does not contain oxygen, in order to not to disturb medium in a methane tank. It is appropriate to use carbon dioxide, which is obtained during subsequent digestion.

In deeply hydrolyzed sewage water sludge, acidogenesis can be accomplished pretty fast [18]. The main final products, which are created during this stage, are long-chain fatty acids; volatile fatty acids, such as propionic acid, butyric acid, etc.; and alcohols. Along with them, carbon dioxide, small amount of hydrogen, and ammonia are released. It is also possible that hydrogen sulfide is created if sulfur is present in the sludge.

Total composition of final products at the end of this stage depends on partial pressure of hydrogen in the capacity. At high partial pressure, vital activity of microorganisms, responsible for this stage, goes up. They produce more ethanol, propionic and butyric acids, and other volatile fatty acids. Because of this, medium pH can drop from 5.2 to 5.0 [9]. At these conditions and at atmospheric pressure, carbon dioxide, which is produced at this stage, is released easily. But methane is not produced almost at all (no more than 4% of carbon dioxide quantity). It is because low pH of the medium inhibits vital activity of methanogenic microorganisms [27]. So, gaseous products of this stage will consist mostly of carbon dioxide with small amount of hydrogen, methane, and with admixture of ammonia and hydrogen sulfide. They can be withdrawn out from the capacity and will not end up in the final biogas.

At the low partial pressure of hydrogen, vital activity of acidogenic microorganisms is inhibited, and, instead of it, acetogenic microorganisms become more active. It leads to the situation when acids, which could have been created otherwise, are not created and are not processed into methane by methanogenic microorganisms. It results in decreasing final methane output. To conduct acidogenic stage more efficiently, it is necessary to maintain higher partial pressure of hydrogen in the capacity by injecting its additional amount into it.

Research shows that the stage of acidogenesis can be accomplished in 10 h if the conditions in the capacity are favorable [7]. The group of acetogenic microorganisms is divided into two types. The first type makes acetate by decomposing long-chain fatty acids, volatile fatty acids, and alcohols. For example, propionic and butyric acids are decomposed according to the next formulas

$$\mathrm{CH3CH2COOH + 2H2O \rightarrow CH3COOH + CO2 + 3H3} \quad (11.4)$$

$$\mathrm{CH3CH2CH2COOH + 2H2O \rightarrow 2CH3COOH + 3H3} \quad (11.5)$$

The second type makes the same acetate by using carbon dioxide and hydrogen according to the next formula

$$4H2 + 2CO2 \rightarrow CH3COOH + H2O \quad (11.6)$$

The second type of microorganisms cannot use all of the carbon dioxide and all of the hydrogen. So, a lot of carbon dioxide and some amount of hydrogen are released as gases.

Vital activities of both types of microorganisms also depend on partial pressure of hydrogen. At low partial pressure, vital activity of the first type becomes more active and of the second group is inhibited. At high partial pressure, the second group becomes more active and the first one is inhibited. The value of hydrogen's partial pressure, when there is equilibrium between vital activities of these two types, is 10 Pa [9]. The first type of microorganisms is more sensitive to increasing the pressure. So it is necessary to keep the partial pressure in the capacity below 10 Pa.

Acetogenesis is necessary to carry out at atmospheric pressure. Then carbon dioxide easily released. It is also necessary to keep oxidation restoration potential of the sludge in the capacity at the level of 330 mV. If this potential goes down, hydrogen is formed actively, which leads to inhibition of the acetogenesis. But at the lower level of this potential, sulfides appear actively in the sludge, which leads to hydrogen sulfide appearing in the gaseous final products of the stage. The group of methanogenic microorganisms (as well as acetogenic) is also divided into two types. First type decomposes acetate to methane and carbon dioxide; second type reduces methane using carbon dioxide. The processes go according to the next formulas

$$CH3COOH \rightarrow CH4 + CO2 \quad (11.7)$$

$$CO2 + 4H2 \rightarrow CH4 + 2H2O \quad (11.8)$$

The stage of methanogenesis requires very strict stabilization of medium characteristics. It is because methanogenic microorganisms are very sensible to them. To have this stage running efficiently, it is necessary to keep pH variations not more than from 7.5 to 8.0. While carrying out the stage at thermophilic conditions, it is necessary to keep temperature variation no more than ± 0.2 °C. Concentration of oxygen must be as low as possible. If it is higher than 0.01 mg per liter, it kills these organisms.

It is possible to improve vital activity of methanogenic microorganisms by adding biostimulants to the sludge. Research shows that adding biostimulant BIOSTIM-SBCH4, which is based on melamine salt of bis(oxymethyl)phosphoric acid, accelerates fermentation process from three to four times and leads to increasing methane concentration in biogas [6, 9].

Methane is produced by two groups of microorganisms in two different ways. The first way is by decomposing acetate (Formula 11.7). At this process, carbon dioxide is produced. It becomes the source of methane production for the second group of microorganisms. They consume carbon dioxide, but for methane producing, they need also hydrogen (Formula 11.8). Research shows that large amount of carbon dioxide remains not consumed because of the lack of hydrogen in the sludge. This is the reason why biogas, obtained using conventional technologies, contains up to 40% of carbon dioxide [13]. To increase the amount of consumed carbon dioxide and consequentially produced methane, it is necessary to add some hydrogen in the sludge at this stage.

Carbon dioxide is soluble in water. Its solubility increases with pressure increasing. Methane, on the other hand, has low solubility, which practically does not depend on pressure. Carbon dioxide, which is produced by the first group of microorganisms and is not consumed by the second group, is usually released as gas. But if the pressure in the capacity is increased, more carbon dioxide dissolves in the water and remains accessible to the second group. Whereas methane continues to be released in gaseous form. So it is necessary to increase the pressure in the capacity, which enables to increase biogas yield and to decrease content of carbon dioxide in the biogas.

11.3 Results and Discussion

The main purpose of the proposed new technology is enhancement of environmental safety and economic efficiency of sewage water treatment processes. The next challenges are met here: diminish harmful impacts of treated sewage waters and their sludges on environment, thereby improving ecology and life conditions, and produce additional alternative energy sources during sewage water treatment and their sludge utilization, thereby enhancing enterprises' energy independence. The proposed sewerage is organized in the way that allows increasing quality of sewage water treatment process and obtaining additional alternative energy sources.

The major idea is that within the bounds of the proposed technology, it is necessary to do the next things. First, it is necessary to organize an additional treatment (posttreatment) of sanitary sewage water, after the water has been treated in conventional treatment facilities, by using it as growth medium for algae cultivation and then producing biofuel out of the algae. Second, it is necessary to organize a process of anaerobic digestion of sanitary sewage water sludges subject to modern scientific concepts, which factors in kinetics of the process.

Technological diagram of the proposed technology is presented on the Fig. 11.1. The technology is functioning in the next way.

Sewage waters from the airport that can be sanitary sewage waters and waters which are close to them by composition go to a conventional sewage water treatment plant. The plant can consist of conventional installations, which apply technology of mechanical, biological, or chemical and biological treatment, depending on sewage water composition. Treated water goes into a photobioreactor for posttreatment. The photobioreactor serves as an installation for using treated in conventional

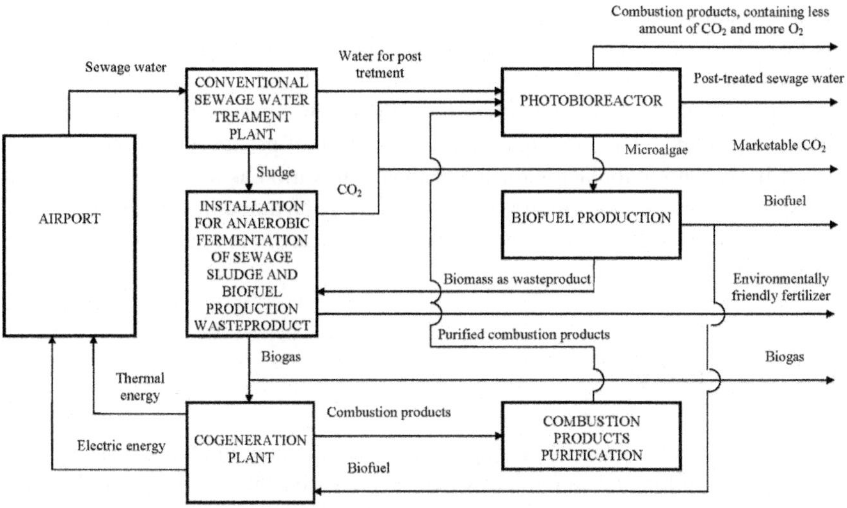

Fig. 11.1 Technological diagram of sewerage organization of an airport

way sewage waters as medium for microalgae cultivation by conversion of solar energy into their biomass. It is proposed to apply microalgae *Botryococcus braunii* for cultivation, as the most suitable culture for motor biofuel production. The installation can be designed as it is described in [22]. It must be placed in an area with enough natural illumination.

To make microalgae cultivation more effective, the following is proposed: using carbon dioxide, obtained during anaerobic digestion of sludge and biomass waste product, conducted according to new technology, and containing carbon dioxide combustion products from cogeneration plant; conducting cultivating process under excessive pressure for increment of carbon dioxide solubility in water; providing possibility to introduce carbon dioxide into the reactor continuously; providing possibility to introduce necessary nutrients for microalgae; and providing continuity of cultivating process. All these are possible while applying the proposed design of installation.

Sewage water, which has been posttreated in the photobioreactor, can be used for technical purposes or can be discharged into a water body.

Cultivated microalgae, unloaded out of the photobioreactor, go to installations for biofuel production. Waste product of the fuel production, which is biomass, is still rich in organic compounds. One of the solutions of utilizing biomass is to ferment it along with sewage water sludge. So the biomass goes to installations for anaerobic fermentation. Produced biofuel can be used at local cogeneration plant or to be sold as marketable product.

Sludges, which are generated during treatment process at conventional treatment facilities, go into installations for anaerobic fermentation. Along with the sludge, biomass as waste product of biofuel production out of microalgae can be placed into the installation and fermented there. The installation can be designed as it is described in [23]. The installation allows obtaining separately biogas with increased methane content and carbon dioxide. Produced biogas can be burned at local cogeneration plant for obtaining heat and electric energy or it can be sold as marketable product. The most reasonable utilization of produced carbon dioxide is using it in the photobioreactor for microalgae cultivation as a source of carbon in photosynthesis. Excessive amount of it, if there is such amount, can also be sold as marketable product. Fermented sludge and biomass can be used as organic fertilizer for agriculture.

11.4 Conclusion

Expected results of organizing sewage water posttreatment in photobioreactors are decreasing of pollutants' amounts, which are released into the environment with sewage waters; obtaining raw materials (microalgae) for producing motor fuel; and reducing amount of carbon dioxide, which is released into the atmosphere.

Expected results of new organization of fermentation process are shortening the process of fermentation approximately from 15 to 3 days, increasing the output of biogas per one volume unit of fermented substance, increasing content of methane in the biogas from 60% to approximately 95%, obtaining marketable carbon dioxide, and obtaining environmentally friendly organic fertilizer, which does not decay and does not contain pathogenic microorganisms.

References

1. Tiruneh, A. T., Fadiran, A. O., & Mtshali, J. S. (2014). Evaluation of the risk of heavy metals in sewage sludge intended for agricultural application in Swaziland. *International Journal of Environmental Sciences, 5*(1), 197–216.
2. Akpor, O. B. (2011). Wastewater effluent discharge: Effects and treatment processes/3rd International Conference on Chemical. *Biological and Environmental Engineering IPCBEE, 20*(2011), 85–91.
3. Alrawi, R., Ahmad, A., Norli, I., & AK, M. (2010). Methane production during start-up phase of mesophilic semi-continues suspended growth anaerobic digester. *International Journal of Chemical Reactor Engineering, 8*(8), 89.
4. Altas, L. (2009). Inhibitory effect of heavy metals on methane-producing anaerobic granular sludge. *Journal of Hazardous Materials, 162*, 1551–1556.
5. Babaev, V. N., Goroch, N. P., & Korinko, I. V. (2011). Power budget of methane formation during mesophilous anaerobic decomposition of organic parts of wastes. *Eastern-European Journal of Advanced Technology, 4*(6), 59–65. (In Russian).
6. Barskiy, E. L., Shandieva, I. O., Savanina, J. V., et al. (2011). Effect of melafen on development of cyanobacteria cultures and green mikroalgae under stress conditions. *Proceedings of the Moscow State University, 1*, 15–20. (In Russian).
7. Danilovich, D. A., Kozlov, M. N., Kevbrina, M. V., & Gusjev, D. V. (2009). Influence of sewage water sludges pretreatment on completeness of methane fermentation processes. *Water: Technologies, Materials, Equipment, Ecology, 2*, 24–26. (In Russian).

8. Gersberg, R. M., Elkins, B. V., Lyon, S. R., & Goldman, C. R. (1986, March). Role of aquatic plants in wastewater treatment by artificial wetlands. *Water Research, 20*(3), 363–368.
9. Gjunter, L. I., & Goldfarb, L. L. (1991). Methanetanks. M.: Strojizdat. 280 p. (In Russian).
10. Healy, A. M., & Cawleyb. (2002). Nutrient processing capacity of a constructed wetland in Western Ireland. *Journal of Environmental Quality, 31*, 1739–1747.
11. Jakovlev, S. V., Karelin, J. A., Zhukov, A. I., & Kolobanov, S. K. (1975). *Sewerage*. Moscow: Strojizdat. 632 p. (In Russian).
12. Kadam, A., Oza, G., Nemade, P., Dutta, S., & Shankar, H. (2008). *Chemosphere, 71*, 975–981.
13. Kovalev, V. V., Ungurjanu, D. V., & Kovaleva, O. V. (2012). Theoretical and practical aspects of improving processes of biogas technology. *Problems of Regional Energetics, 1*, 102–114. (In Russian).
14. Kramariova, U. S. (2011). Prospects of application of tetrasodium salt of ethylene diamine tetraacetate as an extractant of heavy metals from urban sewage water sludges. *Medical Prospects, XVI/4*(11), 127–129. (In Ukrainian).
15. Liao, B. Q., Kraemer, J. T., & Bagley, D. M. (2006). Anaerobic membrane bioreactors. Application and research directions. *Science Technology, 36*, 489–530.
16. Möller, U. (1988). Entseuchung von Klarschlam. Eine Standartbestimmung. *Korrespondenz Fbwasser, 1*, 24–30. (In German).
17. Pavliukh, L. I., & Boichenko S. V. (2012). Application of natural raw plant materials based sorbents in the purification systems of oil contaminated waters of aviation enterprises (Vol. 3, pp. 4.3.48–4.3.50). Proceedings the Fifth World Congress ["Aviation in the XXI-st Century"]. Kiev, 25–27 September.
18. Pereira, M. A., Cavaleiro, A. J., Mota, M., & Alves, M. M. (2003). Accumulation of long chain fatty acids onto anaerobic sludge under steady state and shock loading conditions: Effect on acetogenic and methanogenic activity. *Water Science Technology, 48*, 33–40.
19. Predzimirska, L. M. (2015). Cavitational purification of natural and sewage waters from organic and biological contaminations: aref. of PhD thesis. Ivano-Frankivsk. 21 p. (In Ukrainian).
20. Gaber, S. E., Rizk, M. S., & Yehia, M. M. (2011, March 31). Extraction of certain heavy metals from sewage sludge using different types of acids. *Boikemistry, 23*(1), 41–48.
21. Sorokina, K. N., Jakovlev, V. A., Piligaev, A. V., et al. (2012). Potential of using microalgae as a raw material for bioenergy sector. *Catalysis in Industry, 2*, 63–72.
22. Shamanskyi, S. I. (2015). Continuously working installation for solar energy bioconversion. *Science Intensive Technologies, 2*(26), 115–119. (In Ukrainian).
23. Shamanskyi, S. I., & Boichenko, S. V. (2015). Energy efficient and environmental friendly technology of stabilizing of airline enterprises' wastewater sludges. *East European Journal of Advanced Technology, 5/8*(77), 39–45. (In Ukrainian).
24. Shamanskyi, S. I., & Nestorjak, D. M. (2012). *Bioconversion of solar energy as a perspective direction in alternative energy* (pp. 371–372). "Green Energy". International theoretical and practical conference: 17–19 July 2012, Kyiv.
25. Trakhunova, I. A. (2014). Increasing efficiency of anaerobic treatment of organic wastes in a methane tank with hydraulic agitation on the basis of numerical experiment. aref. of PhD thesis. Kazan'. 19 p. (In Russian).
26. Volova, T. G. (1999). *Biotechnology*. Novosibirsk: Publishing House of Siberian Branch of Russian Academy of Science. 252 p. (In Russian).
27. Ziemiński, K., & Frąc, M. (2012). Methane fermentation process as anaerobic digestion of biomass: Transformations, stages and microorganisms. *African Journal of Biotechnology, 11*(18), 4127–4139.

Chapter 12
Statistical Model Development for Military Aircraft Engine Exhaust Emissions Data

Akhlitdin Nizamitdinov, Yasin Şöhret, Aladdin Shamilov, and T. Hikmet Karakoç

Nomenclature

$\dot{m}$	Mass flow rate [kg/s]
p	Static pressure [Pa]
P	Total pressure [Pa]
P-GAM	Penalized generalized additive model
q	Dynamic pressure [Pa]
T	Temperature [K]

Greek

$x_1, \cdots, x_n$	Knot points
$b_j(x)$	Basis function
$\mathrm{CV}(\lambda)$	Cross-validation method
$f(x)$	Regression function

A. Nizamitdinov (✉)
Department of Statistics, Graduate School of Sciences, Anadolu University, 26470 Eskisehir, Turkey
e-mail: ahlidin@gmail.com

Y. Şöhret
Department of Airframe and Powerplant Maintenance, Süleyman Demirel University, School of Civil Aviation, 32700 Isparta, Turkey

A. Shamilov
Department of Statistics, Faculty of Science, Anadolu University, 26470 Eskisehir, Turkey

T.H. Karakoç
Department of Airframe and Powerplant Maintenance, Anadolu University, Faculty of Aeronautics and Astronautics, 26470 Eskisehir, Turkey

T.H. Karakoç et al. (eds.), *Advances in Sustainable Aviation*,
DOI 10.1007/978-3-319-67134-5_12

$f(\mu_i)$	Link function
GCV(λ)	Generalized cross-validation method
$S(f)$	Minimization function
λ	Smoothing parameter

12.1 Introduction

Energy consumption arising in the last decade leads numerous environmental issues visible or not. Particularly utilization of petroleum-derived fuels is the main source of the carbon emissions. It is a known fact, arise of emitted carbon amount is the basis of global warming, whereas nitrogen oxide emissions cause ozone depletion. Kyoto Protocol and Paris negotiations are milestones for preventing major environmental problems. However, many sectors including transportation, cement, iron, textile, and so on industries are still consuming fossil fuels.

According to the IPCC report [1], aviation is responsible of carbon emission as well as other industries with 3 %. Percentage of aviation emission seems less than others, but it has more influence compared to other industries and sectors. The greenhouse gas emissions induced from other industries affect atmosphere and within under the troposphere, but aircraft-induced greenhouse gas emissions have a significant influence on atmosphere and within above troposphere. Emitted carbon monoxide, carbon dioxide, nitrogen oxides, and unburned hydrocarbons from aircrafts are named as greenhouse gases. Additionally a few number of water vapor and sulfur oxides are emitted from aircrafts. Nitrogen oxides react with other ingredients of the air; meanwhile water vapor, sulfur oxides, and unburned hydrocarbons are inclusive of various microphysical processes in the atmosphere. As a result of the nitrogen oxides' chemical reactions, methane and ozone amount increases. On the other hand, after microphysical process, water vapor, unburned hydrocarbons, and sulfur oxides create contrail and have an influence on amount change of aerosol and water vapor in the atmosphere. Carbon monoxide emissions turn into carbon dioxide in consequence of the chemical reactions. Carbon dioxide emissions formed directly and indirectly are uptaken by the oceans. Besides, contrails and cirrus clouds have a significant influence on radiative forcing. Aircraft-induced cirrus clouds are formed by two ways. Persistent linear contrails spread through wind shear and form cirrus-like cloud structures, commonly named as contrail-cirrus cloud. During cruise phase of the aircraft-induced flight, particulate emissions accumulate in the troposphere layer of the atmosphere and act like cloud. Both of the mentioned cirrus clouds contribute radiative forcing in the upper atmosphere [2, 3]. For this reason many studies associated with aviation-induced emissions have been conducted until today [4–12].

Statistical methods are beneficial for many sciences. Statistic science is essential for, especially, studies carried out on the basis of experiments. Different approaches are applicable for data processing at each step of an experimental study. To understand accuracy, errors (measurement or human factor induced), and uncertainties, basic or advanced statistical methods are commonly used in engineering field.

In the current study, advanced methods of statistics are benefited to extend emission characteristics of a turboprop engine. Additionally, peaks and falls of applied statistical tools are presented with a comparison.

12.2 Theoretical Background of Used Techniques

Regression analysis is a statistical technique for modeling and investigating the relationship between dependent and independent variable. Applications of regression are widely used in almost every field, including economics, engineering, physical and chemical science, biological sciences, management, and the social sciences. Regression analysis may be the most widely used statistical technique.

Many problems in engineering and the sciences involve a study or analysis of the relationship between two or more variables. Most widely used regression technique is linear regression. However, there are many situations where the relationship between variables is not deterministic.

Multivariate regression approach such as generalized additive models, thin plate splines, and penalized generalized additive models have a wide implementation in nonparametric problems.

In this study we used linear multivariate model, smoothing spline, and penalized generalized additive models (P-GAM) for analysis of exhaust emissions data of a turboprop engine. Further we give a short theoretical background of these techniques.

12.2.1 *Smoothing Splines*

Regression spline [13] chooses a basis, defining the space of functions of which f is an element. Choosing a basis amounts to choosing some basis functions, which will be treated as completely known: if $b_j(x)$ is a jth such basis function, then f is assumed to have a representation for some values of the unknown parameters $b_j(x)$ as the following [14, 15]:

$$f(x) = \sum_{j=1}^{q} \beta_j b_j(x) \tag{12.1}$$

Smoothing spline [16] estimate of the function arises as a solution to the following minimization problem: find $\widehat{f} \in C^2(a, b)$ that minimizes the penalized residual sum of squares:

$$S(f) = \sum_{i=1}^{n} \{y_i - f(x_i)\}^2 + \lambda \int_a^b \{f''(x)\}^2 \mathrm{d}x \tag{12.2}$$

for some value $\lambda > 0$. The first term in Eq. 12.2 denotes the residual sum of the squares, and it penalizes the lack of fit. The second term which is weighted by λ denotes the roughness penalty. In other words, it penalizes the curvature of the function f. The λ in Eq. 12.2 is known as the smoothing parameter.

The solution based on smoothing spline for minimum problem in the equation Eq. 12.2 is known as a "natural cubic spline" with knots at $x_1, \cdots, x_n$. From this point of view, a special structured spline interpolation which depends on a chosen value λ becomes a suitable approach of function f in the regression model.

12.2.2 Generalized Additive Models

A generalized additive model [16] is a generalized linear model with a linear predictor involving a sum of smooth functions of covariates. The main structure of the model is as follows:

$$g(\mu_i) = X_i^* \theta + f_1(x_{1i}) + f_2(x_{2i}) + \cdots + f_p(x_{\mathrm{pi}}) \tag{12.3}$$

where $\mu \equiv E(Y_i)$ and Y_i – some exponential family distribution.

In this model Y_i is a response variable, X_i^* is a row of the model matrix for any parametric model components, θ is the corresponding parameter vector, and the f_j are smooth functions of the independent variables, X_i. First, we have to introduce additive models. In additive model each smooth function in Eq. 12.3 can be introduced using a penalized regression spline basis. The spline basis for smooth functions can be represented in this form:

$$f_1(x_1) = \delta_1 + x_1\delta_2 + \sum_{j=1}^{q-2} R\left(x_1, x_{1j}^*\right)\delta_{j+2} \tag{12.4}$$

$$f_2(x_2) = \delta_1 + x_2\delta_2 + \sum_{j=1}^{q-2} R\left(x_2, x_{2j}^*\right)\delta_{j+2} \tag{12.5}$$

where δ_j are the unknown parameters for functions f, q is the number of unknown parameters for f, while x_j^* are the knot location for the function. The smoothness of the function can be measured as the following:

$$\int f_1''(x)^2 \mathrm{dx} = \beta^T S_1 \beta \tag{12.6}$$

$$\int f_2''(x)^2 \mathrm{dx} = \beta^T S_2 \beta \tag{12.7}$$

For Eq. 12.3 the sum of squared error can be found:

$$\sum_{i=1}^{n} (y_i - b_0 - f_1(x_{1i}) - f_2(x_{2i}))^2 + \lambda_1 \int (f_1''(x_1))\, dx_1 + \lambda_2 \int (f_2''(x_2))\, dx_2 \tag{12.8}$$

The parameters β of the model are obtained by minimization of the penalized least-squares objective:

$$\|\mathbf{y}-\mathbf{X}\boldsymbol{\beta}\|^2+\lambda_1\boldsymbol{\beta}^{\mathbf{T}}\mathbf{S}_1\boldsymbol{\beta}+\lambda_2\boldsymbol{\beta}^{\mathbf{T}}\mathbf{S}_2\boldsymbol{\beta} \tag{12.9}$$

Smoothing parameters λ_1 and λ_2 control the weight to be given to the objective of making f_1 and f_2 smooth, relative to the objective of closely fitting the response data. The problem of choosing the smoothing parameter is one of the main problems in curve estimation. If we use fitting curves by polynomial regression, the choice of the degree of the fitted polynomial is essentially equivalent to the choice of a smoothing parameter.\vspace*{-6pt}

12.2.3 Penalized GAM

Marx and Eilers [17] make some significant changes in smoothing spline technique. They made the following two assumptions: First, they assume that $E(y)=\mathbf{Ba}$ where $\mathbf{B}=(B_1(x),B_2(x),\cdots,B_k(x))$ is an $n\times k$ matrix of B-splines and $\mathbf{a}$ is the vector of regression coefficients. They suppose that the coefficients of adjacent B-splines satisfy certain smoothness conditions that can be expressed in terms of finite differences of the a_i. Thus, from a least-squares perspective, the coefficients are chosen to minimize

$$S=\sum\nolimits_{i=1}^{m}\left\{y_i-\sum\nolimits_{j=1}^{n}a_jB_j(x_i)\right\}^2+\lambda\sum\nolimits_{j=k+1}^{n}\left(\Delta^k a_j\right)^2 \tag{12.10}$$

For least-squares smoothing, we have to minimize S in Eq. 12.10. Penalized generalized additive models [18] consider in the form of $(\mu)=\mathbf{Ba}=\boldsymbol{\eta}$, where $\mathbf{B}=(1,\|\mathbf{B}_1\|,\ldots,\|\mathbf{B_p}\|)$ is the $N\times\left(1+\sum_{j=1}^{p}n_j\right)$ regressor matrix and $\mathbf{a}=(\alpha,a_1,\ldots,a_p)'$.

P-splines directly fit GAMs through a slightly modified method of scoring algorithm and avoid the call to back fitting. P-GAM technique essentially eliminates the local scoring algorithm.

12.2.4 Smoothing Parameter

There are a number of different methods to choose smoothing parameter. Probably the most well known is cross-validation and generalized cross-validation. Let $(S_\lambda)_{\mathrm{ii}}$ be the ith diagonal element of S_λ. For smoothing splines the usual cross-validation score function is

$$\mathrm{CV}(\lambda)=\frac{1}{n}\sum\nolimits_{i=1}^{n}\left\{\frac{y_i-\widehat{f_\lambda}(x_i)}{1-(S_\lambda)_{\mathrm{ii}}}\right\}^2 \tag{12.11}$$

Here λ is chosen to minimize CV(λ). The basic idea of generalized cross-validation is to replace the denominators $1-(S_\lambda)_{ii}$ of cross-validation by their average $1-\frac{1}{n}\mathrm{tr}\,(S_\lambda)$, giving generalized cross-validation score function

$$\mathrm{GCV}\,(\lambda)=\frac{1}{n}\frac{\sum_{i=1}^{n}\left\{y_i-\widehat{f_\lambda}\,(x_i)\right\}^2}{\left\{1-\frac{1}{n}\mathrm{tr}\,(S_\lambda)\right\}^2} \tag{12.12}$$

In this equation λ is chosen to minimize GCV(λ). In this study we used generalized cross-validation for selection of smoothing parameter. The performance of the model is related with how close are the prediction values for test data and the observed values. The mean squared error (MSE) is defined as follows:

$$\mathrm{MSE}=\frac{1}{n}\sum\nolimits_{i=1}^{n}(y_i-\widehat{y}_i)^2 \tag{12.13}$$

12.3 Data Acquisition

The engine, under investigation in the current study, is a type of single-shaft turboprop engine. The propeller shaft of the engine is offset above the core engine that is comprised of 14 stage axial-flow compressors, 6 can through-flow combustors congregated in a single annular chamber, and 4 stage turbines. The T56-A-15 turboprop engine is still in use as the power unit of the C-130 Hercules tactical transport aircrafts in the Turkish Air Forces [19–21].

Experimental data used in this study is obtained from a previous report of EPA [19]. Measured quantities during engine test are unburned hydrocarbons, carbon monoxide, carbon dioxide, nitric oxide, nitrogen oxide, smoke, and aldehydes. Measures used in the current study are given in Table 12.1.

12.4 Statistical Data Processing Results

In this study it has used four different models based on dataset in Table 12.1. Models are obtained as follows:

$$\dot{\mathbf{m}}_{\mathbf{CO}}=\dot{\mathbf{m}}_{\mathbf{f}}+\mathrm{RPM}+\mathrm{AFR}+\varepsilon \tag{12.14}$$

$$\dot{\mathbf{m}}_{\mathbf{CO_2}}=\dot{\mathbf{m}}_{\mathbf{f}}+\mathrm{RPM}+\mathrm{AFR}+\varepsilon \tag{12.15}$$

$$\dot{\mathbf{m}}_{\mathbf{UHC}}=\dot{\mathbf{m}}_{\mathbf{f}}+\mathrm{RPM}+\mathrm{AFR}+\varepsilon \tag{12.16}$$

Table 12.1 Experimental emissions data [19]

Test #	RPM	AFR	$\dot{m}_f$ (kg/s)	$\dot{m}_{co}$ (kg/s)	$\dot{m}_{CO_2}$ (kg/s)	$\dot{m}_{UHC}$ (kg/s)	$\dot{m}_{NO_2}$ (kg/s)
1	10,000	76.92308	0.07119	0.0023184	0.2169972	0.00102312	0.000247212
	13,500	133.33333	0.08442	0.0011718	0.2622312	0.00043974	0.00057519
	13,800	44.44444	0.27405	0.0004158	0.8610336	0.00002142	0.002883888
	13,800	52.63158	0.23562	0.0004284	0.7401492	0.0000252	0.002585142
	13,800	83.33333	0.10836	0.0000504	0.3406536	0.0	0.000852768
	13,800	64.51613	0.17514	0.0002898	0.5502798	0.0	0.001771812
	10,000	76.92308	0.07119	0.0023184	0.2169972	0.00102312	0.000247212
	13,500	129.87013	0.08442	0.0011718	0.2622312	0.00043974	0.00057519
2	10,000	80.00000	0.07056	0.0020412	0.2153592	0.00105084	0.000050526
	13,500	128.20513	0.08442	0.0008316	0.2638188	0.00010332	0.000517986
	13,800	45.66210	0.27405	0.001449	0.8578206	0.00052668	0.002975868
	13,800	47.61905	0.23562	0.0003276	0.7402248	0.00005292	0.001968246
	13,800	76.92308	0.10836	0.0005544	0.3398598	0.00000252	0.000609588
	13,800	63.69427	0.17514	0.000252	0.550242	0.00002646	0.001005354
	10,000	90.90909	0.07056	0.0020412	0.2153592	0.00105084	0.000050526
	13,500	135.13514	0.08442	0.0008316	0.2638188	0.00010332	0.000517986
3	10,000	66.66667	0.071064	0.0018522	0.2177028	0.00090972	0.000266868
	13,500	128.20513	0.08442	0.0004536	0.2647512	0.0	0.000283878
	13,800	44.84305	0.27405	0.0006678	0.8607186	0.0	0.002323062
	13,800	45.45455	0.23562	0.0	0.7409052	0.0	0.002087442
	13,800	76.92308	0.10836	0.0005166	0.3398976	0.00000882	0.000530082
	13,800	65.78947	0.17514	0.0002268	0.550368	0.0	0.001402506
	10,000	76.33588	0.071064	0.0018522	0.2177028	0.00090972	0.000266868
	13,500	128.36970	0.08442	0.0004536	0.2647512	0.0	0.000283878
4	10,000	70.92199	0.07119	0.001953	0.2176524	0.00093114	0.000262206
	13,500	140.84507	0.08442	0.0003024	0.2649906	0.0	0.00059913
	13,800	44.76276	0.27405	0.0001008	0.861588	0.0	0.00253008
	13,800	83.33333	0.23562	0.000315	0.7404138	0.0	0.002280348
	13,800	76.92308	0.10836	0.0001134	0.3405654	0.0	0.000821394
	13,800	63.29114	0.17514	0.0005544	0.5498388	0.0000063	0.001589742
	10,000	66.66667	0.07119	0.001953	0.2176524	0.00093114	0.000262206
	13,500	129.87013	0.08442	0.0003024	0.2649906	0.0	0.00059913
5	10,000	76.92308	0.073836	0.0013734	0.2275938	0.00077364	0.000167076
	13,500	133.33333	0.08442	0.0008316	0.2635164	0.00020286	0.000442386
	13,800	44.44444	0.27405	0.0002898	0.8613108	0.0	0.00214578
	13,800	52.63158	0.23562	0.0000882	0.7407792	0.0	0.00192528
	13,800	83.33333	0.10836	0.0002268	0.3403764	0.0	0.000645624
	13,800	64.51613	0.17514	0.0001008	0.5505822	0.0	0.001323756
	10,000	76.92308	0.073836	0.0013734	0.2275938	0.00077364	0.000167076
	13,500	129.87013	0.08442	0.0008316	0.2635164	0.00020286	0.000442386

$$\dot{\mathbf{m}}_{\mathbf{NO_2}} = \dot{\mathbf{m}}_{\mathbf{f}} + \text{RPM} + \text{AFR} + \varepsilon \tag{12.17}$$

Further, we estimated these variables with the nonparametric techniques. The following nonparametric models based on the smoothing splines are:

$$\dot{\mathbf{m}}_{\mathbf{CO}} = f\left(\dot{\mathbf{m}}_{\mathbf{f}}, \text{bs} = \text{"bspl"}\right) + f\left(\text{RPM}, \text{bs} = \text{"bspl"}\right) + f\left(\text{AFR}, \text{bs} = \text{"bspl"}\right) + \varepsilon \tag{12.14'}$$

$$\dot{\mathbf{m}}_{\mathbf{CO_2}} = f\left(\dot{\mathbf{m}}_{\mathbf{f}}, \text{bs} = \text{"bspl"}\right) + f\left(\text{RPM}, \text{bs} = \text{"bspl"}\right) + f\left(\text{AFR}, \text{bs} = \text{"bspl"}\right) + \varepsilon \tag{12.15'}$$

$$\dot{\mathbf{m}}_{\mathbf{UHC}} = f\left(\dot{\mathbf{m}}_{\mathbf{f}}, \text{bs} = \text{"bspl"}\right) + f\left(\text{RPM}, \text{bs} = \text{"bspl"}\right) + f\left(\text{AFR}, \text{bs} = \text{"bspl"}\right) + \varepsilon \tag{12.16'}$$

$$\dot{m}_{NO_2} = f\left(\dot{m}_f, bs = \text{"}bspl\text{"}\right) + f\left(RPM, bs = \text{"}bspl\text{"}\right) + f\left(AFR, bs = \text{"}bspl\text{"}\right) + \varepsilon \tag{12.17'}$$

The third technique that we used is penalized generalized additive models, which based on smoothing term with B-spline functions estimated using Newton divided difference method:

$$\dot{\mathbf{m}}_{\mathbf{CO}} = f\left(\dot{\mathbf{m}}_{\mathbf{f}}, \text{bs} = \text{"ps"}\right) + f\left(\text{RPM}, \text{bs} = \text{"ps"}\right) + f\left(\text{AFR}, \text{bs} = \text{ps}\right) + \varepsilon \tag{12.14''}$$

$$\dot{\mathbf{m}}_{\mathbf{CO_2}} = f\left(\dot{\mathbf{m}}_{\mathbf{f}}, \text{bs} = \text{"ps"}\right) + f\left(\text{RPM}, \text{bs} = \text{"ps"}\right) + f\left(AFR, bs = \text{"}ps\text{"}\right) + \varepsilon \tag{12.15''}$$

$$\dot{\mathbf{m}}_{\mathbf{UHC}} = f\left(\dot{\mathbf{m}}_{\mathbf{f}}, \text{bs} = \text{"ps"}\right) + f\left(\text{RPM}, \text{bs} = \text{"ps"}\right) + f\left(\text{AFR}, \text{bs} = \text{"ps"}\right) + \varepsilon \tag{12.16''}$$

$$\dot{\mathbf{m}}_{\mathbf{NO_2}} = f\left(\dot{\mathbf{m}}_{\mathbf{f}}, \text{bs} = \text{"ps"}\right) + f\left(\text{RPM}, \text{bs} = \text{"ps"}\right) + f\left(\text{AFR}, \text{bs} = \text{"ps"}\right) + \varepsilon \tag{12.17''}$$

These models constructed to find out the determinants that have an effect to emissions. Further Eqs. 12.14, 12.15, 12.16, and 12.17 are named as first, second, third, and fourth models, respectively. Results of estimation of these models with the aid of used techniques are shown further in Table 12.2.

In order to realize what types of relationships are between variables, we plot the scatter matrix of all the variables used in our analysis.

As we can see from Fig. 12.1, most of the variables have a nonlinear effect between dependent and independent variables.

Table 12.2 Estimation results of linear regression model

Dependent variable	Mass flow rates (CO (kg/s))			
Coefficients	Estimate	Std. error	*t* value	Pr(>|*t*|)
(Intercept)	5.836e–03	4.449e–04	13.118	2.79e–15 ***
Revolution per minute (min-1)	−4.572e–07	5.517e–08	−8.287	7.28e–10 ***
Air/fuel ratio (kg air/kg fuel)	6.417e–06	3.241e–06	1.980	0.0554
Mass flow rate (kg.s-1)	2.213e–03	1.615e–03	1.371	0.1789

Significant codes: 0 '***' 0.001 '**' 0.01 '*' 0.05 '.' 0.1 ' ' 1

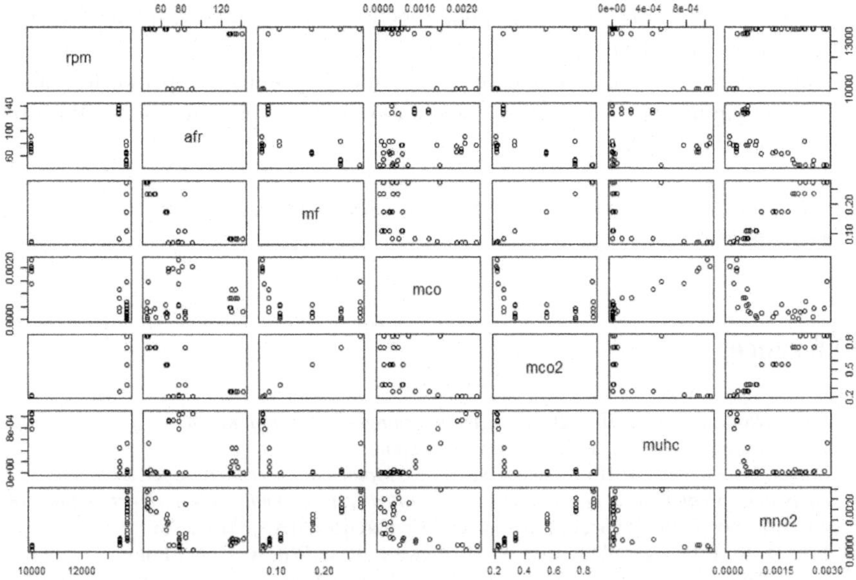

Fig. 12.1 The relationship between variables

First, we have made an analysis of models with multivariate linear regression based in ordinary least-squares technique. From estimation of the models, we found out that not all the independent variables have a linear effect on dependent variable. Further it gives statistical estimation results of the models.

As we can see from Table 12.2, revolution per minute (min-1) is significant variable as its T-statistics and probability value is below the 0.05 critical value. But air/fuel ratio (kg air/kg fuel) and mass flow rate (kgs-1) are not significant as their probability values are outside of the critical value. Results of estimation of these models with the aid of used techniques are shown further in Table 12.3.

As it could be seen from the results of estimation, nonparametric techniques show better performance criteria according to mean squared error value. It is an expected result, because regression techniques based on spline functions have closer approximation to data than the linear regression.

Table 12.3 MSE values of estimated models

	Linear	B-spline	Smoothing spline
First model	9.5e–08	8.6e–08	8.3e–08
Second model	7.1e–07	6.6e–07	6.1e–07
Third model	1.5e–08	1.4e–08	1.1e–08
Fourth model	3.5e–08	3.2e–08	2.9e–08

It is based on construction of spline functions. The smoothness of function can be controlled through changing the degree of spline function. That is, for more wiggle function, it can use higher-degree spline functions. But generally, splines are used with third or fourth degree.

Additionally smoothing splines have one more optimization parameter called smoothing parameter λ. It tunes the smoothing term in spline to make a function better approximate dataset.

Acknowledgments Support provided by Anadolu and Suleyman Demirel Universities is gratefully acknowledged.

References

1. IPCC Working Group-III. (2014). *Climate change 2014: Mitigation of climate change.* Intergovernmental Panel on Climate Change Report.
2. Sohret, Y., Karakoc, T. H., Karakoc, N. (2015). *Mathematical modelling for carbon dioxide equivalent prediction of greenhouse gases emitted from a small scale turbojet engine.* 7th AIAA Atmospheric and Space Environments Conference. AIAA 2015–3326.
3. Lee, D. S., Fahey, D. W., Forster, P. M., Newton, P. J., Wit, R. C. N., Lim, L. L., Owen, B., & Sausen, R. (2009). Aviation and global climate change in 21st century. *Atmospheric Environment, 43*, 3520–3537.
4. Altuntas, O. (2014). Calculation of domestic flight-caused global warming potential from aircraft emissions in Turkish airports. *International Journal of Global Warming, 6*, 367–379.
5. Beck, J. P., Reeves, C. E., de Leeuw, F. A. A. M., & Penkett, S. A. (1992). The effect of aircraft emissions on tropospheric ozone in the Northern Hemisphere. *Atmospheric Environment, 26*, 17–29.
6. Blakey, S., Rye, L., & Wilson, C. W. (2011). Aviation gas turbine alternative fuels: A review. *Proceedings of the Combustion Institute, 33*, 2863–2885.
7. Fan, W., Sun, Y., Zhu, T., & Wen, Y. (2012). Emissions of HC, CO, NOx, CO2, and SO2 from civil aviation in China in 2010. *Atmospheric Environment, 56*, 52–57.
8. Hileman, J. I., Blanco, E. D. R., Bonnefoy, P. A., & Carter, N. A. (2013). The carbon dioxide challenge facing aviation. *Progress in Aerospace Sciences, 63*, 84–95.
9. Beyersdorf, A. J., Timko, M. T., Ziemba, L. D., Bulzan, D., Corporan, E., Herndon, S. C., Howard, R., Miake-Lye, R., Thornhill, K. L., Winstead, E., Wey, C., Yu, Z., & Anderson, B. E. (2014). Reductions in aircraft particulate emissions due to the use of Fischer–Tropsch fuels. *Atmospheric Chemistry and Physics, 14*, 11–23.
10. Ekici, S., Yalin, G., Altuntas, O., & Karakoc, T. H. (2013). Calculation of HC, CO and NOx from civil aviation in Turkey in 2012. *International Journal of Environment and Pollution, 53*, 232–244.

11. Synylo, K., & Duchene, N. (2014). NOx emission model of turbofan engine. *International Journal of Sustainable Aviation, 1*, 72–84.
12. Yilmaz, I., & Ilbas, M. (2012). Investigation of pollutant emissions in aircraft gas turbine engines. *Journal of the Faculty of Engineering and Architecture of Gazi University, 27*, 343–351.
13. Wood, S. N. (2006). *Generalized additive models: an introduction with R. Chapman Hall/CRC.* (Texts in Statistical Science)
14. De Boor. (1978). *A practical guide to splines*. New York: Springer.
15. Dierckx, P. (1993). *Curve and surface fitting with splines*. Oxford: Clarendon Press.
16. Green, P. J., & Silverman, B. W. (1994). *Nonparametric regression and generalized linear models*. London: Chapman and Hall.
17. Marx, B. D., & Eilers, P. H. C. (1996). Flexible smoothing using B-splines and penalized likelihood (with comments and rejoinders). *Statistical Science, 11*(2), 89–121.
18. Marx, B. D., & Eilers, P. H. C. (1998). Direct generalized additive modeling with penalized likelihood. *Computational Statistics and Data Analysis, 28*, 193–209.
19. Vaught, J. M., Johnsen, S. E., Parks, W. M., & Johnson, R. L. (1971). *Collection and assessment of aircraft emissions base-line data turboprop engines (Allison T56-A-15), Final Technical Report. EDR 7200. EPA Contract No: 68-04-0029.* Ann Arbor: Environmental Protection Agency, Office of Air Programs.
20. Balli, O., & Hepbasli, A. (2013). Energetic and exergetic analyses of T56 turboprop engine. *Energy Conversion and Management, 73*, 106–120.
21. Sohret, Y., Kıncay, O., & Karakoç, T. H. (2015). Combustion efficiency analysis and key emission parameters of a turboprop engine at various loads. *Journal of the Energy Institute, 88*(4), 490–499. ISSN 1743-9671.

Chapter 13
The Improvement of Energy-Saving Performance at Ukrainian Airports

Margaryta Radomska, Larysa Chernyak, and Olexandr Samsoniuk

13.1 Introduction

Sustainable development is the process of productive forces harmonization in order to provide the essential needs of all society members, while maintaining the integrity and the gradual restoration of the environment, creating opportunities for balance between the potential and needs of people and natural biocapacity. Ukraine is able to provide sustainable development by the efficient use of all natural resources, structural and technological restructuring of the production, and use of scientific and technological potential of society. The main way to implementation of sustainable development in the state economy is to analyze the impact of each industrial sector on the environment and perform its reformation and upgrading toward limiting negative impacts on the environment and improving rational use of natural resources.

Transport is one of the most important elements of the economic system of each country and at the same time belongs to the group of major environment polluters. Accounting the continuing active expansion of the transport sector, the increasing number of vehicles and maintenance facilities, the transport integrated load on the environment appropriately is growing and raising the severity of the issue of its environmentally sustainable development in the future.

Air transport as a part of the general transport system has certain advantages over ground and water modes of transport: maximum speed of transportation, possibility of reducing path by straightening routes, relatively low dependence on physical and geographical conditions of the area, variety of destinations, etc. These make air traveling all cargo transportation gaining more popularity every year all over the world. On a global scale, passenger air travel is expected to maintain positive growth

M. Radomska (✉) • L. Chernyak • O. Samsoniuk
Institute of Environmental Safety, National Aviation University, Kyiv, Ukraine
e-mail: m_radomskaya@mail.ru; larikch@bigmir.net

T.H. Karakoç et al. (eds.), *Advances in Sustainable Aviation*,
DOI 10.1007/978-3-319-67134-5_13

rates up to 2030, despite a number of challenges faced by the industry (Current Market Outlook 2012–2031 [1] by Boeing Company). Airlines around the world and in Ukraine have to cope with high jet fuel prices and decelerating economic growth. However, these difficult economic conditions are predicted to be offset by growing mobility and communication activity around the planet.

The aviation industry involves many departments, and among them airports and airlines, and aircraft factories are the main.

During the period from 2007 to 2013, Ukraine demonstrated high development rates in aviation sphere: passenger flow has grown 5.1 times, and the number of passenger has grown 4.3 times as compared to 2000. Summarized values of estimated prognosis stated that total volumes of transported passengers from all Ukrainian airports would reach seven to eight million passengers annually by 2020. However, the dramatic changes in political and social life of the country, destruction of territorial integrity, and continuous military conflict together with strong economic recession, including skyrocketing inflation, have canceled the bright prospects of Ukrainian aviation development. The most serious impacts were complete destruction of the newly reconstructed Donetsk International Airport, the third biggest airport in Ukraine, and the loss of the airport "Simferopol," located on the annexed Crimea peninsula.

The system of Ukrainian airports includes 72 aerodromes and 36 active airports. Statistical data on activity of the biggest national airports state that during the last 3 years, the stable growth tendency of air flights volumes as well as number of passengers transported through these airports is observed. The current leaders on the market which are "Boryspil," "Odessa," "Kyiv," "Lviv," "Kharkiv," and "Dnypropetrovsk" have managed to cover the needs of population and international travelers previously provided by missing airports. In 2015 the annual passenger traffic was over ten million. The total volume of passengers using small Ukrainian airports (16 currently operating) was almost 120,000. Considering the situation in the country specialists believe that passenger traffic at these airports will keep stable with potential to reach 120–200 thousand passengers by 2025. The main state airport "Boryspil" is supposed to develop faster, so that the volume of passenger transportation will be 8,300–10,500 thousand passengers per year in 2025, while in smaller airport "Kyiv" (Zhulyany), this number will reach 355–857 thousand passengers annually.

The main problems of aviation development in Ukraine are:

- Obsolescence of aircrafts and shortage of qualified staff
- Airports technical potential noncompliance with current international requirements
- Absence of state support for air transportation development and upgrading
- Loss of control over investments activity at transport sector
- Insufficient level of legal base for air operators certification
- Insufficient modernization of aircrafts park
- Inefficient use of resources by aviation facilities, especially airports

In order to overcome existing problems and strengthen the position of domestic and international aviation on the transport market Ukraine, the government invested into airports modernization and attraction of investments for the development of technologies and improvement of equipment. However, this will not be enough to provide sustainability for Ukrainian airports. The major drawbacks are not regulated environmental impacts and consumption of natural resources, especially power resources at airport.

13.2 Energy Efficiency Background in Ukraine

Energy saving and environmental efficiency are interconnected. The life of humans in modern urban systems is under the constant threat of negative influences, caused by technogenic environment pollution. This pollution is produced mostly by power generating facilities and facilities of fuel consumption. From this point, reduction of power resources consumption is the direct way to improve the environment quality at urban areas. From the other point, implementation of new technologies in the field of power economy optimization is limited with environmental safety of the relevant equipment, materials, alternative fuels, etc. Moreover, the energy efficiency is first of all conditioned by the need to reduce the anthropogenic pressure on the environment and provide the needs of future generations. Transport turns to be the most important and urgent branch for improvement of sustainability due to direct contact and impacts on population.

Energy efficiency is currently one of the main trends of both domestic and global economy. Therefore for Ukraine, which was not properly engaged into the process of energy saving during long period, the trend is now a major motion vector of economic development. An important participant of the energy-saving technologies implementation is transport of all kinds, and especially air. This is because the aviation industry combines two components: intensive impacts on the environment and big volumes of energy consumption, including fuel. Thereby, the purpose of the research is to define the potential for energy conservation at airports of Ukraine and for the improvement of environment quality at the adjoined area.

Acting Ukrainian legislation on energy efficiency is quite imperfect. In order to create sufficient legal basis and ensure the appropriate level of energy efficiency in all sectors of economy production, the Office on Energy Efficiency Regulation has carried out work on preparation and amendment of 25 draft legal acts and provided comments on 40 regulations and legal acts.

Legislation in the field of energy efficiency, as a branch of the national legislation, was initiated by the Law of Ukraine "On energy saving." This law was developed in the midst of the economic crisis of 1994. It was supposed to provide the system of institutional, regulatory, and incentive measures to implement the mode of rational use of fuel and energy resources, but this law, like most others adopted at the time, is not the law of direct action, and most of its provisions are declaratory and have referential character.

At the same time, many scientists have paid attention to the development of theoretical grounds for energy optimization of Ukrainian economy and experience of their application. Thus, the leading specialists on energy efficiency in this field B.S. Stogniy, M.M. Kulik, and V.A. Zhovtyansky have published a fundamental work "Energy Conservation Strategy in Ukraine: Analytical and Reference Materials" in two volumes. The first book summarizes the first experience of public administration in the field of energy saving in Ukraine compared with the achievements in the other countries. The second book contains practical materials primarily related to power management methods in manufacturing, financial, and economic mechanisms of energy conservation, information, and international experience in this field.

Among other important works in the field, there are studies by A. Shcherbina, A.K. Shydlovsky, G.T. Vasyukov, G.M. Kaletnik, I.P. Vasilyev, V.S. Samokhvalova, Korchevoy Yu.P, Maystrenko A.Y, Topal A.I, and Rozen V.P. dealing with management fundamentals of energy conservation, alternative fuels application on transport, environmental efficiency of nonconventional power generating facilities, secondary energy resources, and energy efficiency.

The issues of energy consumption optimization at airports were considered by Rozen [5], Sokolova and Zakharchenko [3], Leschynsky and Konovalyuk [4], Velychko Yu and Kozlov [2], and others. They have highlighted the major problems and opportunities for the improvement of energy efficiency at airports. Thus, according to their experience, the analysis of power management at the world airports, especially those within the same climatic zone with Ukraine, cannot be provided with adequate assessment due to closed access to the information on power consumption. Based on the statistical data on power consumption collected by Zakharchenko V.P. and Sokolova N.P. at Ukrainian airports during 2002–2013, there exists stable trend to growing volume of electricity consumption and considerable variations of power consumption characteristics in time and regions are present. One reason is the lack of clear system of energy economy management.

13.3 The Overview of Energy Economy at Airports

Practically, the major energy consumers in airports are illumination, steam, and heating systems, climate control systems, and comfort provision systems. The specific trait of airport energy economy is the need to support constantly the activity of certain systems, decisive for airport functioning. This is first of all related to the illumination of runways, aprons and terminals, climate control systems provision, power supply of navigation and routing systems, as well as safety control systems, heating of terminals, and fuel supply systems. It is impossible just switch off these systems, as it will ruin all the work at an airport. In much the same way, it is not possible to reduce the intensity of illumination without increasing accidents risks or to change production capacities of an airport without reduction of its economic characteristics. Nevertheless, accurate planning, control, and management of technological processes, their thorough distribution creates opportunities for the reduction of energy consumption. To address the problem of energy losses at airports efficiently, it is necessary to use complex automated systems of power management based on the latest developments from the leading equipment manufacturers such as Danfoss, ABB, Carrier, and Siteco.

Thus, energy-efficient solutions for illumination include installation of daylight reflection systems, allowing redirection of the most of light into the premises during the daytime. Besides the obvious energy savings, it helps create a soft even illumination in the rooms and comfortable for visitors. Although the intensity of this daylight illumination depends on the level of natural light and must be adjusted automatically or supplemented with electric illumination.

Centralized illumination control system also allows the operator to manage apron illumination, lighting at adjacent areas when the platform is occupied with aircraft, boarding on or off is carried out, and other operations are performed. In the airport waiting areas and terminals lighting is automatically scheduled and adjusted to the timetable of arrivals and departures.

Equally interesting solutions can be used to optimize the management of microclimate. The airport interior is excessively heated by the direct sunlight, illumination, various technological equipment, and people within. So, to neutralize the excess heat effectively, the special circuits with variable cooling media flow are used. It is supplied only to those areas where and when sensors record the air temperature rise above the level comfortable for visitors or staff, for example, in the terminal area under daylight. The extracted heat is also a valuable resource and must be reused, which is performed with the systems of heat recuperation. Thus, 85% heat energy, present in the exhaust air, is returned through the energy recovery equipment to the heating system – at present the most efficient energy-saving technology (Boichenko 2014).

Efficient heating system is essential for complete climate control of airport in temperate latitudes. Its stable operation should be provided by two or more independent sources of heat, as this allows running them separately if required level of heat is insignificant, and thus reducing energy consumption for the functioning of heating system. Climate control systems also involve management of ambient parameters in terminal buildings by means of electronics, computer network related to the airport. This allows implementing the principle of preventive climate control, when ventilation and air conditioning system are adjusted automatically to increase or decrease the intensity of work based on the analysis of information coming from the CO_2 sensors and data about passenger traffic in a particular area of the terminal.

Electricity must be also supplied to terminals from several independent energy centers, which airport is connected through several transformer substations. It is necessary for the organization of uninterrupted electricity supply of security, telecommunication, and navigation systems and should not turn off for a second, even in cases of accidents. However, it is possible to find the way to saving costs and resources by using energy-saving equipment in electrical system – the result may achieve 20% reduction in energy losses. This result is provided by the implementation of reactive power compensation, which reduces the total load on transformers and power lines. The use of lighting equipment with electronic adjustment reduces power losses by 10% (Boichenko 2014).

Dispatching and control over the operation of all engineering and information systems in terminals using automated system of supervisory control also contributes to reducing energy losses. Automation control equipment opens the way to reduce the number of necessary operations and terminal personnel, minimize errors caused by human factor, and increase the reliability and safety of all engineering information infrastructure at the airport. In addition to these management functions, centralized controlling system software is able to collect and process data on power consumption, allowing operators to monitor energy use, identify trends, make appropriate decisions, and implement corrective measures to improve energy efficiency, especially in terms of electricity.

13.4 State of the Art in Energy-Saving Solutions for European Airports

The pioneering representatives of the branch in energy-saving solutions are the Heathrow Airport, Great Britain; Munich Airport, Germany; and Vnukovo Airport, the Russian Federation.

Thus, Heathrow Airport has developed complex program on energy conservation, approved by the state government and airport managers. The certain points of the program to improve energy efficiency in Heathrow Airport include regulation of illumination to cover only the areas being used, heat and cold recovery, and power disabling for the equipment not used.

HVAC systems and operations at Munich Airport include the classic management functions for energy saving in combination with a number of specially developed intelligent functions. CPS implements energy optimization program for remote buildings, as well as areas of outputs, including lighting and temperature control. Individual controllers in premises regulate temperature and ventilation in approximately 1,700 rooms throughout the airport, allowing operators to accurately manage energy use and support comfort levels of ambient parameters.

At the Terminal A at Vnukovo airport, all large premises are equipped with the system of daylight reflection, allowing the best use of it in the daytime. Besides the split system of heat generation, the systems of heat recovery provide regulation of microclimate parameters and formation of energy reserve for heating. The intensity of power supply for ventilation and lighting depends on the level of terminals traffic load and number of passengers and services provided (Boichenko 2014).

The main air gate of Ukraine, the Boryspil airport, has begun to implement the energy-saving technology trying to reduce power and natural resources consumption. Thus, easy to operate systems for the daylight reflection are installed in all major halls, allowing minimal use of artificial illumination during the most of daytime. The expected reduction of energy consumption will be 12–15% (Boichenko 2014). Application of such systems has also decent payback period which is said to be 4–5 years at the most.

So, prior to making decisions about the appropriate energy-saving measures, it is necessary to assess the existing consumption and losses of energy. Based on the obtained data, the energy-saving potential of the studied facility is defined. The complex of procedures designed to perform these tasks is called integrated energy audit.

When the object of study is an occupied building, then reducing energy consumption while maintaining or improving human comfort, health, and safety is of primary concern [7]. Beyond simply identifying the sources of energy use, an energy audit seeks to prioritize the energy uses according to the greatest to least cost-effective opportunities for energy savings [7].

Airport energy audit may involve recording various characteristics of the building envelope including the walls, ceilings, floors, doors, windows, and skylights. The audit may also assess the efficiency, physical condition, and programming of mechanical systems such as the heating, ventilation, air conditioning equipment, and illumination. The accuracy of energy estimates is greatly improved when the billing history is available showing the quantities of electricity, natural gas, fuel oil, or other energy sources consumed over a certain period [8].

Some of the greatest effects on energy consumption are user behavior, climate, and age of the facility. The energy audit must therefore include interviews with the staff to understand their patterns of power use when performing technological operations (ASHRAE [9]). The energy audit is used to identify cost-effective ways to improve the comfort and efficiency of airport facilities and, in addition, make it possible to apply for energy efficiency grants from central government.

13.5 The Assessment of Energy Conservation Potential for Ukrainian Airports

With the analysis of European experience for energy conservation at airport facilities, it was established that the following measures will be highly efficient under Ukrainian conditions:

- Installing energy-efficient lighting with motion sensors throughout the airport
- Natural illumination indoors for passengers in terminals
- Double-glass windows and solar-shading devices, providing natural light penetration into the building but minimizing heat received from the sun
- Turn off escalators and baggage lines at night
- Turn off peripheral illumination in daytime

The offered solutions were analyzed as a perspective for a range of airports. The studied objects included four small airports of Ukraine with the most intensive passenger traffic: Zhuliany, Dnipropetrovsk, Zaporizhia, and Ivano-Frankivsk. The information below is taken from the official websites of the studied airports.

Kyiv International Airport (Zhuliany) is one of the two passenger airports of the Ukrainian capital Kiev. It is owned by the municipality of Kiev and located in the southern Zhuliany neighborhood of the city. Aside from facilitating regular passenger flights, Kyiv International Airport is also the main business aviation airport in Ukraine (Official Web page of the Kyiv International Airport [11]).

After Ukraine gained independence in 1991, "Kyiv" airport began receiving international flights from nearby countries, but in 2011, when Wizz Air, the locally pioneering low-cost airline, had moved all its operations to "Zhuliany" from the Boryspil Airport, the new era of around-the-clock flights at the airport started and the passenger traffic increased by 1,520%. The new "A" terminal opened in 2012 and now receives all international and some domestic flights. Projects for expanding Zhuliany's taxiways and aircraft parking lots considered as well, unfortunately, the issues of energy conservation are not sufficiently covered in these plans.

The overall effect of the offered complex of activities will lead to reduction of energy consumption by 19–27%, depending on the intensity of energy conservation opportunities implementation. Even under the minimal scenario, the monetary value of the project will be equal to 627 kW of energy capacity, which is a dramatic improvement. In applied presentation this volume of energy can provide heating for two 16-storey residential buildings. The payback period for the corresponding capital investments will range from 7 to 11 years, but considering the instability on energy resources provision typical for current economic and political situation in Ukraine, the need to invest in energy efficiency improvement turns to be the need of survival importance.

Dnipropetrovsk International Airport is an airport serving Dnipro, the center of the corresponding oblast. It is located 15 km southeast from the city center. The airport is currently owned by its major airline partner Dniproavia. This has resulted in a number of management problems and has slowed the airfield's development as Dniproavia has, on a number of occasions, refused to be forthcoming with the required funds to undertake a comprehensive modernization program. In addition to this, foreign airlines have found it difficult to gain access to Dnipro as a result of Dniproavia's protectionist policies along routes to and from the airport (Official Web page of the Dnipropetrovsk International Airport [12]).

In 2011 the airport's owners initiated a program to develop a new terminal complex. This project envisaged the construction of a large new international terminal, similar in specifications to the newly built terminal at Kharkiv International Airport. However, the construction was soon frozen, and, as of 2016, building work has not progressed beyond the laying of foundations.

Being a part of the Soviet Union, in 1990, the highest level of the airport passenger traffic was registered. Dnepropetrovsk International Airport has noted a slow growth in the number of passengers since year 2010, when the active phase of reconstruction was started.

Among the activities offered for implementation according to our investigations, the Dnipropetrovsk International Airport is in need for the installation energy-efficient lighting; provision of natural illumination indoors for passengers in terminals, double-glass windows, and solar-shading devices; and regulation of active equipment exploitation according to operation's intensity at night and in daytime.

In this case the effect will be the reduction of energy consumption by 16–21%, due to lower capacity of power economy of the airport. As a result of actions implementation, the potential energy savings reach 403 kW. The payback period for the capital investments will be over 10 years, which is an important factor, accounting the instability of economic and political situation in the country and continuous inflation.

Zaporizhia International Airport is the international airport that serves Zaporizhia, Ukraine one of three airfields around the city. The aircraft engine factory Motor Sich has its base here. The traffic volume has increased dramatically over the last 5 years, and it is currently on the 7th place in the country. In 2013, the Zaporizhia International Airport changed owners and became municipal property.

The airport is included in the State Target Program of airports reconstruction for the period to 2023. The development of Zaporizhia air gates involves reconstruction of the sector for domestic airlines, which started in July 2016. Reconstruction of the airport is expected to be completed by the end of 2017. It is also planned to build a new terminal that will include eight check-in counters and areas for separate international and domestic flights (Official Web page of the Zaporizhia International Airport [13]).

The project of airport reconstruction includes all of the mentioned activities and is declared to provide 35–40% higher energy efficiency of airport operations. Our calculations have showed that under the condition of stable growth of passenger traffic the proved savings will be 32% maximum, but this will mean almost 387 kW of power equivalent. The investment will be paid back within 10–15 years.

Ivano-Frankivsk International Airport is 4.4 km by road from the town center. It has maintained border and customs operations since 1992. Its traffic capacity is claimed to be 400 passengers per hour. Officials have made efforts in the past to promote the airport and its relative proximity to the Bukovel ski area, Vorokhta and the Carpathian National Nature Park, and other quiet, spectacular mountain environments.

There are two major runways: the first is used for civil flights, and the second concrete runway is now used by the military as a parking lot, and a large apron (located northwest of the civilian terminal) is still in use by the Ukrainian Air Force. On average, it serves 30,000 passengers (Official Web page of the Ivano-Frankivsk International Airport [14]).

Among the offered measures that are valid for this airport are as follows: energy-efficient lighting with motion sensors, indoor natural illumination in terminals, double-glass windows and solar-shading devices. and turning off peripheral illumination in daytime. Turning off escalators and baggage lines at night is currently performed as the airport is practically not serving flights at night. The result of actions implementation will probably be 152 kW or 13% cost savings. The lower intensity of traffic as compared to other considered objects makes the investment twice less efficient and postpones the probability of energy-saving opportunities implementation.

13.6 Evaluation Methodology for Environmental Benefits of Energy Conservation

Energy saving makes it possible to reduce the pressure both on the energy economy and environment. Environmental effect is not limited merely to the decreased consumption of natural resources. Each saved calorie of heat or kilowatt-hour of electricity also provides significant environmental benefits at all previous stages of energy generation, associated with fuel extraction, enrichment, processing, and transport and production, transportation of electrical and thermal energy to the consumer, and its distribution.

Additionally, aviation is one mode of transportation that, in turn, it is one of many GHG-emitting sectors, generating at a global level over 730 million tons of carbon dioxide per year with an increase of 45% compared to 1990 [10]. It is expected that growth in global air transport will triple aviation carbon dioxide emissions between 1990 and 2050, and that total radiative forcing (global warming) effects will increase fourfold over the same period [10].

Emissions from fuel combustion in aircraft represent from 2% to 4% of the total global GHG inventory [15]. Based on airport emission inventories prepared to date, emissions from non-aircraft airport-related operations represent an additional 0.1–0.3% of the global total [15].

One of the most significant sources of emissions is related to local power generating facilities (which directly undergo the proposed energy-saving improvements) and transportation of employees and passengers to and from the airport, and they may be accounted for elsewhere in "on road" transportation emissions inventories. While the airport contribution can be relatively small, many improvements can still be made, as we have showed. It is important to distinguish between aircraft emissions and emissions directly associated with airport facilities. In practice, airports use a variety of definitions to determine the aircraft emissions contribution. Some scientists base the emissions entirely on the fuel dispensed at the airport. Others count the emissions from aircraft only while their wheels are on the ground; others include the whole landing and takeoff cycle down from and up to an altitude of 900 m. Including the landing and takeoff cycle, taxiing, and APU use, the aircraft emission contribution to an airport CO_2 inventory is typically in the range of 50–80% [15].

Carbon dioxide (CO_2) and water vapor (H_2O) are the most abundant products of jet fuel combustion (emission indices for CO_2 and H_2O are 3.15 kg/kg fuel burned and 1.26 kg/kg fuel, respectively) and conventional fuels combustion at local stationary power facilities [16]. However, both substances have significant natural background levels. Neither current aircraft emission rates nor likely future subsonic emission rates will affect the ambient levels by more than a few percent, unlike power generation, if it continues to grow intensively.

NOx represents the next most abundant airport emission (emission indices range from 5 to 25 g of NO_2 per kg of fuel burned), which is able to affect background levels/ozone levels [16].

Carbon monoxide (CO) emissions are of the same order of magnitude as NOx emissions. Like NOx, CO is a key participant in tropospheric ozone production. However, natural and non-transport anthropogenic sources of CO are substantially larger than analogous NOx sources, thereby reducing the role of aircraft CO emissions in ozone photochemistry to a level far below that of aircraft NOx emissions, while power generation facilities turn to be much more important in the process [16].

Emissions of sulfur dioxide (SO_2) and hydrocarbons depend on the quality of fuel used at power generation facilities at and outside the airport. Their primary potential impacts are related to formation of sulfate and carbonaceous aerosols and acid rains.

Considering the fact, that most of these emissions appear to be in the ground layer of the atmosphere, the excessive heat generation from airport power economy is strongly unfavorable factor, deepening microclimate imbalance on the territory.

13.7 Results and Discussions

Based on the presented considerations, the perspective reduction of energy consumption, for example, at Kyiv International Airport, based on our calculations, will be equivalent to the emissions of solid particles decreased by 9.48 t, sulfur oxide emissions decreased by 12.75 t, and nitrogen oxide emissions decreased by 3.97 t every year. The resulted effect will include lower contribution of airports activity to greenhouse effect enhancement, atmosphere dimming, and intensity of acid rains formation (Table 13.1).

As for the major greenhouse gas generated by airport stationary facilities, which is carbon dioxide, its contribution to microclimate warming at the territory of an airport could be numerically evaluated given the fact that generated on a regular basis depending on the intensity of power production.

Our investigations for Dnipropetrovsk Airport have showed the following results. The potential temperature growth accounting carbon dioxide radiative forcing ranges from 0.4 to 1.2 °C depending on the period of the year and intensity of power generation and consumption (minimal is in June, maximal is in January). The estimation of the cost of generated carbon units equals 304,000 UAH for the period of 2015, which could be deducted from airport incomes. This money could be spent on greening the airport territory to catch some part of CO_2 emissions or for other environmental purposes.

Thus, airports are sources of emissions that affect climate: emissions generated from activities occurring inside and outside the airport perimeter fence associated with the operation and use of an airport, therefore, may represent a significant danger to the health of people living near airports [17]. Airport operators are realizing just how construction, operation, maintenance, and other activities at airport facilities can contribute to the industry's overall impacts on the environment due to operations and energy consumption.

Table 13.1 The potential environmental efficiency of the energy-saving measures at Kyiv International Airport

Technical measures for energy efficiency	Energy conservation, MW/year	Volume of emissions reduction, kg			Volume of wastewaters reduction, t
		Dust	SO_2	NO_x	
1. Improvement of heat energy provision					
1.1. Heating system automation	456.8	1918.4	2580.6	803.9	2283.8
1.2. Building envelope insulation	203	852.6	1147.0	357.3	1015.0
1.3. Windows reconstruction	182.7	767.3	1032.3	321.6	913.5
1.4. Roof insulation	66	277.1	372.8	116.1	329.9
Total	908.4	3815.4	5132.6	1598.8	4542.1
2. Power supply improvement					
2.1. Replacement of lighting equipment at terminals	75.3	316.3	425.6	132.6	376.6
2.2. Replacement of lighting equipment at the airfield	130	546.0	734.6	228.8	650.1
2.3. Automation of power supply	44.3	185.9	250.1	77.9	221.3
Total	249.6	1048.3	1410.2	439.3	1248.0
3. Improvement of climate control system					
3.1. Ventilation system automation	18.9	79.2	106.5	33.2	94.3
3.2. System of air composition control	12.8	53.7	72.3	22.5	64.0
3.3. System of outside air flow provision	11.5	48.1	64.7	20.2	57.3
Total	43.1	181.0	243.5	75.8	215.5
Overall efficiency	1201,1	5044,7	6786,3	2114,0	6005,6

The leading airports of Europe show strong potential of energy conservation, which could be applied in Ukraine for major airports. The investigations show that implementation of the basic set of energy-saving recommendation releases energy capacity enough to provide the needs of residential blocks and leads to reduction of emissions related to greenhouse effect enhancement, atmosphere dimming, and intensity of acid rains formation.

The adoption of measures to prevent, minimize, or mitigate adverse impact of aviation on the environment is the main target of airport stakeholders; today more than ever they are called upon to assess the local air quality at and around the airport, as well as the efficiency of natural resources consumption at airports [18].

Airports should review ground service equipment and ground vehicles and land transport for emissions reduction opportunities. New buildings should employ best practice energy efficiency:

- Underground thermal sinks can be used to enhance heating and cooling efficiencies;
- Combined cooling, heat, and power systems use waste heat from electricity generation to heat the terminal in winter. In summer, absorption cycle refrigeration systems can use the same heat source to generate chilled water to cool the building;
- Smart building technologies can be used to reduce lighting and heating or cooling in unoccupied spaces. Unoccupied escalators can be slowed or paused until people need to use them;
- For large interior spaces in hot climates, thermal stratification can be used to cool occupied areas at floor level while allowing unoccupied space near the ceiling to remain hot;
- In cold climates, new steam plume-suppressing technologies can be used to allow heating plants to be located close to terminal and control tower structures without affecting visibility. This can substantially reduce piping losses and inefficiencies [19].

New and existing buildings should have best practical thermal insulation and glazing: installation of shading or light-filtering films on windows to reduce solar load; modifying and modernizing heating, ventilation, and air-conditioning systems, such as installing variable speed electric motors to reduce air flows when occupancy is low or temperatures are mild; installation of more efficient and long-life light bulbs for both interior and exterior lighting [19].

Operational procedures can also be used to improve energy efficiency: maintenance hangar door opening and closing procedures can be improved to reduce heat loss in winter or heat gain in summer; lighting procedures can be improved to minimize lighting in unoccupied areas or during low occupancy.

Renewable energy should be used, where practicable, to reduce fossil fuel consumption: generating electricity on site with wind turbines, photovoltaic, solar cells; solar hot water heating; using biofuels, hydrogen, and other nonfossil fuels for ground vehicles and support equipment; and using boilers that burn wood pellets or similar forestry or recycled waste material [19].

13.8 Conclusions

The potential development of energy-saving technologies and energy efficiency in all areas of human activity can be compared as a whole with the potential to increase economic performance of the country and its resource base, which is especially important for Ukraine.

Among the barriers to the development of energy saving and energy efficiency in our country is mostly the lack of motivation, including among government authorities, insufficient information support, lack of experience in financing energy efficiency projects, and lack of organization and coordination of implementation. Such a barrier as technology drawback is, to date, substantially removed, including through the investments from the developed countries. Currently the market has a very wide range of energy-efficient equipment, energy-saving materials, and a range of consulting services on energy conservation and efficiency, creating a strong infrastructure base. This is also valid for airports as they are the facilities with extremely high and expensive energy consumption, conditioned in many cases by safety and functionability. Nevertheless, leading airports of Europe show strong potential of energy conservation, which could be applied in Ukraine for major airports. The investigations show that implementation of the basic set of energy-saving recommendation releases energy capacity enough to provide the needs of residential blocks and leads to reduction of emissions related to greenhouse effect enhancement, atmosphere dimming, and intensity of acid rains formation.

References

1. Current Market Outlook 2012–2031. New York: Boeing Market Research.
2. Velychko Yu., Kozlov V. (1996). Power supply of airports. Kyiv International University of Civil Aviation, Kyiv.
3. Zakharchenko, V., & Sokolova, N. (2014). The model of power consumption efficiency management at airports. *Eastern-European Journal of Enterprise Technologies, 5*(71), 9–15.
4. Leshchinskiy, O., Konovalyuk, V., & Sokolova, N. (2014). Model of prognosis of power consumption volumes by the illumination equipment of airport. *Tekhnologicheskiy audit i reservi proizvodstva, vol. 2, 1*(16):27–31.
5. Rozen, V. (2005). Upravlinnya rezhimom elektrospozhivannya promislovogo pidpryemstva [management of industrial enterprises power consumption]. *Promelectro, 6*, 35–41.
6. Boichenko, S., et al. (2014). *Aviation ecology*. Kyiv: National Aviation University.
7. Sourse Book for Energy Auditors/Edited by M.D. Lyberg (1987), Stockholm: International Energy Agency, Swedish Council for Building Research. 693 p.
8. *Mortgage industry national home energy rating systems standards* (2013), Oceanside: Residential Energy Services Network, 257 p.
9. The ASHRAE Standard 90.1. (2010). *Energy standard for buildings except low-rise residential buildings*. Atlanta: US Department of Energy.
10. Khodayari, A., et al. (2013). Intercomparison of the capabilities of simplified climate models to project the effects of aviation CO_2 on climate. *Atmospheric Environment, 75*, 321–328.
11. Kyiv International Airport. (2011). http://iev.aero. Accessed 10 Sept 2016.
12. Dnipropetrovsk International Airport. (2017). http://dnk.aero. Accessed 10 Sept 2016.
13. Zaporizhia International Airport (2013). http://avia.zp.ua. Accessed 10 Sept 2016.
14. Ivano-Frankivsk International Airport (2017). http://airport.if.ua/. Accessed 10 Sept 2016.
15. ICAO Environment report 2010: Aviation and climate change. Montréal: International Civil Aviation Organization, 2010.
16. Joyce, E., Penner, D. H., Griggs, D. J., Dokken, D. J., & McFarland, M. (Eds.). (1999). *Aviation and the global atmosphere, intergovernmental panel on climate change*. New York: Cambridge University Press.

17. *Guidance Manual: Airport greenhouse gas emissions management* (2009). Washington: ACI World Environment Standing Committee.
18. Smale, R., Krahe, M., & Johnson, T. (2012). *Aviation report market based mechanisms to curb greenhouse gas emissions from international aviation*. Gland: WWF International.
19. Airports Council International. (2009). *Policy and recommended practices handbook* (7th ed.). Montreal.

Chapter 14
Ship Emissions and Human Health Relationship: A Theoretical and Numerical Investigation in Asyaport

M. Suner and E. Yalcin

14.1 Introduction

Sector basis studies show that the biggest emissions source in the world is transportation vehicles [1]. Among these transportation vehicles, ships are the biggest emitters. The amount of emissions that a ship emits and the rate at which ships consume fuel depend on many factors such as different navigation phases i.e., cruising, maneuvering, hotelling, and the time spent in these phases, the type of engine, the fuel, etc. [2]. During the burning of fuel, pollutants and greenhouse gases are emitted in the forms of CO_2, SO_2, nitrogen oxides, CO, hydrocarbons (HC), CH_4, nonmethane hydrocarbons (NMHC), and particulate matter (PM) [3] and increasing fuel consumption is directly proportionate to the amounts of emissions in the abovementioned forms. It is well known that emissions have a lot of impact on people. It has been stated in some remarkable studies that ship emissions are directly related to rising mortality in coastal regions and cause 60,000 deaths from various causes [4–6]. Moreover, Garbaccio et al. have pointed out that if carbon emissions decrease by 5% every year, premature deaths will be reduced by 3.5% to 4.5% [7].

Ocean-going merchant vessels are the main players with over 90% market share in transportation of goods, and the shipping sector has been growing on average by 5% every year over the last three decades [8]. According to various estimates, global fuel consumption by ships was predicted to be over 300 million tons in 2007. During the burning of the abovementioned amount of fuel, approximately 1046 million tons of CO_2 is emitted. These CO_2 emissions constitute 3.3% of global emissions [9].

Ship emissions are directly related to fuel consumption and in a survey conducted by Ballou et al., a tool was used to display this relationship. In their paper, the authors focus on optimization of vessel operations and fuel consumption

M. Suner (✉) • E. Yalcin
Maritime Faculty, Istanbul Technical University, Tuzla, Istanbul, Turkey
e-mail: msuner@itu.edu.tr

T.H. Karakoç et al. (eds.), *Advances in Sustainable Aviation*,
DOI 10.1007/978-3-319-67134-5_14

based on vessel operational dynamics and environmental parameters by using the abovementioned tool [10]. Besides the studies on optimization of ship operations and fuel consumption, some studies are concerned with energy-saving types of equipment and ship designs. The study by Hochkirch and Bertram [11] is one of them. Hochkirch and Bertram emphasized that energy saving is possible with electronically controlled pumps; heating, ventilation, and air conditioning (HVAC) systems; and energy-efficient lighting. In order to save energy or decrease emissions, additionally, exhaust gas and hot coolant may be used [12]. In terms of ship design, various studies discuss different hull resistance reduction technologies, prime movers, and ship propulsion for the best implementation [13]. New approaches in ship design may be practicable for new ships, although it may be difficult to apply them to ships that have already been constructed [14]. According to Corbett et al., speed optimization is the other way to decrease ship emissions in short-run simulations. However, as speed optimization provides lower ship emissions, it increases transportation costs because of the lower speed and additional ships required [15]. In parallel with advances in ship emissions reduction technologies, the International Maritime Organization (IMO) has issued a policy related to shipping emissions. Although the IMO's emissions policy has vital importance within long-run simulations for reducing emissions from ships, the results of the investigation conducted by Harrould-Kolieb et al. demonstrate that this policy does not work well enough in short-run simulations [12]. It is expected that the IMO emissions management policy and other regional and national institutions and organization(s) will play a central role in issuing a number of regulations, which are listed below, for decreasing health and environmental risks in the long run:

- International Convention for the Prevention of Pollution from Ships (MARPOL)—Annex VI: Prevention of Air Pollution from Ships
- European Union (EU) Legislation on Air Quality Standards and Limitations Related to Ship Exhaust Gas
- Emission Control Area (ECA) and Limitations in the ECA (Baltic Sea, North Sea, English Channel, North America and US Caribbean)
- Regulations of National Organizations such as the US Environmental Protection Agency (EPA)

The main objective of these policies and regulations is to minimize the potential impact of emissions on living things by reducing emission rates in highly populated coastal areas. In order to reduce emission rates, the limit of sulfur content and particulate matter in emissions has been changed by mass from 4.5% to 0.1%, depending on navigation phases and areas, through the abovementioned rules and regulations over the years. This series of changes is summarized in Table 14.1.

The limitations of emissions are not only regulated internationally and regionally but also by national legislation. An article of the Turkish Constitution of 1982 identifies the role of countries as follows: "everyone has the right to live in a healthy and balanced environment. It is the duty of the State and citizens to protect the environment" [17]. In conformity with this article, Turkish legislative authority has been addressed by the Regulation on Reduction of Sulfur Ration on Some

Table 14.1 Changing sulfur and particulate matter emissions limitations of marine fuel over the years

Emission	Limit outside an ECA	Limit inside an ECA
Sulfur	3.50% m/m after January 1, 2012	–
Particulate matter	0.50% m/m after January 1, 2020	0.10% m/m after January 1, 2015

Source: International Maritime Organization, accessed on May 10, 2016 [16]
ECA emission control area

Type of Fuel Oil. This regulation covers the emissions limitations from ships in Turkish territorial waters in two classifications, i.e., nonstop ships in marine pollution control areas of the Republic of Turkey, and ships in inland waterways, including during hotelling. Articles 6 and 7 of the regulation explain the limits of sulfur content in using fuels, which are 1.50% m/m for ships navigating in Turkish territorial waterways and 0.1% m/m for ships using Turkish inland waterways and berths.

Although it is thought that sulfur amounts in ship emissions can be minimized with supporting rules and regulations, there are significant long and short-term health risks in highly populated coastal regions. These sulfur ratios and emissions reduction technologies alone are not enough to deal with health risks from ships during port approach and maneuvering. To draw attention to the long- and short-term health risks in changing sulfur and particulate matter limitations of marine fuel, the authors conducted a comprehensive analysis in different navigation phases. For this aim, hotelling and port approach of a ship were selected for analysis because they include all navigation phases from cruising to hotelling. However, the authors thought that if the initial port emissions rate was minimized in this analysis, the ship emissions could be determined more objectively. These views led the authors to select a green, environmentally friendly, and less polluted port. In this sense, one of the best ports in Turkey for all necessities of this analysis is Asyaport, in a densely populated area in Tekirdag.

14.2 Characteristic of Asyaport and the Tekirdag Region

Tekirdag is one of the most prominent provinces in Turkey, with a strong industry network, transportation facilities, and hinterland. With port investments in recent times, the city is shaping up to be potential transfer center for ships between the Black Sea, the Marmara Sea, and Far East destinations. The Asyaport Container Port is one of the main ports located in the Tekirdag region. The port started to conduct operations with nine ships per week (four mother ships and five feeder ships) in the year 2015, and it was estimated that the port would handle 2.5 million TEU (twenty-foot equivalent unit) containers/year in 2016 and the following years, with operation and handling of approximately 11 ships per week in its berths [18]. As well as the operational capacity of the port, the port authority manages

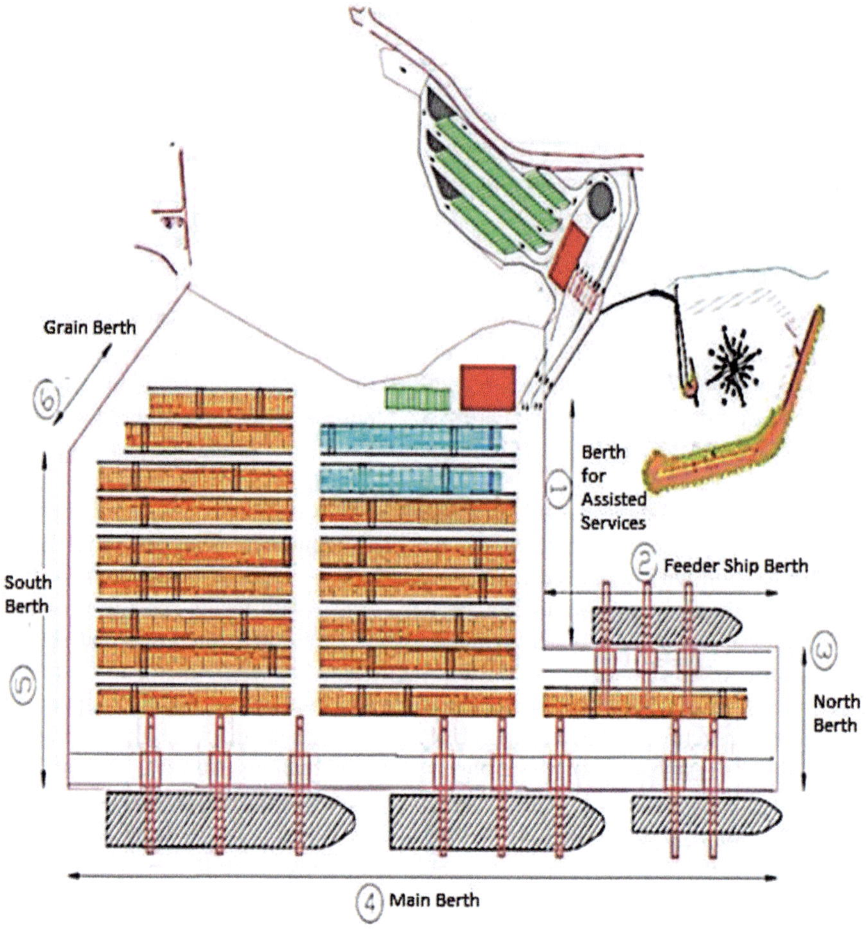

Fig. 14.1 General plan of Asyaport

the environmental effects of port operations. For this, the port is equipped with a solar power–generating system as an environmentally friendly approach. All electrical energy needs of the port have been met via this system. To minimize emissions, inland transportation vehicles use liquified natural gas (LNG) as a fuel. This environmentally friendly management approach has been granted a Port Environmental Review System (PERS) certificate by the European Sea Ports Organization (ESPO) [19] (Fig. 14.1).

14.3 Types of Research

The types of research used in this study are theoretical and numerical research. The theoretical research addresses chemical reactions in the burning of marine fuel. It also investigates fuel emissions and human health risk relationships according to US EPA reference doses. The numerical research involves statistical analysis of emissions and health risks by considering the theoretical research findings and different scenarios.

14.4 Theoretical Perspective of Research

14.4.1 Chemical Reactions in Combustion

The majority of the energy used is from fossil sources. These energy sources include hydrogen, carbon, and sulfur. In use, these energy resources are generally defined as hydrocarbons (C_nH_{2n+2}). These hydrocarbons are burned with air. In combustion, with the effect of nitrogen in the air, oxides (nitrogen oxides (NOx), sulfur oxides (SOx), and carbon oxides) are formed. These oxides significantly and adversely affect the health of the environment and living things, and even cause deaths. As is known, sulfur oxides cause damage to the human respiratory system and eyesight. Also, they cause harm to machines in many ways. Sulfur oxides discharged into the atmosphere react with oxygen and moisture and create sulfuric acid. Nitrogen oxides combine with oxygen, causing toxic effects. Nitrogen oxides combine with moisture in the lungs and form small amounts of nitric acid as a result [20].

As is known, emissions ensue from fuel combustion, and combustion is an exothermic chemical reaction. There are two kinds of combustion: complete combustion and incomplete combustion. In complete combustion, the general form of H_2, C, and S burning reactions can be written as Eqs. 14.1–14.3 [21]:

$$\mathrm{hkgH_2} + \frac{h}{4}\mathrm{molO_2} = \frac{h}{4}\mathrm{molH_2}O \tag{14.1}$$

$$\mathrm{ckgC} + \frac{c}{12}\mathrm{molO_2} = \frac{c}{12}\mathrm{molCO_2} \tag{14.2}$$

$$\mathrm{skgS} + \frac{s}{32}\mathrm{molO_2} = \frac{s}{32}\mathrm{molS_2O} \tag{14.3}$$

In these equations, H_2, C, and S express their percentages by weight in the fuel, and they are as given in Table 14.2.

Combustion of fuel with sulfur creates SO_3. Fuel combines with water, which is formed in combustion, and this results in formation of sulfuric acid. The corrosive

Table 14.2 Percentages of H, C, and S by weight in fuel and an alternative fuel, liquified natural gas (LNG)

Type of fuel	%H	%C	%S
Fuel oil	12–18	82–86	0–3.5
LNG	4–6	12–17	0

Table 14.3 Excess air coefficients subject to the type of fuel and type of engine [21]

Type of fuel	Type of engine	Air coefficient
Fuel oil	High-speed diesel machine	1.30–1.70
	Low-speed diesel machine	1.70–2.00
LNG	Low-speed	1.30–1.50
	High-speed	1.50–1.70

LNG liquified natural gas

effect of the sulfuric acid may be controlled by a suitable arrangement of the burning of lubricating oil and the local temperature. CO is a toxic, odorless, and heavy gas. This gas consists of 1 mole of carbon and 1 mole of oxygen. Internal combustion engines cannot provide the high oxygen content of the air necessary for combustion, and transformation to CO_2 happens in a short time. So effective combustion does not occur, and the amount of CO in exhaust gas increases. In combustion, the excess air coefficient is 1.8, the sulfur ratio is 2.5, and the lower heat value of the fuel is 42,676.6 kJ/kg (fuel). The excess air coefficient changes according to the type of fuel and the type of engine. The air coefficients are listed in Table 14.3.

When the theoretical amounts of available oxygen and air are used, the required minimum amount of oxygen is 10.520625 mol (O_2)/kg (fuel) and the required quantity of air is 2,606,011 kg (air)/kg (fuel) in the light of the following equations [22]:

$$\text{Lo1} = \frac{1}{0.21}\left(\frac{s}{32} + \frac{h}{4} + \frac{c}{12} - \frac{o}{32}\right)\text{mol/kg} \tag{14.4}$$

$$\text{Lo2} = 24, 4.\alpha.\text{Lo1}\,[\text{mol (air)}/\text{kg (fuel)}] \tag{14.5}$$

The amount of flue gas can be written as Eq. 14.6, and the amount of gas (in kilograms) can be written as Eq. 14.7.

$$M_g = \alpha.\text{Lo1} + \frac{o}{32} + \frac{h}{4} \tag{14.6}$$

$$W_g = 1 + \text{Lo2} \quad [\text{kg (gas)}/\text{kg (fuel)}] \tag{14.7}$$

In the poor oxygen level, the production of water continues. However, the forms of carbon monoxide and carbon occur instead of carbon dioxide. This is called incomplete combustion.

14.4.2 Reference Doses for Health Risks of Emissions

The air consists of 78% N, 21% O_2, and 1% other gases (CO_2, argon, etc.). Although the human phenomenon has changed the balance of nature with activities of humankind, the natural balance of these gases is crucially important for humans and the environment. The natural balance of these gases is a reference for air quality. Nowadays, air quality and emissions are a major problem for countries overall. To deal with this problem, several countries and institutions have established standards. A few of them determine reference doses for health risks of emissions as well as standards. The most well known is the National Ambient Air Quality Standards (NAAQS) set by the US EPA, based on the Clean Air Act. NAAQS considers reference doses for some pollutants as given in Table 14.4. Besides these pollutants, the other forms of emissions are CO_2, NOx, SOx, PM, etc. PM occurs from emissions directly, and the other PM sources are NOx and SOx. The increasing percentage of CO_2 in the atmosphere causes a series of health risks. For example, with 3% CO_2 composition in the atmosphere, people's breathing becomes deeper and more frequent. With increases, this effect changes radically.

Table 14.4 National Ambient Air Quality Standards (NAAQS) reference doses

Pollutant	Standard value	
CO		
8-h average	9 ppm	10 μg/m³
1-h average	35 ppm	40 μg/m³
NO_2		
Annual arithmetic mean	0.053 ppm	100 μg/m³
SO_2		
Annual arithmetic mean	0.030 ppm	80 μg/m³
24-h average	0.140 ppm	365 μg/m³
3-h average	0.500 ppm	1300 μg/m³
$PM_{2.5}$		
Annual arithmetic mean		15 μg/m³
24-h average		65 μg/m³
PM_{10}		
Annual arithmetic mean		50 μg/m³
24-h average		150 μg/m³

Source: US Environmental Protection Agency (1991), NAAQS
PM particulate matter

14.5 Numerical Perspective of Research

14.5.1 Fuel Consumption Rate of Ships and Assumptions

The ships usually use fuel energy for the main engine, auxiliary engine, and generators. The required fuel rate is in direct proportion to their capacity, and their capacity generally changes according to ship size. It is seen that there are two types of ships—mother and feeder ships—when analyzing ship characteristics at Asyaport. These ships have maximum capacities of 14000–18000 TEU for the mother ships and 1000–2000 TEU for the feeder ships. The ships of these capacities have approximate maximum engine power as given in Table 14.5.

Although Table 14.4 expresses the maximum engine rate, the ships do not use this maximum engine load in practice. In several studies, approximate engine load has been assumed in different navigation phases as the following:

- Cruising: main engine load 80%, auxiliary engine 70–80%, generators 75%
- Maneuvering: main engine load 40%, auxiliary engine 70–80%, generators 75%
- Hotelling: only generators 75% [22]

The Asyaport authority estimates that 11 ships/week were given service within the 2016 year. With the assumption of four mother ships and seven feeder ships/week (a total of 11 ships) being handled in Asyaport, and the duration of each navigation phase of the different ship types being as given in Table 14.6, the authors calculated emissions by considering chemical reactions during combustion per fuel oil kilogram and the approximate engine load for each navigation phase.

14.5.2 Findings

The theoretical research indicates that the percentages of pollutants in the overall emissions differ according to the fuel type and the sulfur limitation policy. The percentage of each pollutant has been obtained as given in Table 14.7, in the light of the theoretical research considering different sulfur policies, excess air coefficients, and alternative fuel (LNG) use.

Table 14.5 Approximately engine power of ships according to ship type

Ship type	Main engine	Auxiliary engine	Generators
Mother (14000–18000 TEU)	50,000 kw	14,000 kw	4 × 1000 kw
Feeder (1000–2000 TEU)	7000 kw	600 kw	2 × 500 kw

TEU twenty-foot equivalent units

Table 14.6 Approximate times for ship operations

Ship type	Cruising	Maneuvering	Hotelling
Mother	1 h	2.5 h	30 h
Feeder	1 h	1.5 h	15 h

Table 14.7 Percentage emissions changes with different sulfur limitations and fuel types

	Fuel oil (for an excess air coefficient of 1.8)			LNG (for excess air coefficients as shown below)	
	S: 3.4[a]	S: 1.5	S: 0.1	1.5	1.3
CO_2	7.37	7.47	7.54	6.82	7.8
O_2	8.98	8.99	8.99	6.56	4.5
SO_2	0.11	0.049	0.003	–	–
SO_3	0.17	0.075	0.00494	–	–
N_2	76.04	76.07	76.09	74.03	73.3
NO	0.0905	0.0905	0.0906	0.088	0.087
NO_2	0.00905	0.00905	0.00906	0.0088	0.0087
SOx + NOx	0.38	0.22	0.11	0.097	0.096

NOx nitrogen oxide, *SOx* sulfur oxide
[a]*S* sulfur limitation

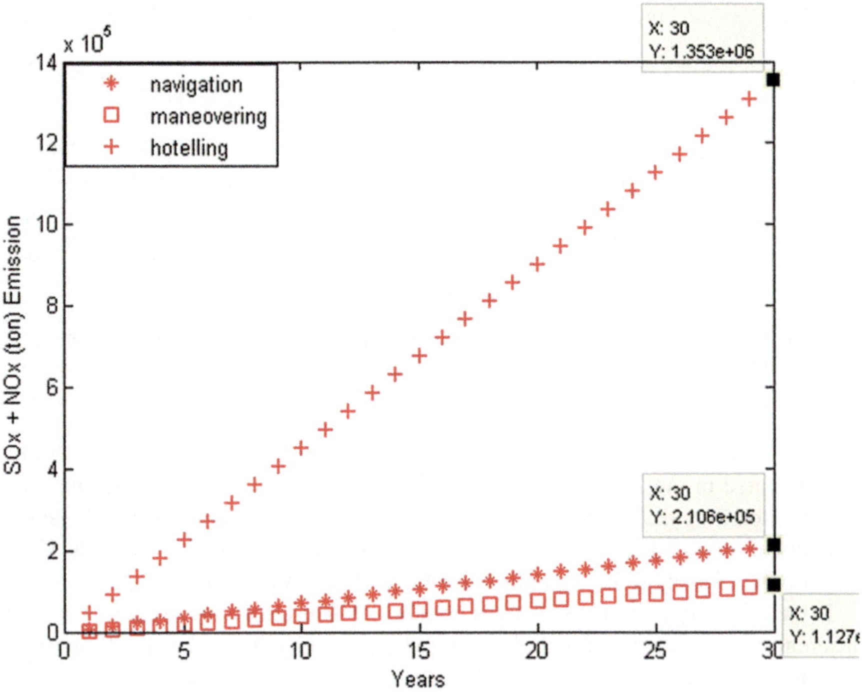

Fig. 14.2 Sulfur oxide (SOx) and nitrogen oxide (NOx) emissions changes in the different navigation phases of mother ships

After calculation of the pollutant percentages, four scenarios were developed for the numerical analysis of emissions and the survey of Asyaport. In the first one, changes in SOx and NOx emissions in the different navigation phases of mother ships was examined by taking into account the 0.1% sulfur limitations for ships using Turkish inland waterways and berths. The results are given in Fig. 14.2, which shows that emissions increase in all navigation phases for mother ships.

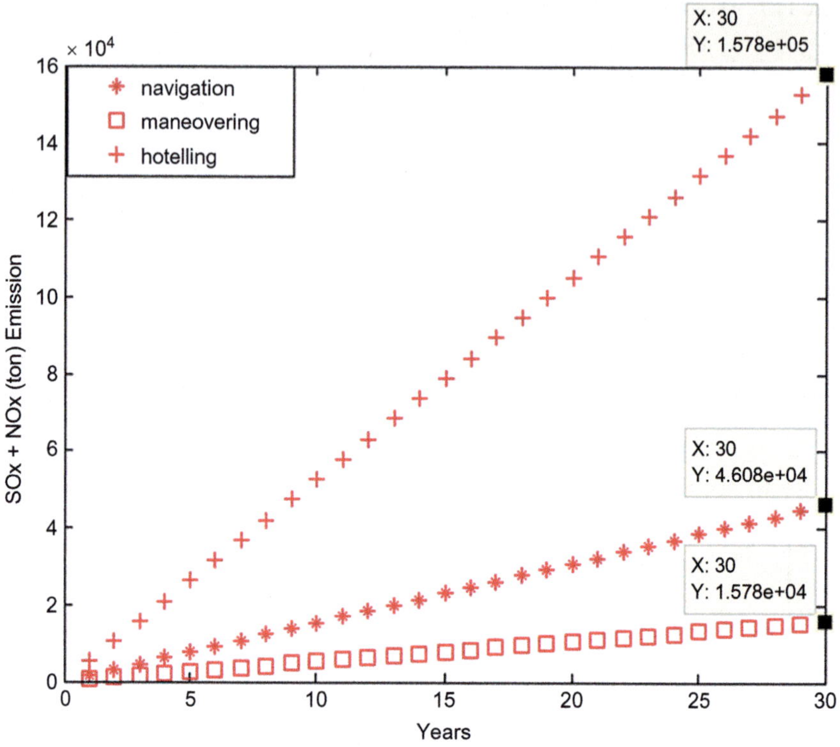

Fig. 14.3 Sulfur oxide (SOx) and nitrogen oxide (NOx) emissions changes in the different navigation phases of feeder ships

In the second scenario, the same process is applied to feeder ships, and the results are illustrated in Fig. 14.3. As can be seen from Figs. 14.2 and 14.3, the emission rates for hotelling have increased dramatically. However, it should be noted that the amounts of emissions during cruising or navigation are present for only 1 h. If we increase the cruising time, the emission rates will increase rapidly.

In the following scenarios, changes in SOx and NOx emissions in the presence of different sulfur content during cruising have been calculated based on the limitations determined by rules and regulations. The results of these calculations are presented in Figs. 14.4 and 14.5.

Changes in SOx, NOx, SO_2, SO_3, NO, and NO_2 emissions with a 3.4% ratio of sulfur content during cruising have been calculated based on the limitations determined by rules and regulations. The results of these calculations are presented in Figs. 14.6 and 14.7.

Changes in CO_2 emissions in the different navigation, maneuvering, and hotelling phases have been calculated based on the limitations determined by rules and regulations. The results of the calculation are presented in Figs. 14.8 and 14.9.

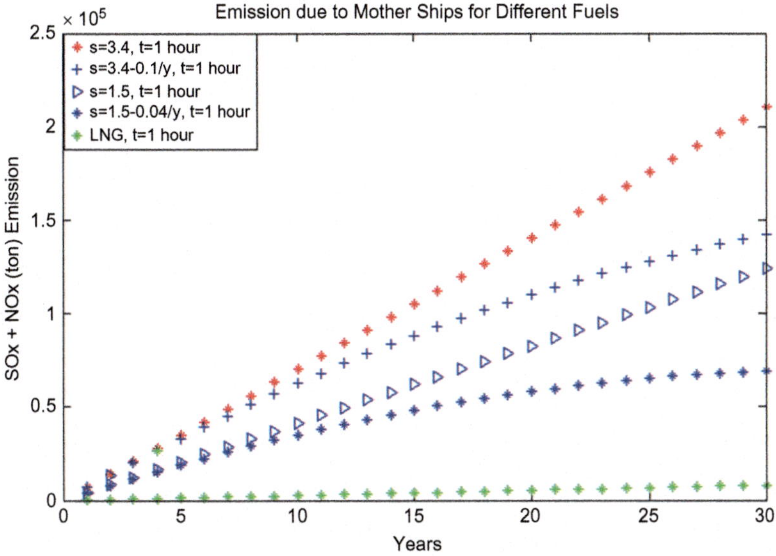

Fig. 14.4 Sulfur oxide (SOx) and nitrogen oxide (NOx) emissions changes with different sulfur content and in the case of alternative fuel use (liquified natural gas (LNG)) during cruising of mother ships

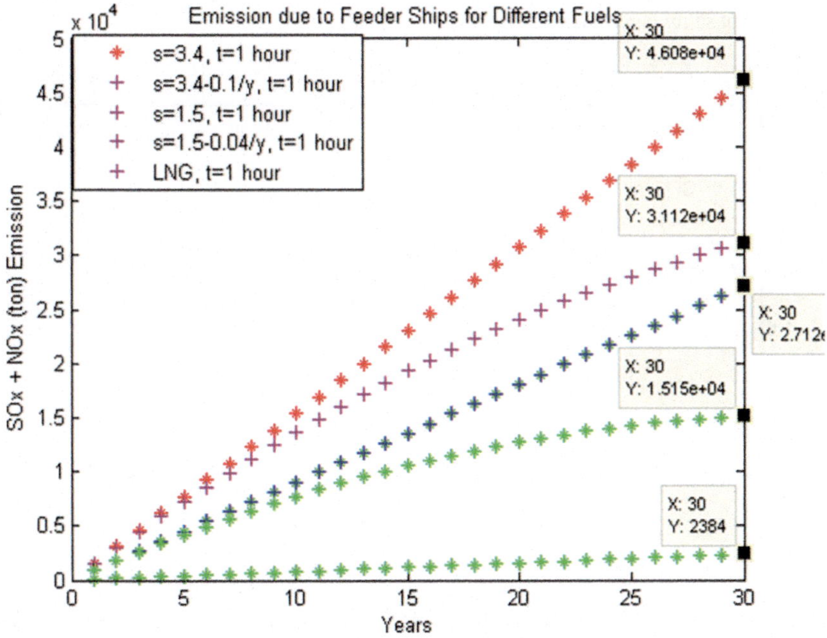

Fig. 14.5 Sulfur oxide (SOx) and nitrogen oxide (NOx) emissions changes with different sulfur content and in the case of alternative fuel use (liquified natural gas (LNG)) during cruising of feeder ships

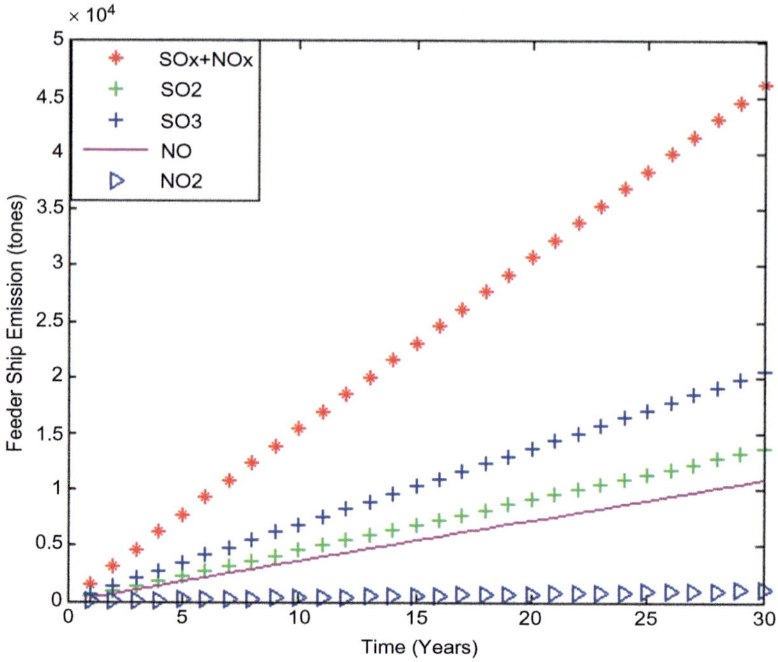

Fig. 14.6 Sulfur oxide (SOx), nitrogen oxide (NOx), SO_2, SO_3, NO, and NO_2 emissions changes with a constant sulfur percentage during cruising of feeder ships

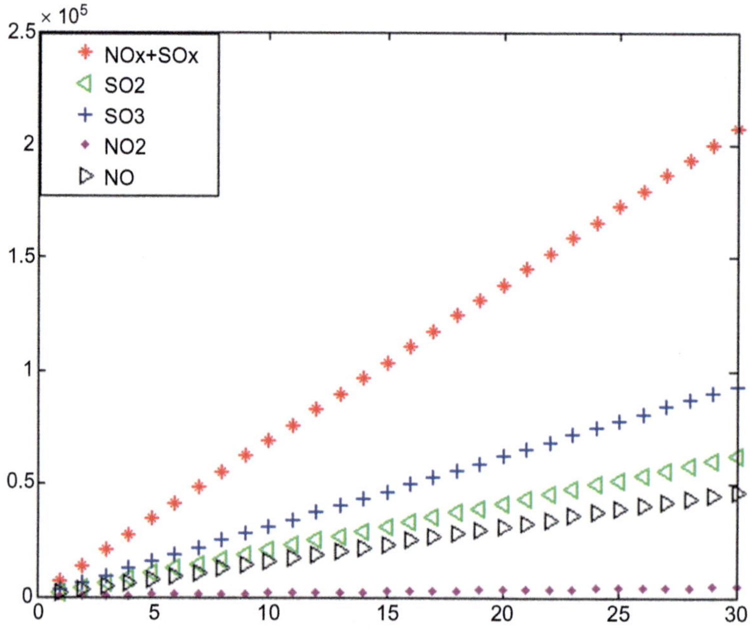

Fig. 14.7 Sulfur oxide (SOx), nitrogen oxide (NOx), SO_2, SO_3, NO, and NO_2 emissions changes with a constant sulfur percentage during cruising of mother ships

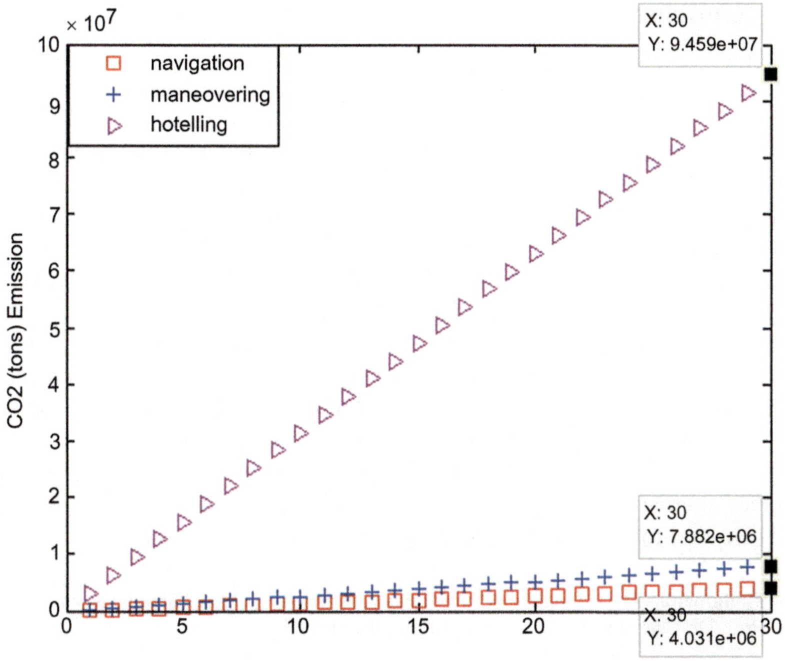

Fig. 14.8 CO_2 emissions changes in different navigation phases of mother ships

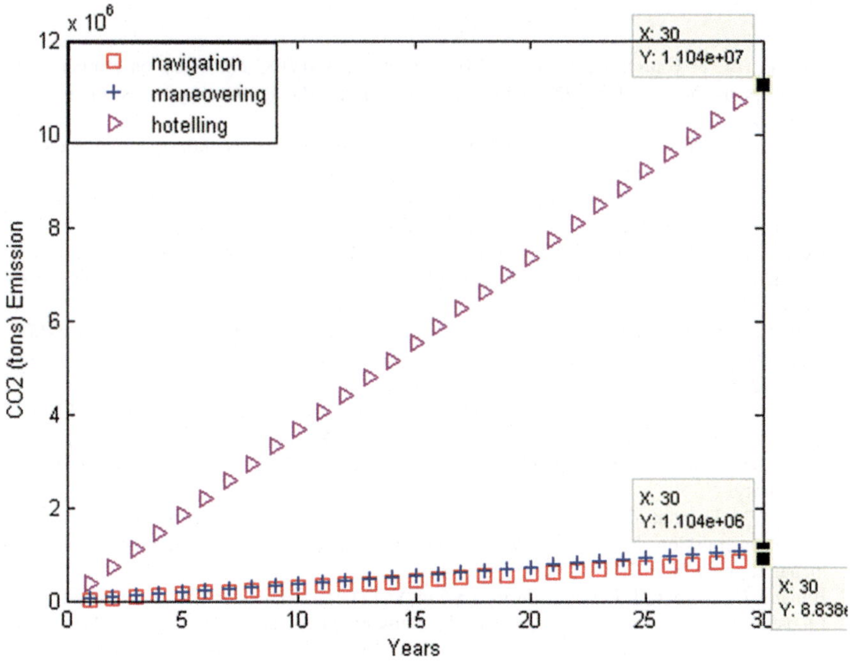

Fig. 14.9 CO_2 emissions changes in different navigation phases of feeder ships

14.6 Conclusions

Total SOx and NOx emissions in the next 20 years have been estimated to be between 118,100 and 146,439 tons for feeder ships and between 1,059,495 and 11,117,261 tons for mother ships in Asyaport. It is estimated that the grand total of SOx and NOx emissions will be 1,249,278 tons for a total of 11 ships/week including both mother and feeder ships. These findings prove that limitations in sulfur content are not enough on their own to eliminate health and environmental risks. Even if ports themselves are completely environmentally friendly or have zero emissions, the ships themselves will continue to reduce air quality. As is seen by the example of Asyaport, it is possible to reach critical levels (based on NAAQS reference doses) of air quality for human health and the environment in the short term. Unlike traditional marine fuel oils, use of LNG as fuel in ships can help humankind to improve air quality in the long term. Additionally, scrubbers, international engine modifications, humid aid motors, selective catalytic reduction, and other alternative energy sources such as solar power, biofuels, and fuel cells could be useful for emissions-reducing policies in the near future.

Acknowledgments The authors thank Chief Eng. Muhammet Sandikci, Fahrettin Kucuksahin and Assoc. Prof. Dr. Cengiz Deniz for commenting on the manuscript.

References

1. Laurent, A., & Hauschild, M. Z. (2014). Impacts of NMVOC emissions on human health in European countries for 2000–2010: Use of sector-specific substance profiles. *Atmospheric Environment, 85*, 247–255.
2. Deniz, C., & Durmusoglu, Y. (2008). Estimating shipping emissions in the region of the sea of Marmara, Turkey. *Science of Total Environment, 390*, 255–261.
3. Schrooten, L., et al. (2009). Emissions of maritime transport: A European reference system. *Science of the Total Environment, 408*(2), 318–323.
4. Healy, R. M., O'Connor, I. P., Hellebust, S., Allanic, A., Sodeau, J. R., & Wenger, J. C. (2009). Characterisation of single particles from in-port ship emissions. *Atmospheric Environment, 43*, 6408–6414.
5. Corbett, J. J., Winebrake, J. J., Green, E. H., Kasibhatla, P., Eyring, V., & Lauer, A. (2007). Mortality from ship emissions: A global assessment. *Environmental Science and Technology, 41*, 8512–8518.
6. Lubick, N. (2007). Death from shipping. *Environmental Science and Technology, 41*, 8206.
7. Garbaccio, R. F., Ho, M. S., & Jorgenson, D. W. (2000). *Modelling the health benefit of carbon emissions reductions in China*. Cambridge, MA: Kennedy School of Government Harvard University.
8. ICCT (The International Council on Clean Transportation). (2007). Air pollution and greenhouse gas emissions from ocean-going ships: Impact, mitigation options and opportunities for managing growth, p. 7.
9. Buhaug, Ø., Corbett, J. J., Endresen, Ø., Eyring, V., Faber, J., Hanayama, S., Lee, D. S., Lee, D., Lindstad, H., Markowska, A. Z., Mjelde, A., Nelissen, D., Nilsen, J., Pålsson, C., Winebrake, J. J., Wu, W., & Yoshida, K. (2009). *Second IMO GHG study*. London: IMO (International Maritime Organization).

10. Ballou, P., Chen, H., & Horner, J. D. (2008). Advanced methods of optimizing ship operations to reduce emissions detrimental to climate change. Oceans 2008. April 8–11 2008, Kobe Japan.
11. Hochkirch, K., & Bertram, V. (2010). *Engineering options for more fuel efficient ships*. Montreal: GreenTech.
12. Harrould-Kolieb, E., & Savitz, J. (2010). Shipping solutions: Technological and operational methods available to reduce CO2, *Oceana*. http://oceana.org/reports/shippine-solutions-technological-and-operational-methods-available-reduce-co2. Accessed on 19 Sept 2017.
13. Vergara, J., McKesson, C., & Walczak, M. (2012). Sustainable energy for the marine sector. *Energy Policy, 49*, 333–345.
14. Kontovas, C., & Psaraftis, H. N. (2011). Reduction of emissions along the maritime intermodal container chain: Operational models and policies. *Maritime Policy & Management, 38*(4), 451–469.
15. Corbett, J. J., Wang, H., & Winebrake, J. J. (2009). The effectiveness and costs of speed reductions on emissions from international shipping. *Transportation Research Part D: Transport and Environment, 14*(8), 593–598.
16. http://www.imo.org/en/OurWork/Environment/PollutionPrevention/AirPollution/Pages/Sulphur-oxides-%28SOx%29-%E2%80%93-Regulation-14.aspx. Accessed on 10.05.2016.
17. Coskun, A., Turkey, O. Y., & Velioglu, N. (2011). Air pollution regulations in Turkey and harmonization with the EU Legislation. *iForest- Biogeosciences and Forestry, 4*, 181–185.
18. http://www.denizhaber.com.tr/dev-gemiler-asyaport-limaninda-agirlaniyor-haber-65040.htm. Accessed on 01.08.2016.
19. http://www.asyaport.com/tr-TR/news/asyaport-turkiyenin-ilk-ecoport-pers-sertifikali-limani-oldu/283/News.aspx. Accessed on 12.05.2016.
20. Gurel, O., & Bicen, S. (2013). Exhaust emission calculations, Istanbul Technical University Maritime Faculty Marine Engineering Thesis, Istanbul (In Turkish).
21. Küçükşahin, F. (2013). *Çevre ve Gaz Emisyonları*. İstanbul: Birsen Yayınevi.
22. Ucar, O. F. (2014). Investigation of exhaust gases produced by vessels called at Samsun port and their environmental impacts, republic of Turkey Ministry of Transport, Maritime and Communication, Maritime Specialist Thesis.

Chapter 15
Smart Protective Clothing for Aircraft Crew

Emel Onder, Ezgi C. B. Noyan, Sena C. Duru, Cevza Candan, Selcuk Paker, and Rafet Sayar

15.1 Introduction

In recent years, electromagnetic radiation has become the fourth most serious source of public pollution in addition to noise, water and air pollution. Electromagnetic interference (EMI) is a well-known problem in commercial and scientific electronic instruments, antenna systems and military electronic devices [1]. Wireless communication links have been used worldwide for some years, particularly in radio, television broadcast stations, mobile phones, radar and microwave systems [2]. There is an increasing sensitivity and interference between digital devices, which not only reduces the device performance but has also required for a safe and healthy environment [3]. Limits of electromagnetic emissions from these devices should be set in order to reduce the possibility of interference. Many countries pass some regulations for the producers of electrical and electronic equipment to satisfy the expectations about limiting the electromagnetic radiation [4, 5]. Electromagnetic interference shielding effectiveness (EMI SE) level of 30 dB is considered acceptable for most applications and attenuates 99.9% of electromagnetic radiation. In some military and aerospace applications, 80–100 dB SE is required, due to rising electronic integration and the need for the effective working of these instruments [6].

E. Onder (✉) • E.C.B. Noyan • S.C. Duru • C. Candan
Faculty of Textile Technologies and Design, Istanbul Technical University, İstanbul, Turkey
e-mail: onderem@itu.edu.tr; bozez@itu.edu.tr; cimilli@itu.edu.tr; candance@itu.edu.tr

S. Paker
Faculty of Electrical and Electronics, Istanbul Technical University, İstanbul, Turkey
e-mail: spaker.@itu.edu.tr

R. Sayar
Turkish Republic Land Forces, İstanbul, Turkey
e-mail: rftsyr@gmail.com

T.H. Karakoç et al. (eds.), *Advances in Sustainable Aviation*,
DOI 10.1007/978-3-319-67134-5_15

Another important issue is the fact that flying at higher altitudes and latitudes results in an increased exposure to cosmic radiation compared to the radiation on Earth. This is the result of galactic cosmic rays (GCR) and secondary particles produced by interactions of the GCR with the atmosphere [7]. The cascades of secondary cosmic radiation consist of an electromagnetic component – X-ray and gamma radiation – and of electrically charged and uncharged subatomic particles, which are created when the primaries enter the atmosphere and interact with its constituents. The intensity of the radiation components, travelling downwards in the atmosphere, is a function of their energy, and it generally depends on altitude. From sea level to a maximum at 20–25 km altitude, the radiation intensity grows almost exponentially [8, 9]. At current cruising altitudes of airliners, this corresponds to 300 times greater radiation fluxes than on the ground. The exposure increases in parallel with latitude of the flight path, such that polar routes are exposed to radiation most [7–9].

Cosmic radiation in aviation has been a concern since the 1960s, and annual effective dose limits of 20 mSv are recommended for crewmembers and 1 mSv for the public and averaged over a 5-year period by the International Commission on Radiological Protection (ICRP) as a measure of the stochastic health risk due to the ionizing radiation. A 2 mSv limit on the accumulated dose over 9 months of pregnancy is also recommended [10, 11]. According to the Basic Safety Standards Directive of European Union members, two thirds of 20 mSv or 6 mSv is an action level for aircrew for flights above 8 km. This level is designed as a guidance level for the dose control intervention and for a medical assessment and tracking [11]. Results taken for several decades by Air France show that aircraft crew generally receive 3–4 mSv $year^{-1}$ for 750 boarding hours. Compliance with the trigger level of 6 mSv $year^{-1}$ is achieved by route selection. Work schedules can be developed for pregnant pilots to enable the dose to the foetus to be kept below 1 mSv [12].

The in-flight dose has been associated with increases in crewmember cancer incidence and mortality with doses of radiation estimated to yield a 5.5 % increase in cancer risk per Sv absorbed. One study showed that airline pilots receive a greater annual dose than industrial, medical or even nuclear workers. However, epidemiological researches have not presented conclusive evidence of an increase in cancer cases for radiation doses below 100 mSv [10, 11].

Textiles demonstrate many advantages as clothes and are unique in several aspects. The building stones of textile materials are fibres or filaments. Fibres are available in a broad range of materials, such as natural or synthetic, biocompatible, biodegradable and optical or electroconductive. They can differ in terms of length, fineness, cross-sectional shape, surface roughness, etc. Fibres of various types are employed to form yarns, from which fabrics, including three-dimensional structures, can be constructed.

With the help of special treatments (finishes) such as antimicrobial finish, textile materials may combine advanced multifunctionality with traditional textile properties. Moreover, the laundering machines available are well developed for easy care of textiles. Furthermore, textiles can be manufactured relatively fast at reasonable costs. These characteristics present many opportunities for a number of novel applications of textiles.

The intelligent character of a textile material can be introduced at fibre, fabric and/or finishing level. It is even possible to connect independent appliances with a textile structure [13]. As nano-, bio- and information technologies and biomaterials have evolved, there are many possibilities for enhanced functionalities within textiles, from new fibre/fabric structures to the integration of wearable electronic assemblies into textiles. Bio- and polymer chemistry, for example, enables computer processor miniaturization, so that new forms of textile sensors, actuators and some other smart textiles for functional and wearable computing can be produced. In addition to that, conductive yarns can be woven, knitted and even embroidered into electronically enabled textiles to provide soft textile interfaces.

The adoption of sophisticated technologies to textiles is surely not limited to wearable computers. Thanks to the great deal of research to improve the functional properties of textiles using current and emerging technologies, the innovative textile products find applications in the fields such as environmental protection, safety and protection, health and energy. Some of the most common technologies which are developed for specific medical and textile applications may include phase change materials and microencapsulation, light-emitting polymers and diodes, nanotechnologies, thermo-chromic materials, shape memory alloys and polymers, photovoltaics and solar cells, radio-frequency identification (RFID) tags, photoluminescence, microelectromechanical systems (MEMS), piezoelectric resistance, photochromic materials, biomaterials and plasma technologies [14].

In parallel with the developments outlined above, the textile industry has also made considerable advances in the production of multifunctional value-added products. These functional properties are beyond the aesthetic and decorative ones, and they may be defined as "smart properties" since the very properties enable the textiles to act according to an external stimulation. The stimulus and the response may have an electrical, chemical, magnetic or thermal origin. Depending on the extent of intelligence, smart textiles can be divided in three subgroups: (a) passive smart textiles which can only sense the environment, (b) active smart textiles which can sense the stimuli from the environment and also react to them and (c) very smart textiles which can adapt their behaviour to the circumstances [13, 15–17].

Textiles, which can be metalized to obtain electrically conductive surfaces, have broad application in the field of industrial, military and civil shielding [18–22]. Some researchers and industrial companies have provided new sorts of electromagnetic shields either to isolate a space from external electromagnetic radiation or to avoid unwanted emission of electromagnetic waves radiated by internal sources [23, 24]. Apart from the advanced methods, e.g. ionic plating, cathode sputtering and vacuum metallization, the conventional surface production methods (i.e. weaving, knitting and braiding) are also utilized to develop shielding textile materials. Generally speaking, the conductive textiles – which are coated with aluminium, copper, silver and nickel – are significant types of materials for preventing EMI [22, 23, 25]. Using special yarns, such as silver-plated yarns, it is possible to decrease the strength of electromagnetic field or waves influencing human health seriously [26]. Silver-coated yarns offer good electrical conductivity and shielding with good corrosion resistance, antistatic, antibacterial, odour-inhibiting property [1, 27].

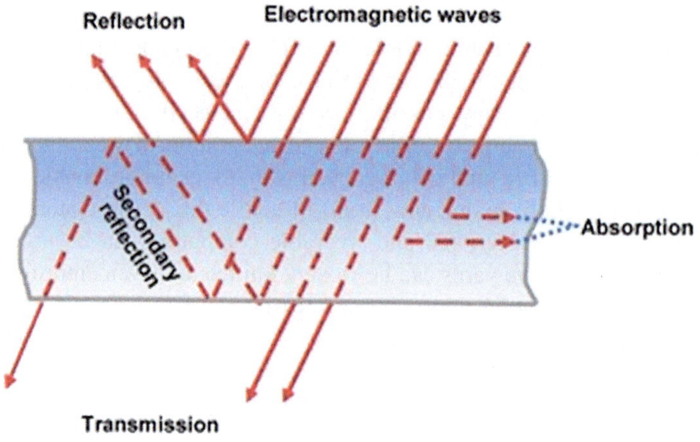

Fig. 15.1 Schematic representation of shielding phenomena [30]

Electromagnetic interference shielding effectiveness (EMI SE) is described as the attenuation of an electromagnetic wave produced by its passage through a shield and is expressed in decibel (dB). EMI SE is calculated according to Eq. 15.1.

$$SE = 20 \log \frac{E_i}{E_t} \tag{15.1}$$

where E_i is the incident electrical field strength and E_t is the transmitted electric field strength of electromagnetic wave [28, 29]. EMI SE shows the ability of a surface to reduce the propagation of electromagnetic fields by reflection, absorption and/or multiple reflection mechanisms (see Fig. 15.1). For a knitted fabric, shielding effectiveness is dependent on the signal amplitude and frequency and affected by different fabric parameters such as course and wale densities, yarn linear density, fabric structure, fabric thickness and shape of shield discontinuities.

Smart materials that provide thermal equilibrium between body and environment attract attention over the last decade. Thermally responsive polymers (TRP) undergo remarkable changes in their swelling behaviour and network structure when the small temperature changes in environment [31]. Phase change materials (PCMs) can absorb or release great amount of latent heat during phase transitions, and hence they may offer significant potential for the thermal management applications [32]. When a textile is incorporated with thermally active materials such as TRP and PCM, it not only provides thermal regulation function, but also an improved thermal capacity is added to its present passive insulation characteristic. These textiles can absorb heat from human body or environment at high-activity levels and/or in hot climate conditions. Conversely, they redistribute and release heat in cold environments.

In the present work, we designed spacer weft-knitted fabrics composed of wool and silver-plated nylon yarns for the purpose of underwear clothing – which has an electromagnetic shielding and good thermal property – of air aircrew personnel.

15.2 Material and Method

The experimental work deals with combining two functions, a good thermal response and an electromagnetic shielding effect, in one fabric with the use of smart polymers, wool and functional yarns. To do so, highly conductive silver-plated polyamide 6.6 single and 2-ply yarns, namely, SHIELDEX®, were purchased from Statex, and 100% worsted wool yarns were supplied from Yünsa A.Ş. These yarns were used to produce two types of spacer knitted fabrics (S1 and S2) on Shima Seiki flat knitting machine of E18, with the parameters given in Table 15.1. Figure 15.2 shows one of the knitted samples.

To improve the thermal performances of the wool/PA-silver knitted fabrics, two types of environmentally friendly, biodegradable and biocompatible polyol-based organics, one TRP and one PCM, were chosen for their thermal characteristics matching the dynamic heat management requirements next to the skin. These technical grade organics were purchased from Sigma-Aldrich and employed without further purification.

Table 15.1 Fabric properties

Knitted sample	Inlay yarn	Face yarn	Back yarn	Wales cm^{-1}	Courses cm^{-1}	Fabric weight ($g\ m^{-2}$)
S1	Nm 30/1 wool	110/34 dtex one-ply silver-plated nylon liner resistance: < 30 Ω/cm	Nm 30/1 wool	Face: 6	Face: 9	208.8
				Back: 6	Back: 10	
S2	Nm 30/1 wool	110/34 dtex two-ply silver-plated nylon liner resistance: < 100 Ω/m	Nm 30/1 wool	Face: 7	Face: 7	254.4
				Back: 6	Back: 9	

Fig. 15.2 Wool/PA-silver knitted spacer fabric

The triblock copolymer poly(ethylene oxide)-poly(propylene oxide)-poly(ethylene oxide), PEO-PPO-PEO, was used as thermo-responsive polymer (TRP) due to its well-known reversible sol-gel and hydrophilic/hydrophobic transition at nearly body temperature, which makes this polymer attractive for textile and other applications. This TR polymer is a non-ionic surfactant and soluble in water at low temperatures (usually below 15°C), and it melts at 56°C.

In addition, PEG1000 was used as phase change material (PCM) to maintain thermoregulating property in the fabric. Poly(ethylene glycols) (PEGs), also referred to as poly(oxyethylenes), consist of linear dimethyl ether chains with hydroxyl ending groups, $HO–CH_2–(CH_2–O–CH_2–)_n–CH_2–OH$, and accordingly possess dual feature of water solubility and organic solubility. Polyethylene glycols (PEGs) refer to high-latent heat capacities as PCMs in addition to many other advantages such as thermal stability, ecologically friendly property, non-toxicity, non-corrosiveness, biodegradability, easily availability and inexpensiveness.

Prior to TRP and PCM applications, impurities were removed by washing fabric samples in an aqueous solution of non-ionic surfactant at 30°C for 1 h. The polymer solutions were applied to the wool/PA-silver knitted samples (S1 and S2) via stepwise coating process combined with intermediate and final drying.

In the first step, 30% aqueous solution of the thermo-responsive triblock copolymer (TRP) were prepared and held in water bath at 4°C for 48 h to dissolve it in water homogenously. The fabric samples were then immersed in TRP solution at the liquor ratio of 1:10 for 30 min. TRP-treated samples were held in water bath again at 4°C for 2 h before they were processed on a laboratory-type foulard machine. Squeezing proceeded at 10 °C in the foulard machine so that the pressure and the speed of the rollers were set to be 2 bars and 1.5 m min^{-1}, respectively. Coated fabrics were dried at 85°C for 15 min for fixation and then conditioned for the next 24 h. TRP-coated fabrics were coded as S1-1 and S2-1.

In the second step, two pieces of TRP-coated samples were impregnated with 20% aqueous solution of PEG1000 at the ambient temperature and squeezed by applying a slight pressure via roller. Coated fabrics were dried at 85°C for 15 min for fixation and then conditioned for the next 24 h. TRP- and PCM-coated samples were coded as S1-2 and S2-2.

Structural characterizations, electromagnetic shielding properties and thermal responses of the samples were studied. Fourier transform infrared (FTIR) analyses were conducted using a PerkinElmer Spectrum 100 FTIR spectrometer equipped with a universal attenuated total reflection (ATR) accessory in the range 4000–450 cm^{-1} and a resolution of 4 cm^{-1}. The coaxial transmission line method described in ASTM D 4935–99 was employed to test the electromagnetic interference shielding effectiveness of the samples in the frequency range 0–8000 MHz, conducted by means of Rohde & Schwarz FSHIB-Spectrum Analyser.

Differential scanning calorimeter (DSC) analyses were carried out using PerkinElmer DSC 4000 under nitrogen atmosphere where the test specimens were heated and cooled within a temperature interval ranging from 15 to 75°C at 3°C min^{-1}. Second heating and cooling cycles were considered to eliminate the thermal histories of the samples. Temperature scans were run under the same conditions for both knitted samples and the coating material.

In order to characterize thermo-responsive behaviour of TRP, 1% of aqueous solution of the triblock copolymer was prepared, and particle size change was determined via heating and cooling cycles at temperatures between 20 and 40 °C using Malvern Nanosizer [33].

15.3 Results

15.3.1 Results of Structural Analyses

The comparative Fourier transform infrared (FTIR) graphs in Fig. 15.3 showed that all of the coated samples had typical peaks of their corresponding reference fabrics (S1, S2) and applied polymers.

The peaks at 3276–3282 cm^{-1} are associated with NH stretching bands of wool. The peaks at 1630–1640 cm^{-1} are assigned to the C=O stretching of amides I and II, as well as to the N–H scissoring of amide I, and the peaks at 1515–1516 cm^{-1} are labelled as the bending deformation peak of C–N–H bond related to the wool/nylon fabrics [34]. The broad bands observed at 3410–3414 cm^{-1} in IR spectra correspond to stretching vibrations of both structural and free OH groups. Similarly, the band due to the in-plane bending vibrations of the OH groups appeared at 1630–1640 cm^{-1}.

FTIR results of PEO-PPO-PEO and PEG1000 are quite similar to each other because of the similarities in chemical groups. The CH_2 bending band at 1466 cm^{-1} and the CH rocking band at 1341–1342 cm^{-1} of the PEG and PEO groups appeared as distinct peaks in the spectra of knitted fabric samples [35]. The strong bands around 1089–1100 cm^{-1} are assigned to C–O stretching vibration band of PEO block and PPO block of the triblock copolymer, respectively [36]. The bands at 1242–1241 cm^{-1}, 963 cm^{-1} and 842–841 cm^{-1} corresponding to the C–O–C stretching vibrations of ether groups also appear in spectra of fabric samples.

Consequently, wavenumbers and intensities of the transmission bands observed in spectra of the fabric samples verified that both thermo-responsive polymer (PEO-PPO-PEO) and organic phase chance material (PEG1000) could be successfully transferred into the wool/PA-silver knitted fabric structures in relation to the prepared recipes and the applied coating methods. Thus the knitted fabrics could be achieved to have a composite structure with the enhanced properties.

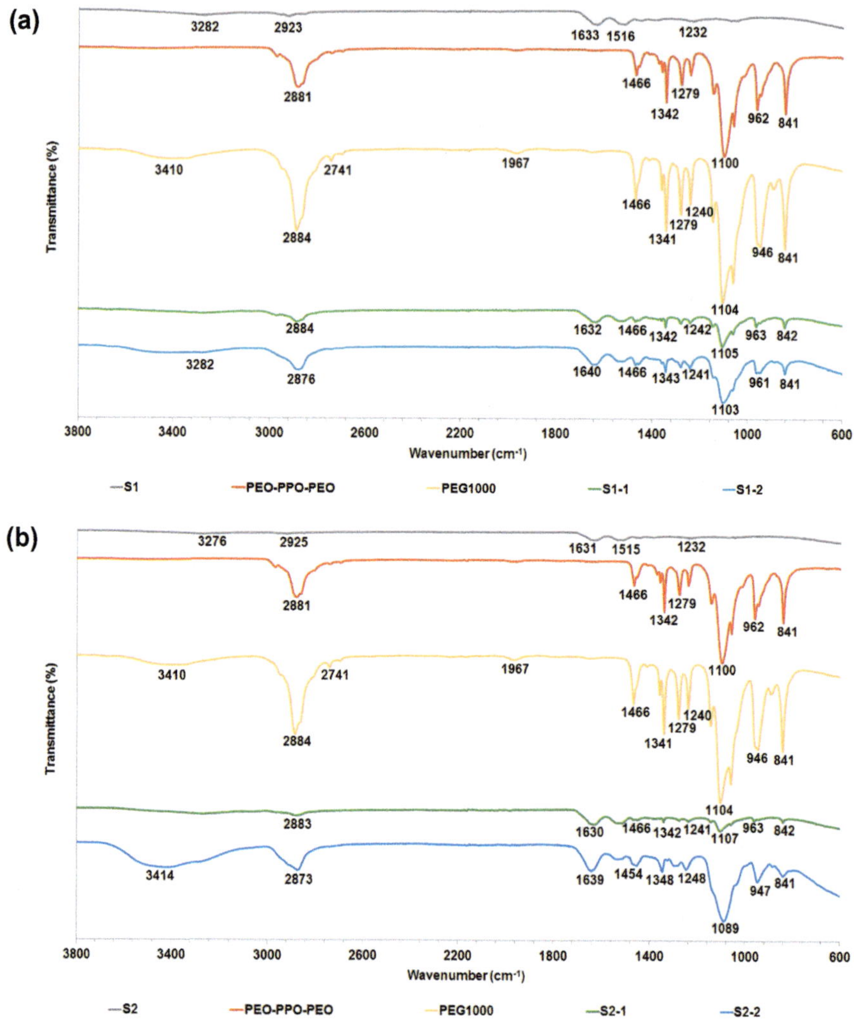

Fig. 15.3 (**a**, **b**) Schematic FTIR results of the samples

15.3.2 *Results of EMI SE*

As may be seen from Table 15.2, both of the knitted spacer fabrics developed for the study, S1 and S2, demonstrated Faraday shielding effect against electromagnetic waves at frequencies between 1 and 8000 MHz. The lighter knitted fabric (208.8 g.m^{-2}) made from wool and silver-plated single nylon yarn, S1, performed EMI SE of 22.2 dB at 1–300 MHz and 10.5 dB at 301–2400 MHz on average. The sample S1-1, including TRP, performed more or less the same shielding effectiveness (SE) in a wider spectrum, say, 28.2 dB at 1–300 MHz and

Table 15.2 Electromagnetic interference (EMI) shielding results of the samples

	1–300 MHz			301–2400 MHz			2401–8000 MHz		
Sample	Min dB	Max dB	Av. dB	Min dB	Max dB	Av. dB	Min dB	Max dB	Av. dB
S1	14.6	59.2	22.2	6.1	22.1	10.5	0.0	26.2	8.5
S2	32.5	61.6	41.2	23.7	36.4	29.2	0.0	33.4	19.2
S1-1	22.4	43.8	28.2	16.9	25.6	21.3	0.0	28.9	23.5
S2-1	37.2	59.2	46.6	10.1	42.3	25.5	3.0	64.5	29.1
Al foil	46.4	86.2	57.3	37.2	85.9	56.2	6.1	85.4	47.0
Wool jumper	0.0	4.5	0.6	0.0	0.5	0.1	0.0	4.1	0.3
Cotton T-shirt	0.0	15.2	2.6	0.0	0.6	0.1	0.0	6.4	0.9

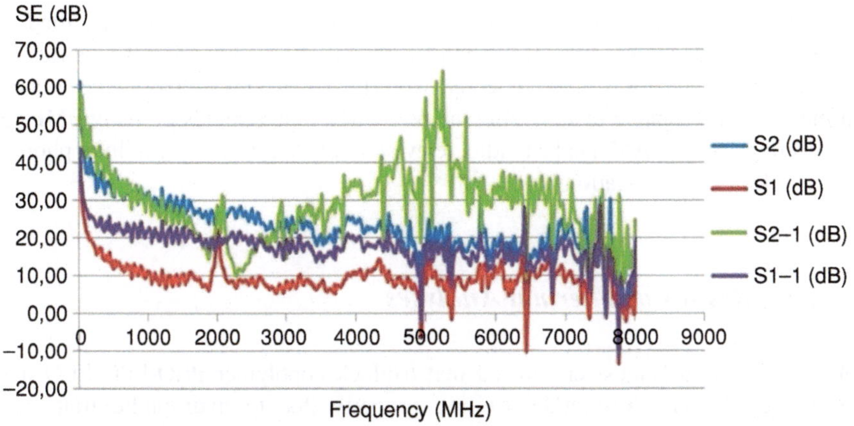

Fig. 15.4 EMI SE results of wool/PA-silver knitted fabrics

21.3 dB at 301–2400 MHz on average. SE values of both samples showed significant decreases in 2401–8000 MHz band; however, SE of the sample S1-1 remained slightly higher than that of its counterpart S1 (see Table 15.2 and Fig. 15.4). The heavier knitted fabric (254.4 $g.m^{-2}$) made from wool and silver-plated plied nylon yarn, S2, revealed a better of EMI SE, 41.2 dB at 1–300 MHz and 29.2 dB at 301–2400 MHz on average, and S2-1, its TRP-treated counterpart, showed a similar EMI shielding performance, say, 46.6 dB at 1–300 MHz and 25.5 dB at 301–2400 MHz on average. The mean SE values at 2401–8000 MHz band were obtained as 19.2 dB for S2 and 29.1 dB for S2-1.

To compare and evaluate EMI SE performances of the fabric samples, two types of daily clothing and an aluminium sheet were chosen as reference material and tested under the same conditions. Wool jumper and cotton T-shirt did not show a protection against electromagnetic waves. Al foil performed EMI SE of 57.3 dB at 1–300 MHz, 56.2 dB at 301–2400 MHz and 47 dB at 2401–8000 MHz on average (see Table 15.2). Therefore, EMI SE levels attained with the samples S1 and S1-1 were only 39% and 49% of the corresponding SE value of Al foil, respectively, at 1–300 MHz, although EMI SE values of the samples S2 and S2-1 were obtained as high as 72% and 81% of that of Al foil, respectively, at the same frequency band. All knitted samples performed EMI SE corresponding to one third to one half the values of Al foil at 301–2400 MHz, and further they sustained their SE performances in a range from 18% to 62% of that of Al foil at 2401–8000 MHz. This was the quite evidence for the considerable EMI SE performances of the knitted fabrics.

These results also indicated that the knitted fabric parameters, say linear densities of the yarns, wale and course densities, types of raw material as well as the thicknesses and porosities of the structures had influences on the EMI shielding functions of the samples. The continuity of the silver-plated yarn on the face ensured the electrical conductivity and allowed the achievement of EMI SE in the fabrics. TRP and PCM applications did not have any adverse effect on EMI SE of the fabrics; conversely, slight rises in SE values were observed due to the filled-in pores. Furthermore, this very effect can be increased by employing the fabrics as two or more plied form. Consequently, wool/PA-silver knitted fabrics developed in the study are capable of satisfying the protection requirements of many EMI SE applications such as AM and FM radio, television broadcast stations, cellular phones and wireless communication links.

15.3.3 Results of Thermal Analyses

Measurements in Nanosizer showed that triblock copolymer PEO-PPO-PEO has reversible thermal swelling/de-swelling property due to hydrophilic transition starting at 22–24°C and ending at nearly 32–34°C (see Fig. 15.5). During heating, the particle sizes in aqueous solution increased rapidly as a result of increased hydrophilic attraction between triblock copolymer and water molecules and, thus, increased number of hydrogen bonds between them at 20–24°C. Particles' diameters attained a maximum of 1100 nm and remained around 600–800 nm up to 32–34°C. Beyond this point, molecules of the copolymer were subjected to a drastic

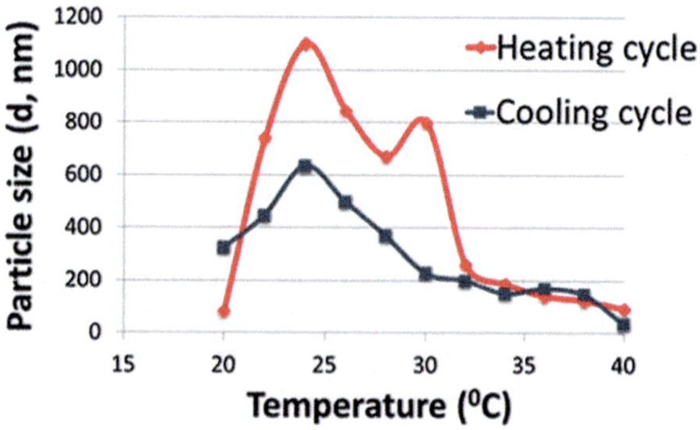

Fig. 15.5 Particle size distribution of 1% TRP solution upon heating and cooling at temperatures between 20 and 40°C using Malvern Nanosizer [33]

Table 15.3 DSC results of the coating materials

	2nd heating				2nd cooling			
	Phase transition temperatures (°C)				Phase transition temperatures (°C)			
TR/PCM polymer	T_{onset}	T_{peak}	T_{end}	ΔH (Jg^{-1})	T_{onset}	T_{peak}	T_{end}	ΔH (Jg^{-1})
PEO-PPO-PEO	51	59	62	133	36	31	27	125
PEG1000	30	38	41	160	26	22	14	159

volumetric change with the additional increase in temperature; consequently, the particle diameter was reduced to 100 nm on average at around 40°C which was attributed to the de-swelling property due to the increased polymer–polymer attractions in this region and repelling of water molecules from the copolymer structure [33]. On the other hand, it was shown that the hydrophilic transition was possible in two ways since the copolymer particles changed their behaviour and tended to attract water molecules again at about 34°C during cooling from 40 to 20°C as indicated by their increased sizes.

This ability of hydrophilic transition and volumetric change against the ambient temperature changes in between 20 and 40°C implies that PEO-PPO-PEO molecules, which are in the gel form and adhered to the knitted fabrics with physical bonds, can react against the temperature changes in surrounding and able to show hydrophilic behaviour and increased moisture-holding capacity next to the skin under mild temperature conditions. These thermo-responsive polymer molecules are also able to assist in air and moisture transfer through the fabric with their much smaller sizes at hot temperatures, which is critical to avoid heat stress generated in the body and to ensure breathability.

Table 15.3 presents the results of DSC analyses obtained for two types of coating materials. The triblock copolymer (PEO-PPO-PEO) absorbed 133 Jg^{-1} in between 51 and 62°C during heating and released 125 Jg^{-1} in between 36 and 27°C during cooling and exhibited a substantial heat-storing/heat-releasing behaviour like a PCM; thus, it provided the dual feature of thermo-response and phase change. However, the thermo-responsive behaviour of PEO-PPO-PEO was more prominent in this application than its latent heat of fusion since the melting temperature of it (56–59°C) was far from the considered ambient temperature interval of 20–40°C. The heat enthalpy changes of PEG1000 during heating and subsequent cooling were also quite large, 160 Jg^{-1} in between 30 and 41 °C during heating and 159 Jg^{-1} in between 26 and 14°C during cooling, accordingly suitable to produce a variety of new heat storage and release products. Enthalpy changes of solid–liquid PCMs are characterized by the phase transitions of the ordered structures in solid phase towards the disordered structures in liquid phase at certain temperature intervals during heating and by phase transitions of the disordered liquid phases to the ordered solid structures at intervals shifted towards relatively lower temperatures during cooling.

Figure 15.6a, b indicates that untreated wool/PA-silver knitted samples, S1 and S2, did not show any enthalpy change, while S1-1 and S2-1 gave the endothermic

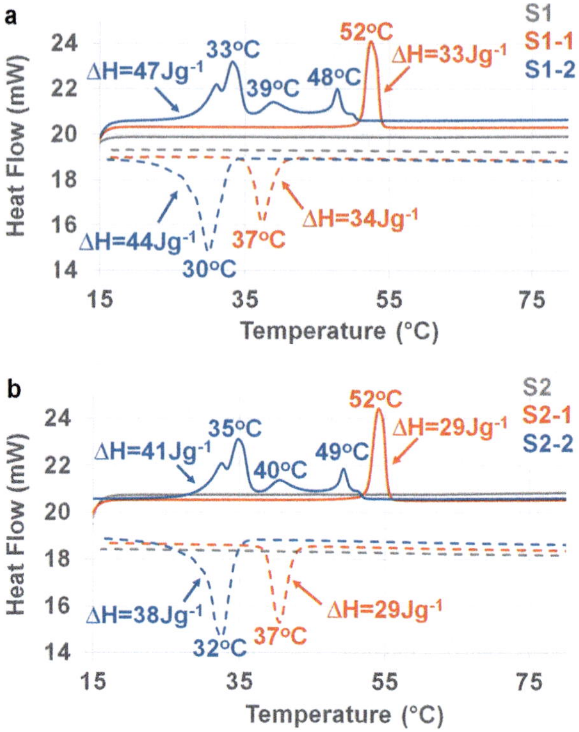

Fig. 15.6 DSC graphs of wool/PA-silver knitted fabrics:. **a**. S1, S1-1, S1-2; **b**. S2, S2-1, S2-2 for the second heating-cooling cycle, at the rate of 3 °C min^{-1}

heat changes of 33 and 29 Jg^{-1}, respectively, at a peak temperature of 52°C during their heating in their DSC analyses. The same amounts of exothermic enthalpies, 34 Jg^{-1} and 29 Jg^{-1}, respectively, observed at a peak temperature of 37°C during cooling. These sharp enthalpy changes and phase change temperatures were mainly coincided with those of the triblock copolymer. Furthermore, the freezing point depressions during cooling arose from solidification difficulty of the disordered molecules of the polymer as well as from the impurities included in composite structures [37]. Thus, TRP-coated knitted samples, S1-1 and S2-1, are able to store or release considerable amount of heat at certain temperature intervals, as mentioned above, in addition to their thermo-stimulated dynamic behaviour related to hydrophilic transitions and volumetric changes of triblock copolymers in between 20 and 40°C.

As shown from Fig. 15.6a, b, the heat-storing capacities of S1-2 and S2-2 were measured as 47 and 41 Jg^{-1}, respectively, and their phase transitions appeared as multiple peaks in a range from 33 to 49 °C during heating, reminding some variations in chain lengths of PEG1000 and PEO-PPO-PEO molecules due to the interactions between their chemical groups. S1-2 and S2-2 performed exothermic enthalpies of 44 Jg^{-1} with a single peak at 30 °C and 38 Jg^{-1} with a single peak at 32 °C, respectively, during solidification. Thus, TRP-/PCM-coated knitted samples, S1-2 and S2-2, performed a better of heat absorption or heat release than their

TRP counterparts. Moreover, these enthalpies occurred during phase changes at temperatures coinciding with the body comfort temperature zone, which brings the advantage of keeping the microclimate next to the skin constant as long as possible in relation to the level of heat exchange for a given set of conditions, e.g. ambient temperature, metabolic heat, fabric insulation, etc. [37]. Therefore, two different transition characteristics, say hydrophilic and phase change transitions, provided by TRP and PCM applications into the knitted fabrics are expected to be both functional with changing temperatures in a micro environment.

15.4 Conclusions

The work focused on the development of wool/PA-silver knitted spacer fabrics having not only good thermal property but also electromagnetic shielding. For this purpose, in addition to the wool yarns as well as silver-plated functional yarns, a temperature-responsive triblock copolymer (PEO-PPO-PEO) and a phase change material (PEG100) were applied on to the spacer knitted fabrics using a stepwise coating process. FTIR results presented all typical peaks of wool and other components in the composite structure and verified that both of the polymers (PEO-PPO-PEO and PEG1000) could be successfully transferred into the wool/PA-silver knitted fabric structures in relation to the prepared recipes and the applied coating methods.

EMI shielding results revealed that the structures developed can be employed especially in the design of underwear for aircraft crew. All types of wool/PA-silver knitted fabrics demonstrated shielding effects against electromagnetic waves at frequencies between 1 and 8000 MHz in general. The achieved EMI SE values were quite remarkable for S2-1, say 46.6 dB at 1–300 MHz, 25.5 dB at 301–2400 MHz and 29.1 dB at 2401–8000 MHz, which were 81%, 45% and 62% of the corresponding SE values of Al foil at the same frequency intervals, respectively. The knitted fabrics having SE in excess of 20 dB are reasonably suitable for many textile applications.

These wool/PA-silver knitted fabrics are quite promising about providing a protection for the aircraft crew when they are subjected to electromagnetic radiation due to the operated systems in the cockpit for low- and high-frequency communication, altitude, positioning, navigation, etc., with an increased intensity during taking off and landing operations. This attempt given in this study can be enhanced by several applications and/or combinations of the materials having shielding effect against electromagnetic waves at higher frequencies in the spectrum and ionizing radiation and by varying structural parameters of the textiles to contribute the achievement of the required protection.

The application TRP and PCM further improved the knitted structures to have the abilities of thermo-response and heat exchange with the surrounding at 20–40 °C. The reversible hydrophilic transition character of PEO-PPO-PEO was proven via particle size measurements conducted in dependence of temperature variation in Nanosizer. DSC analyses showed that TRP-/PCM-coated knitted samples were able to store and release heat up to 47 and 41 Jg^{-1}, respectively, during heating and subsequent cooling. Therefore, coating the knitted fabrics with TRP and PCM is quite encouraging to enhance the thermal properties of such advanced structures and should be modified to cover alternative incorporation and fixing techniques. The analysis of such complex thermal behaviour also requires making some thermal comfort tests and modelling.

Acknowledgements We thank Yünsa A.Ş for their support providing worsted wool yarns.

References

1. Jiang, S. X., & Guo, R. H. (2011). Electromagnetic shielding and corrosion resistance of electroless Ni–P/Cu–Ni multilayer plated polyester fabric. *Surface and Coating Technology, 205*(17–18), 4274–4279.
2. Roh, J. S., Chi, Y. S., Kang, T. J., & Nam, S. (2008). Electromagnetic shielding effectiveness of multifunctional metal composite fabrics. *Textile Research Journal, 78*(9), 825–835.
3. Zhang, C. S., Ni, Q. Q., Fu, S. Y., & Kurashiki, K. (2007). Electromagnetic interference shielding effect of nanocomposites with carbon nanotube and shape memory polymer. *Composites Science and Technology, 67*(14), 2973.
4. Genuis, S.J. (2008). Fielding a current idea: exploring the public health impact of electromagnetic radiation. *Public Health, 22*(2), 113–124.
5. Wessapan, T., Srisawatdhisukul, S., & Rattanadecho, P. (2011). The effects of dielectric shield on specific absorption rate and heat transfer in the human body exposed to leakage microwave energy. *International Communications Heat Mass, 38*(2), 255–262.
6. Al-Saleh, M. H., & Sundararaj, U. (2009). Electromagnetic interference shielding mechanisms of CNT/polymer composites. *Carbon, 47*(7), 1738–1746.
7. Matthiä, D., Schaefer, M., & Meier, M. M. (2015). Economic impact and effectiveness of radiation protection measures in aviation during a ground level enhancement. *Journal Space Weather Space Climate, 5*(A17), 6. http://www.swsc-journal.org/articles/swsc/pdf/2015/01/swsc140044.pdf. Accessed August 2016.
8. Koops, L., Sizmann, A. (2014). Relevance of cosmic radiation exposure for hypersonic flight. Deutscher Luft- und Raumfahrtkongress, Document ID: 340206, http://www.dglr.de/publikationen/2015/340206.pdf. Accessed August 2016.
9. http://www.dglr.de/. Accessed May 2017.
10. Alvarez, L. E., Eastham, S. D., & Barrett, S. R. H. (2016). Radiation dose to the global flying population. *Journal of Radiological Protection, 36*(1), 93–103.
11. Bennett, L. G. I., Lewis, B. J., et al. (2013). A survey of the cosmic radiation exposure of air Canada pilots during maximum galactic radiation conditions in 2009. *Radiation Measurements, 49*(2), 103–108.
12. Desmaris, G. (2015). Cosmic radiation in aviation: Radiological protection of air France aircraft crew. In: ICRP 2015 Proceedings, pp. 64–74.
13. Langenhove, L. V. (Ed.). (2007). *Smart textiles for medicine and healthcare materials, systems and applications*. UK: Woodhead Publishing Limited and CRC Press LLC.
14. Black, S. (2007). *Trends in smart medical textiles: Smart textiles for medicine and healthcare* (pp. 3–26). Cambridge: Woodhead Publishing Ltd..
15. Pan, N., et al. (Eds.). (2011). *Functional textiles for improved performance, protection and health*. UK: Woodhead Publishing Limited.
16. McCann, J., et al. (Eds.). (2009). *Smart clothes and wearable technology*. UK: Woodhead Publishing Limited and CRC Press LLC.

17. Smith, W. C. (Ed.). (2010). *Smart textile coatings and laminates*. UK: Woodhead Publishing Limited and CRC Press LLC.
18. Lu, M., Xie, R., et al. (2016). Enhancement in electrical conductive property of polypyrrole-coated cotton fabrics using cationic surfactant. *Journal of Applied Polymer Science, 133*(32), 43601.
19. Wei, Q., Yu, L., et al. (2008). Preparation and characterization of copper nanocomposite textiles. *Journal of Industrial Textiles, 37*(3), 275–283.
20. Ersoy, M. S., & Onder, E. (2014). Electroless silver coating on glass stitched fabrics for electromagnetic shielding applications. *Textile Research Journal, 84*(19), 2103–2114.
21. Kim, S. H., Oh, K. W., & Bahk, J. H. (2004). Electrochemically synthesized polypyrrole and Cu-plated nylon/Spandex for electrotherapeutic pad electrode. *Journal of Applied Polymer Science, 91*(6), 4064–4071.
22. Lu, Y., Jiang, S., & Huang, Y. (2010). Ultrasonic-assisted electroless deposition of Ag on PET fabric with low silver content for EMI shielding. *Surface and Coating Technology, 204*(16–17), 2829–2833.
23. Koprowska, J., Pietranik, M., Stawski, W. (2007). New type of textiles with shielding properties. *Fibres & Textiles in Eastern Europe* 12 (3):39–42.
24. Perumalraja, R., Dasaradan, B. S., et al. (2009). Electromagnetic shielding effectiveness of copper core-woven fabrics. *Journal Text I, 100*(6), 512–524.
25. Jiang, S. X. (2011). Electromagnetic shielding and corrosion resistance of electroless Ni-P/Cu-Ni multilayer plated polyester fabric. *Surface & Coatings Technology, 205*(17–18), 4274–4279.
26. Varnaite, S. (2010). The use of conductive yarns in woven fabric for protection against electrostatic field. *Materials Science, 16*, 133–137.
27. Statex Products. http://statex.de/index.php/en/shieldex-2/produkte/. Accessed 19 Aug 2016.
28. Geetha, S., Kumar, K. K. S., et al. (2007). Synergetic effect of conducting polymer composites reinforced e-glass fabric for the control of electromagnetic radiations. *Composites Science and Technology, 70*(6), 1017–1022.
29. Satheesh Kumar, K. K. (2005). Freestanding conducting polyaniline film for the control of electromagnetic radiations. *Current Applied Physics, 5*(6), 603–608.
30. Maity, S., Singha, K., et al. (2013). Textiles in electromagnetic radiation protection. *Journal Safety Engineering, 2*(2), 11–19.
31. Hu, J., Meng, H., et al. (2012). A review of stimuli-responsive polymers for smart textile applications. *Smart Materials and Structures, 21*(5), 21–23.
32. Onder, E., & Sarier, N. (2015). Thermal regulation finishes for textiles. In R. Paul (Ed.), *Functional finishes for textiles* (pp. 17–78). Amsterdam: The Textile Institute & Woodhead Publishing, Elsevier Ltd.
33. Uludag, B. (2014). Development of thermo-responsive and thermo-regulated smart textiles, Unpublished MSc Thesis. İstanbul: ITU.
34. Yao, J., Liu, Y., et al. (2008). Characterization of secondary structure transformation of stretched and slenderized wool fibers with FTIR spectra'. *Journal of Engineered Fibers and Fabrics, 3*(2), 22. https://www.researchgate.net/publication/44567421 _Characterization_of_Secondary_Structure_Transformation_of_Stretched_and_Slenderized_Wool_Fibers_with_FTIR_Spectra. Accessed August 2016.
35. Onder, E., Sarier, N., et al. (2013). Ultrasound assisted solvent free intercalation of montmorillonite with PEG1000: A new type of organoclay with improved thermal properties'. *Thermochimica Acta, 566*, 24–35.
36. Sánchez, M. T., Pariente, J. P., Alvarez, C. M. (2008). Dynamics of PEO-PPO-PEO block copolymer aggregation and silicate mesophase formation monitored by time resolved ATR-FTIR spectroscopy. Studies in Surface Science and Catalysis 174: 333–336., https://doi.org/10.1016/S0167-2991(08)80210-8. Accessed 19 Aug 2016.
37. Sarier, N., & Onder, E. (2007). The manufacture of microencapsulated phase change materials suitable for the design of thermally enhanced fabrics. *Thermochimica Acta, 452*, 149–160.

Chapter 16
Advantages and Future of Electric Propulsion in UAVs

Melih Yildiz and T. Hikmet Karakoç

16.1 Introduction

An unmanned aerial vehicle (UAV) can be defined as "remotely piloted or autonomous aircraft, which can carry camera, communication equipment and other payloads" [1]. For operations under extreme conditions where a human pilot is too risky or unnecessary, UAVs have proved to be a viable platform in a wide range of applications [2]. As the operation costs of UAVs are low compared to satellites and aircraft, both military and civilian uses of UAVs are increasing.

UAVs come in all shapes and sizes for wide range of applications. Although UAVs are built in different shape and size depending on the application, autonomous and remote operations are common properties for every UAV.

During and just after World War II, towed banners were used for target practice. Those banners were later replaced by towed gliders, and then during the 1960s, they became remote-piloted drones. In the 1970s, chip-scale circuits and servomotors became available which allowed civilian use and model airplane enthusiasts to be able to build their remote-piloted UAVs. During these years, remotely piloted unmanned helicopters were also built, but as their payload capability was very low, no military or commercial applications were developed. During the Vietnam War, as they requested to fly over antiaircraft systems, high losses of military aircraft and pilots were experienced. As a result of losses, unmanned aircraft was seemed to be a key solution for such military roles.

M. Yildiz (✉)
School of Civil Aviation, Atilim University, Ankara, Turkey
e-mail: melih.yildiz@atilim.edu.tr

T.H. Karakoç
Department of Airframe and Powerplant Maintenance, Anadolu University, Faculty of Aeronautics and Astronautics Eskisehir, Turkey
e-mail: hkarakoc@anadolu.edu.tr

T.H. Karakoç et al. (eds.), *Advances in Sustainable Aviation*,
DOI 10.1007/978-3-319-67134-5_16

Development of high-quality and light digital cameras and real-time long-range communication systems made it possible to develop reconnaissance versions of these unmanned air vehicles or remotely piloted vehicles. Such UAVs were widely used during the Gulf War for reconnaissance missions.

Today a wide range of military applications including attack roles are being developed even with autonomous capabilities. Such capabilities like autonomy also found a base in the civilian market such as controlling the long electric energy lines, pipelines, etc.

16.2 Applications and Requirements

Payload and endurance optimization is always a problem in the design of UAVs. UAVs also suffer the weight and size limitations for the onboard systems and payloads as in for the aircraft. This limitation creates a handicap for small-size UAVS. As electronics and energy storage technologies emerge, electronic components get smaller, which makes it possible for avionic equipment to be smaller in size, weight, and energy usage [3]. On the other hand, payloads also get lighter which makes it easier for a design engineer.

Traditionally, UAVs used ICE engines for propulsion and primary loads which have a thermal efficiency of around 30% [4]. Although hydrocarbon fuels have high-energy density compared to other energy storage alternatives, more efficient power plant designs are sought in order to achieve more endurance with less weight.

For small UAVs, electric propulsion is more desirable because of weight, design, and system simplicity and reliability considerations. Also, quiet operation is a plus for electric propulsion which makes surveillance operations easier where UAVs are preferred widely. Electric motors are capable of operating nearly 98% efficiency [5] which, compared to ICE, make it more preferable.

Market research reports that the electric power is optimal for most future UAV needs, regarding increasing demand for electric power for onboard systems. Using electric power also makes it easier and cost-effective for autonomous applications. Besides, payloads, especially complex systems like cameras, telemetric sensors, etc., are demanding more electric power which is a requirement for UAV to generate and/or store electric energy on board.

US DoD defines the future UAVs as systems that easily accept technological improvement capabilities [6]. This means more complex electronics systems are expected to be integrated for complex missions in the future. Also, airspace traffic is getting crowded where UAVs are required to be integrated. There are continental studies for safe and efficient integration of UAVs in the air spaces. The studies show the necessity of more complex electronic navigation systems will be integrated into UAVs [1–6].

Besides higher energy efficiency, electric propulsion provides the advantage of an ability to use renewable energy sources and more transportable generators like solar cells, fuel cells, ultra-capacitors, and batteries. Some experiments show that in cruise phase, only solar cell is enough for powering the electric motor [7].

Civilian-based UAV design and development are mainly in academic research and hobby activities. As small UAVs are easy and low cost to develop and integrate, the interest for small UAVs is globally high. This also opens a window for a search for commercial use of UAVs. It is estimated that by 2020, military market will reach 11.4 billion USD, where the commercial market and hobby market will be around 6.4 and 4.4. billion USD, respectively [8].

As a good example of the UAV application benefits, glacier research studies were conducted by satellite images up till now, which the researchers shall have to wait for satellite passing and image transfers. Using UAV in research minimized the cost of data acquisition, and researchers are able to work and acquire image according to their schedule [9].

Further, in the future, swarms of UAVs may take place in wide areas of use for both civilian and military. Swarm applications provide the ability to share payloads and functions among the UAVs which are deployed in cooperation.

16.3 Technology

The subsystems used in unmanned vehicles are very similar to those of manned aircraft. The main difference is an absence of the human life support and safety systems in UAVs.

Propulsion technology is the leading actor in the aviation industry. Development of electric motors is promising to provide lower weight and higher power ratios for aviation propulsion systems. This is expected to provide higher performance per unit battery energy stored on board. Recent production technologies made it possible to develop integrated gearing and electric motor in the same case. Besides the weight savings, this integration can reduce the cost of the integration and maintenance functions of the UAVs [14].

The electric propulsion system of a typical UAV consists of the following components: propeller, electric motor, energy storage and/or source, driver, wiring, and connectors. The design of electric propulsion system for UAV requires the cooperation of various disciplines such as the propeller's aerodynamic and structural properties, electrical system, and the vehicle itself [10, 13–16].

For the structural subsystem, 3D printers are also providing support for civilian research and thus making it easy for academicians to enter into this field. New production technologies not only offer to manufacture lightweight products but also provide faster prototyping capabilities, which lead to more efficient designs to be tested in short time.

Power electronics is another enabler technology [10], as smaller and cheaper power electronic components provide more control over electric energy. Electric energy is still expensive to store and control in higher levels. Extensive relays and batteries are required which makes the use of more complex subsystems like battery management systems depending on the nature of the mission the UAV is used for.

As the civilian market develops, more UAV products enter into the market making it more competitive and forcing companies to innovate new functions and capabilities. A company offers a UAV which has over 110 min of endurance with 21.5 kg MTOW using all electric propulsion [11] which can be counted as a result of the abovementioned studies.

Research shows that in the future it is expected that the non-electric-powered UAVs tend to be military only, leaving the civilian market to electric versions [12] because of higher energy need during flight phases.

Electrical energy storage shows a great advance in capacity and availability, following the use of lithium in the batteries. The energy storage capacity became over 200 Wh/kg, which is the most essential element in electric UAVs for storing onboard electric energy [10, 15, 16].

16.4 Conclusion

UAVs provide lower operating costs and flexibility compared to aircraft and satellites. This leads to the wide use of UAVs, especially for remote sensing and surveillance applications.

Market research reports that the electric power is optimal for most future UAV needs. Autonomous applications are easier to integrate and cost-effective when electric is used as a primary power source. Besides, new payloads are demanding electric power which is a requirement for UAV to generate and/or store electric energy on board during flight.

Besides higher energy efficiency, electric propulsion gives an advantage of using renewable energy sources and more transportable generators like solar cells, fuel cells, ultra-capacitors, and batteries. These even not enough for a full flight can be used as a range extending and supporting energy sources. Electric propulsion also makes it easy for the integration of electronic control systems.

Electric motors with integrated gearing as a result of emerging production capabilities also provide more compact and low-weight power-pack solutions. Also using of 3D printers for manufacturing is another enabler for the rapid development of UAVs.

References

1. Unmanned Aerial Vehicles (UAV). online, http://www.fas.org/irp/program/collect/uav.htm
2. Jane's unmanned aerial vehicles and targets. online, http://juav.janes.com/public/juav/index.shtml
3. Logan, M.J., Chu, J., Motter, M.A., Carter, D.L., Ol, M., Zeune, C. Small UAV research and evolution in long endurance electric powered vehicles. In: AIAA infotech aerospace 2007 conference and exhibit, Rohnert Park, CA, USA, 7–10 May, 2007.

4. Bhatia, A., Mendiratta, A., Vaish, M. (2010). Comparison of proposed six stroke internal combustion engine with four stroke engine using ideal cycle. In: proceedings of the 2nd International conference on mechanical electronics engineering (ICMEE2010), Kyoto, Japan; 1–3 August 2010.
5. Energy-efficient electric machines, online, http://www.csiro.au/science/ElectricMachines.html
6. US DoD, Unmanned systems integrated roadmap, FY2013–2038. (2013) .
7. Torabi, H.B., Sadi, D., Varjani, A.Y. (2011). Solar power system for experimental icle (UAV); Design and fabrication. In: 2nd Power Electronics, Drive Systems and Technologies Conference.
8. Statista. http://www.statista.com/statistics/431717/global-uav-market-size-by-application/
9. Bhardwaj, A., Sam, L., Akanshka, Javier F., Kumar, R. (2015). UAVs as remote sensing platform in glaciology: Present applications and future prospects, Remote sensing of environment.
10. Bourdon, J., Asfaux, P., Etayo, A.M. (2015). Review of power electronics opportunities to integrate in the more electrical aircraft. In: 2015 international conference on electrical systems for aircraft, railway, ship propulsion and road vehicles (ESARS) (Vol. 2015–May, pp. 1–6). IEEE. https://doi.org/10.1109/ESARS.2015.7101425
11. UAV Factory, http://www.uavfactory.com/product/69
12. Harrop, P., Harrop, J. (2015). Electric drones: Unmanned aerial vehicles (UAVs) 2015–2025. ID Tech Exchange, Automotive & Electric Vehicles Report, Cambridge, MA.
13. Kayton, M. (2003). One hundred years of aircraft electronics. *Journal of Guidance, Control and Dynamics, 26*, 193.
14. Sehra, K.A., Whitlow, W.J. (2004). Propulsion and power for 21st century aviation, progress in aerospaces.
15. Zhang, H., Saudemont, C., Robyns, B., Petit, M. (2008). Comparison of technical features between a More Electric Aircraft and a Hybrid Electric Vehicle. In: 2008 IEEE vehicle power and propulsion conference, VPPC 2008 (pp. 1–6). https://doi.org/10.1109/VPPC.2008.4677663
16. Fong, K., Galloway, S. J., Harrington, I., Burt, G. M. (2009). Aircraft electrical systems - coping with harmonics for changing power demands. In: Proceedings of the 44th international universities power engineering conference (UPEC), 2009 (pp. 883–887). New York.

Chapter 17
Greenization Factor of a Turbojet Engine

Yasin Şöhret and T. Hikmet Karakoç

17.1 Introduction

Man-made environmental issues are the main concern in the present time. Particularly energy utilizing sectors are the major reason of this threat. Power generation by consumption of fossil fuels leads to environmental pollution (including air, water, and soil pollution) and global warming. Many reports and researches presented until today reveal the significance of efficient power generation systems [1–5]. For this reason it is required to improve power generation and energy utilization systems with respect to environmental issues.

Transportation is an irreplaceable sector for humankind as it provides time-saving, comfortable, and high-level life standard. Depending on recent advances and demand of people, air transportation sector is the most growing and commonly preferred transportation option worldwide. However, increasing demand to air transportation leads to growth of aircraft fleet. As a result of that, energy consumption of the aviation sector and its environmental impact yield more trouble for the future of the earth and humankind [6–10].

Aviation is known to be responsible of emitted greenhouse gases with 3 % of total transportation emissions. In spite of this low rate compared to other sectors, the impact of the aviation-induced emissions plays a vital role for environmental impact on upper layers of the atmosphere [11–16]. Thus, many researches aiming

Y. Şöhret (✉)
Department of Airframe and Powerplant Maintenance, Süleyman Demirel University, School of Civil Aviation, Isparta, Turkey
e-mail: ysohret@gmail.com

T.H. Karakoç
Department of Airframe and Powerplant Maintenance, Anadolu University, Faculty of Aeronautics and Astronautics Eskisehir, Turkey
e-mail: hkarakoc@anadolu.edu.tr

T.H. Karakoç et al. (eds.), *Advances in Sustainable Aviation*,
DOI 10.1007/978-3-319-67134-5_17

to reduce environmental impact of aviation can be found in the accessible literature [17–24]. The main goal of these and all other recent researches is achieving more green and sustainable aviation and air transportation. Each study deals with the topic from different aspects.

The current study aims to introduce greenization factor determination methodology for aerospace propulsion systems with an exemplification. Differently from previous studies, greenization factor is defined for a propulsion system for the first time.

17.2 Definition of the Greenization Factor

The greenization factor is an assessment indicator for energy conversion systems and power generation systems to understand their environmental impact whereas it reveals how to achieve more green system. In this framework greenhouse gases emitted from the system are considered together with energy consumption and power generation. Herein, the current and improved systems are evaluated under operating considerations separately. Latterly greenization factor is used for comparison. In the end greenization impact of the improvements is revealed [25]. Greenization factor any energy system is defined by El-Emam et al. [25] as follows:

$$\mathrm{GF} = \frac{\mathrm{EI}_{\mathrm{ref}} - \mathrm{EI}_{\mathrm{green}}}{\mathrm{EI}_{\mathrm{ref}}} \tag{17.1}$$

Herein, EI represents the environmental impact of the system under investigation, and it is found by [25]:

$$\mathrm{EI} = \frac{\dot{m}_{\mathrm{GHG}}}{\dot{E}_{\mathrm{out}}} \tag{17.2}$$

Environmental impact of a system is defined to be the ratio of emitted total greenhouse gases (kg CO_2) to the useful energy output of the system (kWh). For a propulsion system, Eq. 17.2 is derived into:

$$\mathrm{EI} = \frac{\dot{m}_{\mathrm{GHG}}}{\dot{F}} \tag{17.3}$$

It is known fact that the useful energy output of a propulsion system finally is the thrust force. So, thrust is expressed as:

$$\dot{F} = \dot{m}_{\mathrm{in}}\left[(1+f)\,V_{\mathrm{out}} - V\right] + A_{\mathrm{out}}\left(P_{\mathrm{out}} - P_{\mathrm{in}}\right) \tag{17.4}$$

17.3 Exemplifier Application

The evaluated propulsion system to exemplify the methodology is a turbojet engine. The emission measurement of the engine is explained in details in a former study of authors [16]. To measure greenhouse gases emitted from the turbojet engine, a sampling probe, analyzer, data recorder, and monitoring system are employed. The sampling probe is made of stainless steel and it is a refractor. Also 12 orifices are found on the sampling points of the probe. Absorbed sample exhaust gas is transported from the probe through the analyzer into heated transfer line. The measured emission data is summarized in Table 17.1.

Table 17.2 summarizes the CO_2 equivalents of the emitted greenhouse gases for 100-year time horizon. The conversion of greenhouse gases to CO_2 equivalent is done according to the following statements:

Table 17.1 Greenhouse gas data of the evaluated turbojet engine [16]

		Greenhouse gas emission			
Operating condition	Engine load (min^{-1})	CO (ppm)	CO_2 (vol. %)	UHC (ppm)	NO_x (ppm)
Low-speed ground idle (1)	10,000	380	2.65	331	18
High-speed ground idle (2)	13,500	66	1.35	3	37
Takeoff (3)	13,800	34	3.03	3	91
Normal (4)	13,800	32	3.68	3	138
Near flight idle (5)	13,800	36	1.55	3	18
Reverse (6)	13,800	41	4.33	3	106
Low-speed ground idle (7)	10,000	340	2.57	351	23
High-speed ground idle (8)	13,500	57	2.00	3	32

Table 17.2 CO_2 equivalent of the emitted greenhouse gases from the turbojet engine

Operating condition	Engine load (min^{-1})	Greenhouse gas emission (kg CO_2)
Low-speed ground idle (1)	10,000	98.11
High-speed ground idle (2)	13,500	187.53
Takeoff (3)	13,800	894.43
Normal (4)	13,800	801.9
Near flight idle (5)	13,800	264.35
Reverse (6)	13,800	549.25
Low-speed ground idle (7)	10,000	98.114
High-speed ground idle (8)	13,500	187.53

Table 17.3 CO_2 equivalent of the emitted greenhouse gases from the improved turbojet engine

Operating condition	Engine load (min^{-1})	Greenhouse gas emission (kg CO_2)
Low speed ground idle (1)	10,000	96.85
High speed ground idle (2)	13,500	182.17
Take-off (3)	13,800	875.06
Normal (4)	13,800	785.47
Near flight idle (5)	13,800	258.98
Reverse (6)	13,800	539.50
Low speed ground idle (7)	10,000	97.02
High speed ground idle (8)	13,500	182.45

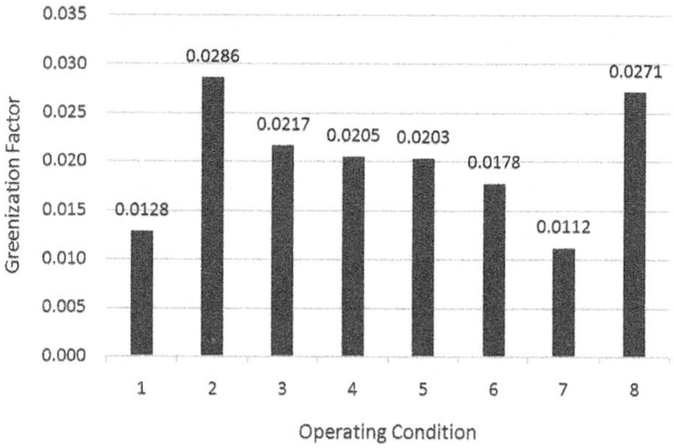

Fig. 17.1 Greenization factor of the evaluated turbojet engine at different operating conditions

$$1\text{kg NO}_x = 310\text{kg CO}_2 \quad (17.5)$$

$$1\text{kg CH}_4 = 21\text{kg CO}_2 \quad (17.6)$$

In the current study, emission measurement for the improved turbojet engine is performed according to the methodology explained in Ref. [16], and the CO_2 equivalent of the emitted greenhouse gases from the improved turbojet engine is given in Table 17.3.

Calculated greenization of the evaluated turbojet engine at different operating conditions using Eq. 17.1 is plotted in Fig. 17.1. As shown in Fig. 17.1, greenization factor of the engine for low-speed ground idle, high-speed ground idle, takeoff, normal, near flight idle, reverse, low-speed ground idle, and high-speed ground idle are found to be 0.013, 0.029, 0.022, 0.020, 0.020, 0.018, 0.011, and 0.027, respectively.

17.4 Conclusion

The current chapter presents the calculation methodology of the greenization factor of a turbojet engine. In the end of the study, the following implications may be addressed:

- The highest greenization factor of the turbojet engine is found to be 0.0286 at high-speed ground idle.
- The lowest greenization factor is determined for low-speed ground idle operating conditions with the value of 0.012.
- The greenization factor indicates the emission reduction rate after improvement of the propulsion system.

The current study can be beneficial for the researchers interested in aerospace, energy, and environmental sciences.

References

1. Wang, Q. (2014). Effects of urbanisation on energy consumption in China. *Energy Policy, 65*, 332–339.
2. Omri, A. (2013). CO 2 emissions, energy consumption and economic growth nexus in MENA countries: Evidence from simultaneous equations models. *Energy Economics, 40*, 657–664.
3. Nejat, P., Jomehzadeh, F., Taheri, M. M., Gohari, M., & Majid, M. Z. A. (2015). A global review of energy consumption, CO 2 emissions and policy in the residential sector (with an overview of the top ten CO 2 emitting countries). *Renewable and Sustainable Energy Reviews, 43*, 843–862.
4. Bilgen, S. (2014). Structure and environmental impact of global energy consumption. *Renewable and Sustainable Energy Reviews, 38*, 890–902.
5. Li, F., Song, Z., & Liu, W. (2014). China's energy consumption under the global economic crisis: Decomposition and sectoral analysis. *Energy Policy, 64*, 193–202.
6. Gwilliam, K. M., & Geerlings, H. (1994). New technologies and their potential to reduce the environmental impact of transportation. *Transportation Research Part A: Policy and Practice, 28*(4), 307–319.
7. Nagurney, A., Qiang, Q., & Nagurney, L. S. (2010). Environmental impact assessment of transportation networks with degradable links in an era of climate change. *International Journal of Sustainable Transportation, 4*(3), 154–171.
8. Zanetti, A., Sabatini, R., & Gardi, A. (2016). Introducing green life cycle management in the civil aviation industry: The state-of-the-art and the future. *International Journal of Sustainable Aviation, 2*(4), 348–380.
9. Bauer, C., Hofer, J., Althaus, H. J., Del Duce, A., & Simons, A. (2015). The environmental performance of current and future passenger vehicles: Life cycle assessment based on a novel scenario analysis framework. *Applied Energy, 157*, 871–883.
10. Messagie, M., Boureima, F. S., Coosemans, T., Macharis, C., & Mierlo, J. V. (2014). A range-based vehicle life cycle assessment incorporating variability in the environmental assessment of different vehicle technologies and fuels. *Energies, 7*(3), 1467–1482.
11. Brunelle-Yeung, E., Masek, T., Rojo, J. J., Levy, J. I., Arunachalam, S., Miller, S. M., & Waitz, I. A. (2014). Assessing the impact of aviation environmental policies on public health. *Transport Policy, 34*, 21–28.

12. Irvine, E. A., Hoskins, B. J., & Shine, K. P. (2014). A simple framework for assessing the trade-off between the climate impact of aviation carbon dioxide emissions and contrails for a single flight. *Environmental Research Letters, 9*(6), 064021.
13. Ekici, S., Yalin, G., Altuntas, O., & Karakoc, T. H. (2013). Calculation of HC, CO and NOx from civil aviation in Turkey in 2012. *International Journal of Environment and Pollution, 53*(3–4), 232–244.
14. Şöhret, Y., Yazar, I., & Karakoç, T. H. (2016). Using some performance parameters to predict exhaust gas emissions of a turboprop engine: Adaptive neuro-fuzzy method. *International Journal of Sustainable Aviation, 2*(1), 1–14.
15. Şöhret, Y., Kıncay, O., Karakoç, T. H. (2017). An environment-friendly engine selection methodology for aerial vehicles. *International Journal of Green Energy*, doi:10.1080/15435075.2017.1324788.
16. Sohret, Y., Karakoc, T. H., Karakoc, N. (2015). Mathematical modelling for carbon dioxide equivalent prediction of greenhouse gases emitted from a small scale turbojet engine. In 7th AIAA Atmospheric and Space Environments Conference, p. 3326.
17. Hepbasli, A. (2016). Proposing an exergy management system standard for establishing exergetically green aviation. *International Journal of Sustainable Aviation, 2*(4), 271–283.
18. Unlu, D., & Hilmioglu, N. D. (2016). Potential fuel bioadditive ethyl levulinate production for the aviation fuel. *International Journal of Sustainable Aviation, 2*(4), 338–347.
19. Bicer, Y., & Dincer, I. (2016). A comparative life cycle assessment of alternative aviation fuels. *International Journal of Sustainable Aviation, 2*(3), 181–202.
20. Platzer, M. F., & Sarigul-Klijn, N. (2016). Carbon-neutral jet fuel production from seawater. *International Journal of Sustainable Aviation, 2*(2), 101–110.
21. Dincer, I., & Acar, C. (2016). A review on potential use of hydrogen in aviation applications. *International Journal of Sustainable Aviation, 2*(1), 74–100.
22. Greene, D. L. (1992). Energy-efficiency improvement potential of commercial aircraft. *Annual Review of Energy and the Environment, 17*(1), 537–573.
23. Huang, R., Riddle, M., Graziano, D., Warren, J., Das, S., Nimbalkar, S., & Masanet, E. (2016). Energy and emissions saving potential of additive manufacturing: The case of lightweight aircraft components. *Journal of Cleaner Production, 135*, 1559–1570.
24. Şöhret, Y., Ekici, S., Altuntaş, Ö., Hepbasli, A., & Karakoç, T. H. (2016). Exergy as a useful tool for the performance assessment of aircraft gas turbine engines: A key review. *Progress in Aerospace Sciences, 83*, 57–69.
25. El-Emam, R. S., Dincer, I., Zamfirescu, C. (2017). Greenization factor as a sustainability measure for energy systems. In Energy solutions to combat global warming (pp. 735–751). Cham: Springer International Publishing. http://www.springer.com/gp/book/9783319269481?wt_mc=ThirdParty.SpringerLink.3.EPR653.About_eBook.

Part IV
Advances in Electronics and Control Technologies

Chapter 18
Integration of Baro-Inertial-GPS Altimeter via Complementary Kalman Filter

Alberto Mañero Contreras and Chingiz Hajiyev

18.1 Introduction

Since Earth's centre is the most extended inertial coordinate frame taken by the scientist society, "where am I?" is probably the most asked question in all of our lifetime. Indeed, it does not matter where the origin is located. We all know how to orientate ourselves in relation to other places, or in other words, through relative positions. Navigation is a really ancient art. History talks by itself and there are several examples throughout human era. In the early times, mariners used to discover new coastlines pushed by their braveness sailing open oceans. They used the North Star as a compass, the same stars and heaven we use today for space navigation. With the advent of the magnetic compass, sextant and chronometers for timing, it was finally possible to determine them both precisely latitude and longitude at the sea.

Air navigation has passed through similar evolutions and revolutions. The First World War was the catalyst for rapid advances in aviation. Since early twentieth century, a wide range of radio-based navigation systems was developed and they were determinants and widely used since World War II. After these milestones, new navigation systems appeared, such as inertial navigation systems (INS), which were thought impossible to implement due to their unstable behaviour. Nevertheless, it was made practical by Charles Draper and is today the backbone of military

A.M. Contreras
Escuela Técnica Superior de Ingeniería, University of Sevilla, Camino de los Descubrimientos s/n, 41096 Sevilla, Spain
e-mail: amcamc92@gmail.com

C. Hajiyev (✉)
Faculty of Aeronautics and Astronautics, Istanbul Technical University, Maslak, 34469 Istanbul, Turkey
e-mail: cingiz@itu.edu.tr

T.H. Karakoç et al. (eds.), *Advances in Sustainable Aviation*,
DOI 10.1007/978-3-319-67134-5_18

navigation and weapon delivery systems. Today we are in the middle of a new technological revolution: the global positioning system (GPS) (United States). Accurate positioning on real time regardless of location, weather or time of day is equal to power on the worldwide equation. Nowadays, several countries are developing and improving similar navigation systems such as GALILEO (European), GLONASS (Russian), DORIS (French), BeiDou (Chinese) or IRNSS (Indian).

The Kalman filter (KF) based integrated navigation system is widely used nowadays [2, 4, 7, 12]. Integrated navigation systems combine the advantages of both autonomous and external sources and eliminate the drawbacks of these systems. Hence, as a result of integration performance of the navigation system gets better, reliability characteristics are improved, and naturally, complexity of the system and the cost increase [1].

The INS are used in the aerospace and maritime widely, providing position and velocity information of the object. As known [8] the measurement errors of the INS increase over time. The mathematical model of these errors is nonlinear and navigation errors accumulate over time. The INS errors is divided into two parts: horizontal errors and vertical channel errors. Error in the vertical channel leads to the error in altitude measurements. The external sources measurements can be used in conjunction with the INS measurements in order to increase the INS vertical channel accuracy.

The barometrical measurements are more accurate than those ones tracked by GPS. In [10] is shown that the GPS integrity monitoring can be improved by non-standard baro-altitude aiding, due to the huge errors of the GPS vertical channel. The non-standard baro altitude is obtained by the on-line integration of the atmosphere hydrostatic equation, rather than the currently used solution based on a standard day lapse rate. Horizontal pressure gradient estimations can be included by employing cross track wind and ground velocity from the INS. The considerable part of the altitude error of barometric altimeter causes by the errors in pressure and temperature, sensors and some small residual atmospheric errors. The continuous calibration of barometric altimeter can be performed via Kalman filter using altitude measurements from GPS.

Integration of the INS, GPS, barometric and radar altimeters by using extended Kalman filter (EKF) is performed in [3]. This integrated altimeter provides sufficiently good accuracy of the altitude measurements.

Kalman filter based integrated GPS, baro-inertial loop aided strap down INS and radar altimeter are developed in [11]. In this study, Kalman filter is designed for the optimal combination of the GPS, INS and radar altimeters. As a result, the shortcomings of the independent navigation systems are eliminated, and the best performance of each altimeter combined through the Kalman filter.

The integrated navigation system, including the INS and radio altimeters, is presented in [6]. The integration is performed via an indirect Kalman filter. The Kalman filter to estimate the vertical channel parameters uses the error models of these altimeters. In this study, the robust KF algorithm against measurement faults is proposed. In the robust filter, the abrupt faults are detected and eliminated using statistical tests and instead of the estimation value, the extrapolation value of filter is used.

The fault tolerant integrated navigation system based on the radar and inertial altimeters is proposed in [5]. In that integrated system, the inertial altimeter is the main source of the altitude information, and radar altimeter gives discrete altitude data to support the inertial measurements. A nonlinear adaptive Kalman filter (NAKF) with the filter gain correction is developed for this integration scheme. The proposed NAKF based on the determination of the posterior probability of the normal operation of system, given for current measurement. This probability is suggested to be calculated using the posterior probability density of the normalized innovation sequence of Kalman filter at the current estimation step. In the proposed filtration algorithm, the filter gain is corrected by multiplying with the mentioned posterior probability, which plays the role of the weight coefficients to the innovation vector. As a result, faults in the estimation system are corrected by the system, without affecting the good estimation behaviour.

The paper [13] presents a multi-sensor INS/GPS/Baro-altimeter integrated navigation system. The main goal of this research was to adapt the error gain and damp coefficients in the 3rd order vertical channel damping loop in order to improve the altitude and the vertical velocity accuracy by using the EKF with a control signal. The adaptation of the coefficients was done with nonlinear functions. In the INS, the altitude error diverges exponentially, especially in low-cost sensors. To suppress divergence of the altitude error, a GPS receiver and a baro-altimeter are utilized in this paper. The integration is done with an EKF.

In this study three different integrated systems have been studied: integrated baro-inertial (IBI), integrated inertial-GPS (IIG) and integrated baro-inertial-GPS (IBIG). IBI and IIG navigation systems have been designed using a complementary Kalman filter (CKF). However, the fused IBIG altimeter gives better altitude measurements than both sources separately. This is the most relevant and essential outcome of this research. The results using the flight data show that the baro-altimeter augmented by IIG navigation system can provide more reliable and accurate navigation solutions. The data acquired by the sensors pass through local filters and then arrive to the integration module, where they are fused in a Kalman filter (optimal Kalman filter if faulty cases are not taken into consideration and robust Kalman filter otherwise) before providing to the user the estimation of the parameters of interest. Those results are shown in Sect. 18.4.

18.2 Optimal Kalman Filter

A Kalman filter is most probably the most famous computer algorithm in order to bounding and updating navigation outputs such as position, velocity and attitude. The Kalman filter is a linear filter with respect to measurements. The differential equations of the INS operation are nonlinear; however, the equations are linearized in order to be valid for the error equations. The first widely known important Kalman filter application was for the Apollo moon flight. From that time, a lot of improvements have been developed in this field and these ones have also been implemented in different important missions and aircraft navigation systems. The model of Kalman filter in its most known universal form is shown along this section. This chapter will focus on the optimal Kalman filter since faulty cases are not taken into consideration throughout this research.

The state space model used throughout this study is expressed with the following equations:

$$X(k) = \phi(k|k-1)X(k-1) + G(k|k-1)U(k-1) \tag{18.1}$$

$$Z(k) = H(k)X(k) + V(k) \tag{18.2}$$

where $X(k)$ is the state space vector of the system, $\phi(k|k-1)$ is the system transition matrix (being k the k-iteration), $G(k|k-1)$ is the noise transition matrix, $U(k-1)$ represents the system noise, $V(k)$ is the measurement noise and $H(k)$ represents the measurement matrix. The state vector components are given by

$$X^T = [\Delta H_I(k)\ \Delta W_z(k)\ \Delta a_z(k)\ \Delta g(k)\ B(k)] \tag{18.3}$$

where $\Delta H_I(k)$ represents the altitude error, $\Delta W_z(k)$ expresses the vertical speed error, $\Delta a_z(k)$ is the vertical acceleration error, gravitational error is given by $\Delta g(k)$ and finally $B(k)$, which represents the bias error. The following statements are the final expressions of the optimal Kalman filter equations to be used in the calculations. The equation of the estimation value on the iteration k is

$$\hat{X}(k|k) = \hat{X}(k|k-1) + K(k)\Delta(k) \tag{18.4}$$

being $K(k)$ the Kalman gain matrix and $\Delta(k)$ represents an innovation sequence defined in the following lines taken from [5]. The variable $\hat{X}(k|k-1)$ estimates the extrapolation value of the space state vector specified above in Eq. (18.3). The expressions of these three variables are

$$\hat{X}(k|k-1) = \phi(k|k-1)\hat{X}(k-1|k-1) \tag{18.5}$$

$$\Delta(k) = Z(k) - H(k)\hat{X}(k|k-1) \tag{18.6}$$

$$K(k) = P(k|k-1)H^T(k)[H(k)P(k|k-1)H^T(k) + R(k)]^{-1} \tag{18.7}$$

where $R(k)$ is the covariance matrix of measurement noise. On the other hand, matrixes P represent covariance matrixes of estimation and extrapolation errors, which respectively have the following form

$$P(k|k) = [I - K(k)H(k)]P(k|k-1) \tag{18.8}$$

$$P(k|k-1) = \phi(k|k-1)P(k-1|k-1)\phi^T(k|k-1) + G(k|k-1)Q(k-1)G^T(k|k-1) \tag{18.9}$$

being $Q(k-1)$ the covariance matrix of the system noise. The software used to implement these algorithms has been MATLAB.

18.3 Integrated Altimeter Design

First of all, it is needed to illustrate the error models of the inertial, barometric and GPS altimeters. Every single estimation model has an unavoidable error. In this section these error models are presented for the case of barometric, inertial and GPS altimeters.

18.3.1 Error Models

The measurement model of the barometric altimeter used in this research includes the bias error of the first order Markov process and a random white Gaussian noise as stated in [13], and they are expressed in the next equations

$$h_B = h_T + B + \nu_B \tag{18.10}$$

$$\dot{B} = -\frac{1}{\tau}B + w \tag{18.11}$$

being h_B the altitude measured from the barometer, h_T is the true height, measurement noise is represented by ν_B, τ is the correlation time of the bias error, w means the driving white Gaussian noise or the bias error and B is the bias error mentioned before.

Having numerous outputs from different channels, the INS has a complex error model. In this research, only the vertical channel of the INS is considered. Although the main parameter of study is the altitude, it is also interesting to consider other fluctuating values such as vertical speed W_z, vertical acceleration a_z and the gravitational acceleration g. The linear, discrete error model of the vertical INS channel is given by Zhukovskiy and Rastorguev [14]:

$$\Delta H_I(k) = \Delta H_I(k-1) + \Delta t \Delta W_z(k-1) \tag{18.12}$$

$$\Delta W_z(k) = \Delta W_z(k-1) + \frac{2g\Delta t}{R_I}\Delta H_I(k-1) + \Delta t \Delta a_z(k-1) + \Delta t \Delta g(k-1) \tag{18.13}$$

$$\Delta a_z(k) = \Delta a_z(k-1) - \Delta t \alpha \Delta a_z(k-1) + \Delta t U_{\Delta a_z}(k-1) \tag{18.14}$$

$$\Delta g(k) = \Delta g(k-1) - \Delta t \beta_g \Delta g(k-1) + \Delta t U_{\Delta g}(k-1) \tag{18.15}$$

where $U_{\Delta a_z}$ and $U_{\Delta g}$ are white Gaussian noises with zero mean, α and β_g are terms of the correlation period, being Δt the discrete time and $R_I = R_0 + H$, where H is the flight altitude and R_0 is the radius of the Earth. For the case of GPS error model, a Gaussian random white noise error has been taken, with zero mean and without bias.

18.3.2 Scheme 1: IBI Altimeter

INS are ubiquitously employed to provide position and velocity information. However, navigation information from INS gets degraded over time as INS errors are nonlinear and accumulate over time. The errors can be divided as horizontal errors and error in vertical channel. Error in the vertical channel directly contributes to error in altitude estimations [4]. Measurements from other external sources are used in conjunction with the INS measurements, to keep the error growth in the vertical channel within bounds. In this section, altitude measurements from the barometric altimeter and the INS measurements are fused with an optimal Kalman filter to obtain an accurate estimation of the flight altitude. While most of the references use a Kalman filter to directly estimate the state variables of system and its derivatives, it is common in INS to instead use a CKF which operates only on the errors in those primary state variables [5]. The scheme used for acquiring a precise altitude throughout IBI altimeters is depicted in Fig. 18.1 [9]. The central processing unit (CPU) block is necessary because the barometrical altimeter measures from sea level while the inertial system measures the relative altitude from an initial alignment location. In a system where these two navigators are combined, this difference could easily be eliminated via CPU. In order to be consistent with the theory explained before it is important to note that the measurement matrix for this system is

$$H_{\mathrm{IBI}}(k) = [1\ 0\ 0\ 0\ -1] \tag{18.16}$$

The scheme shown in Fig. 18.1 symbolizes the acquiring process of vertical measurements when inertial barometrical altimeter is used.

18.3.3 Scheme 2: IIG Altimeter

There are a lot of cases where IIG altimeters are used. Indirect Kalman filtering technique is used for integrating two mentioned navigation systems. Using IIG

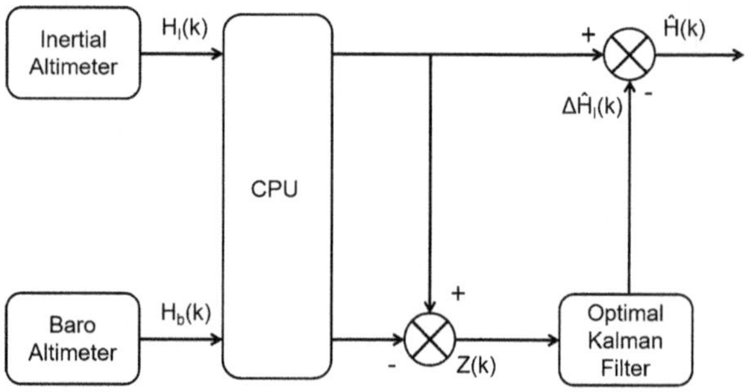

Fig. 18.1 Integrated baro-inertial altimeter scheme

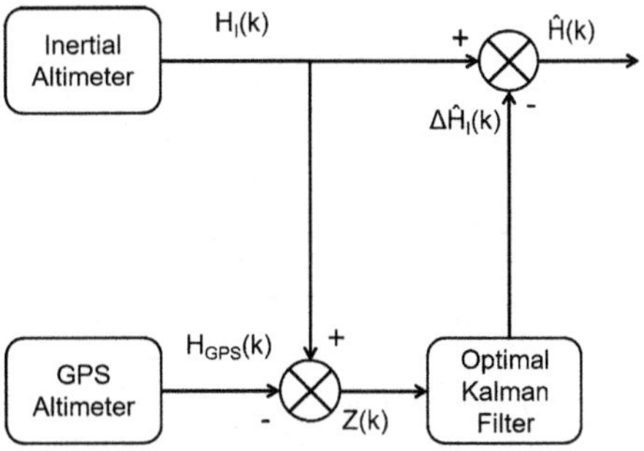

Fig. 18.2 Integrated inertial-GPS altimeter scheme

altimeter a smoother position and velocity estimations can be calculated that can be provided at a sampling rate faster than the GPS receiver. This also allows an accurate estimation of the aircraft attitude (roll, pitch and yaw) angles. In general, GPS/INS sensor fusion is a nonlinear filtering problem, which is commonly approached using a Kalman Filter. The scheme used for implementing this integrated system is depicted in Fig. 18.2. The CPU block is no longer necessary because both measurements are taken from the same reference. The measurement matrix of the system for this case is

$$H_{\mathrm{IIG}}(k) = [1\ 0\ 0\ 0\ 0] \tag{18.17}$$

In this case, the last value is zero because GPS bias error is not taken into consideration for this integrated system.

18.3.4 *Scheme 3: IBIG Altimeter*

The IIG system is designed using a loosely coupled integration architecture that incorporates the GPS navigation solution. The GPS typically has a poor accuracy in the vertical channel due to satellite geometry and atmospheric effects. Even worse, the vertical channel is vulnerable to faults due to its inappropriate dynamic model coupled with the frequent loss of satellites when the aircraft would bank. From the bibliography, huge GPS faults can be observed in the vertical axis, causing the IIG navigation system to become unstable. By augmenting the barometric altimeter data, the vertical channel can be stabilized under GPS fault conditions. The results using the flight data show that the baro-altimeter augmented IIG navigation system can provide more reliable and accurate navigation solutions under high maneuvering environments. The scheme used for this solution is shown in Fig. 18.3 [9]. As shown, the data acquired by the sensors pass through local filters and then arrive to the integration module, where they are fused in an optimal Kalman filter before providing to the user the estimation of the parameters of interest.

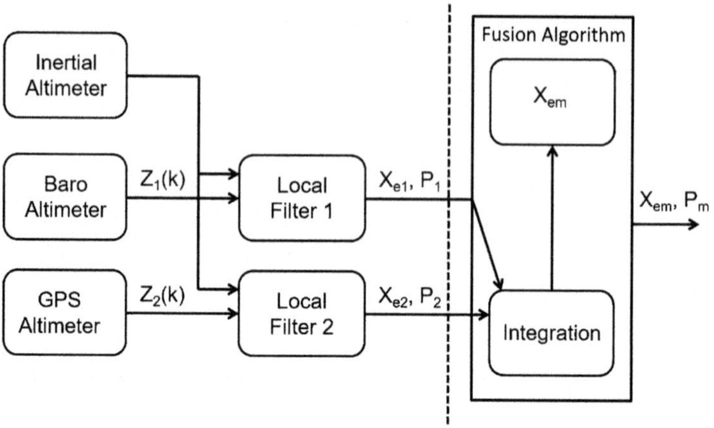

Fig. 18.3 Integrated baro-inertial-GPS altimeter scheme

The algorithm used for fusing the data in the IBIG altimeter is

$$\Delta H_{\mathrm{IBIG}} = \frac{P_1(1,1)\Delta H_{\mathrm{IIG}}(k) + P_2(1,1)\Delta H_{\mathrm{IBI}}(k)}{P_1(1,1) + P_2(1,1)} \tag{18.18}$$

where $P_1(1,1)$ and $P_2(1,1)$ represent a relation with the altitude variable in the covariance matrix of estimation error for IIG and IBI altimeters, respectively. Hence, the altitude estimation is given by

$$\hat{H}_{\mathrm{IBIG}}(k) = \Delta H_I(k) - \Delta\hat{H}_{\mathrm{IBIG}}(k) \tag{18.19}$$

The measurement matrix $H(k)$, covariance matrix of the measurement noise $R(k)$ and measurement vector $Z(k)$ are described above for each case.

18.4 Simulations

The simulation section of this study is mainly based on the error models of the different navigation sources of the integrated altimeters. The simulations have been run on MATLAB using the parameters shown in Table 18.1.

The results shown in this section are basically all the outputs of the state space vector for each integrated system when an optimal Kalman filter is used.

Table 18.1 Simulations parameters

Variable	Value	Units	Definition
Δt	0.5	s	Discrete time
R_0	6378150	m	Radius of the Earth
H	1000	m	Flight altitude
g	9.81	m/s^2	Gravitational acceleration
β	0.01	s^{-1}	Term of the correlation period
β_g	0.005	s^{-1}	Term of the correlation period
α	0.001	s^{-1}	Term of the correlation period
σ_W	0.0001	m/s	Vertical speed white noise mean
$\sigma_{\Delta_{a_z}}$	0.00002	m/s^2	Vertical acceleration white noise mean
σ_{Δ_g}	0.00002	m/s^2	Gravitational acceleration white noise mean
σ_B	1	m	Bias white noise mean
σ_{INS}	3	m	Inertial system white noise mean
σ_{GPS}	2.23	m	GPS white noise mean

18.4.1 IBI Altimeter Simulation Results

In this section, we focus on the space vector outputs of the IBI altimeter. Specifically, altitude error, vertical speed error, vertical acceleration error, gravitational error and baro-altimeter bias error are presented in this order.

In the Fig. 18.4 can be observed how the system tracks the error accurately. This output has a capital importance because since the altitude error is calculated accurately, the actual altitude can be calculated precisely. It is also very interesting in order to refine the data acquired by sensors since real altitude is estimated out of it. Same happens with the rest of the outputs. Vertical speed error is depicted in Fig. 18.5, where the positive and negative peek values are 5 and −4 m/s approximately. Those peek values are more damped in the case of vertical acceleration error, shown in Fig. 18.6.

Even more dumped are the measurements depicted in Fig. 18.7, which correspond to the gravitational acceleration error. In Fig. 18.8, baro-altimeter bias error estimation results given by IBI navigation system can be observed. Finally and most importantly, Fig. 18.9 depicts the evolution of the INS errors versus IBI navigation system errors. The altitude error values of the inertial system grow through time, as known from bibliography. However, the integrated system values dump that error and it reaches a reliance fringe with still some error which will be improved throughout other algorithms defined ahead. In Table 18.2, mean errors of all outputs are presented in order to check the quality of the IBI navigation system.

As known from bibliography, inertial error grows through time while IBI altimeter damps that error achieving values around zero.

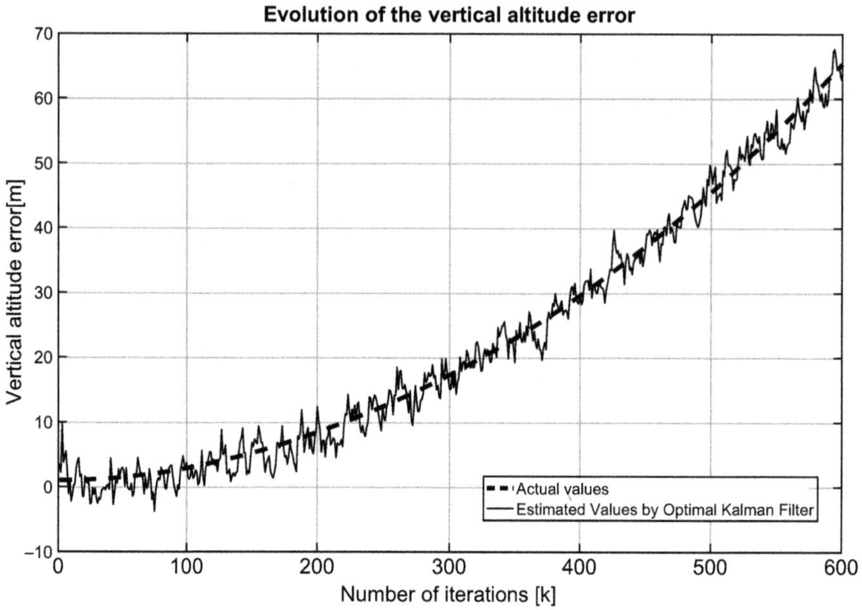

Fig. 18.4 Evolution of the altitude error given by IBI altimeter

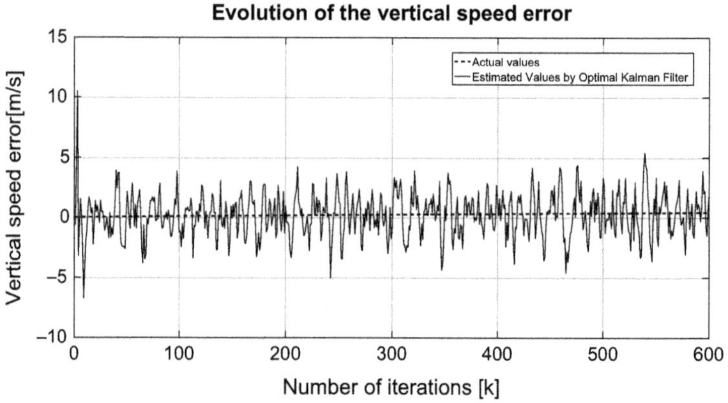

Fig. 18.5 Vertical speed error calculated via IBI altimeter

18.4.2 IIG Altimeter Simulation Results

Similar results have been calculated when the IIG altimeter is implemented. It is assumed that there is not any GPS bias error when IIG altimeter is implemented. In Figs. 18.10 and 18.11 the altitude and vertical speed estimation errors when IIG navigation system was used are shown. Again, the system tracks the altitude error

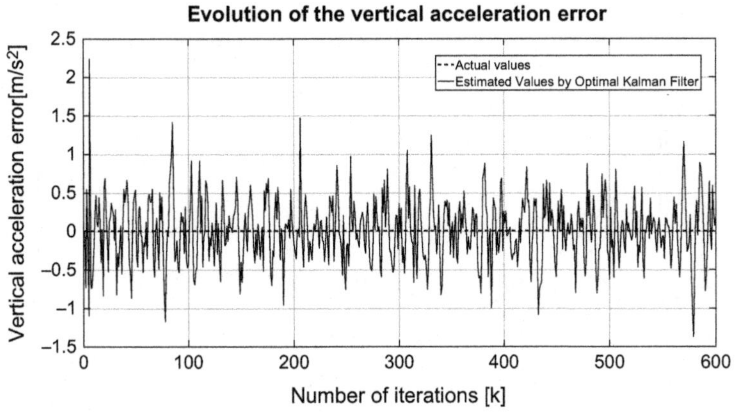

Fig. 18.6 Vertical acceleration error given by IBI altimeter

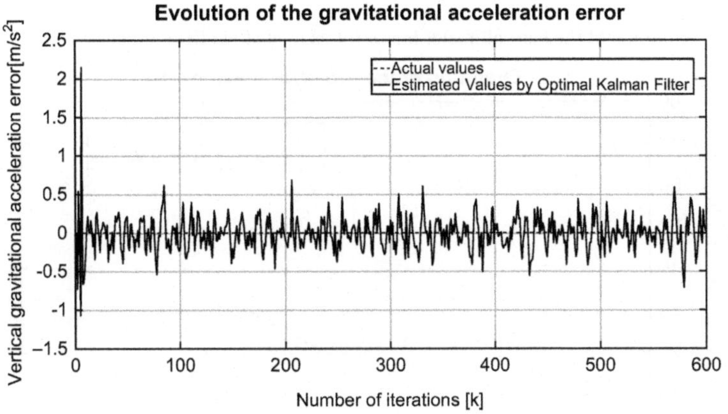

Fig. 18.7 Evolution of the gravitational acceleration error estimated via IBI altimeter

very well. Vertical acceleration error and gravitational error estimation results when IIG navigation system was used are depicted in Figs. 18.12 and 18.13, respectively. Finally, in Fig. 18.14, the evolution of the INS errors versus IIG navigation system errors is shown. As stated before, all these errors will be higher than the ones calculated in the previous section, and it can be compared in Table 18.3, where mean error values have been calculated for this integrated system.

18.4.3 IBIG Altimeter Simulation Results

In this section, the main output of this fused integrated system is shown. Due to the visual similarity with previous results, only the evolution of the altitude error

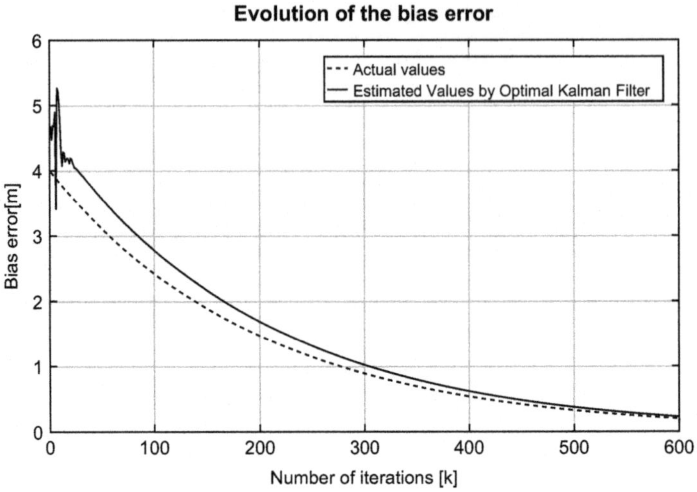

Fig. 18.8 Evolution of the baro-altimeter bias error estimated via IBI altimeter

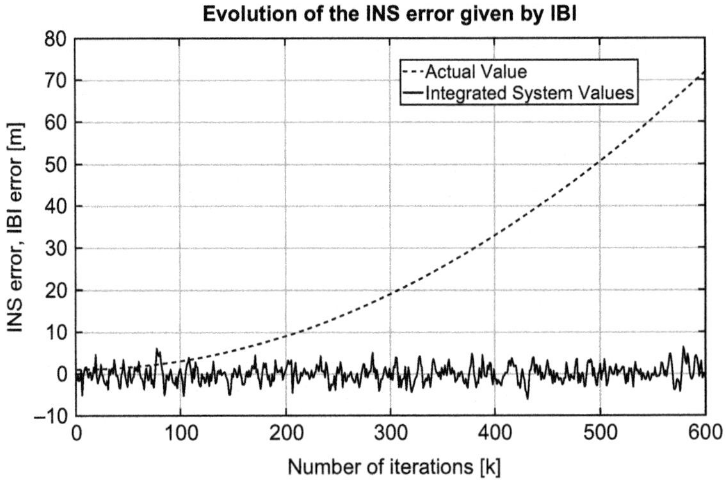

Fig. 18.9 Evolution of the INS error versus IBI altimeter errors

is presented. This result is depicted in Fig. 18.15. The fluctuations for IIG altimeter errors are the largest ones for the three altimeters. The second largest errors are acquired by IBI altimeter and finally, the best results are obtained by the fused IBIG altimeter, as expected. In Table 18.4 are shown the mean errors for all the cases, verifying the visual results.

Table 18.2 Mean error values given by IBI (integrated baro-inertial altimeter)

Variable	Value	Units	Definition
ΔH_I	1.63	m	Altitude error
ΔW_z	1.36	m/s	Vertical speed error
Δa_z	0.33	m/s^2	Vertical acceleration error
Δg	0.17	m/s^2	Gravitational error
B	0.12	m	Bias error

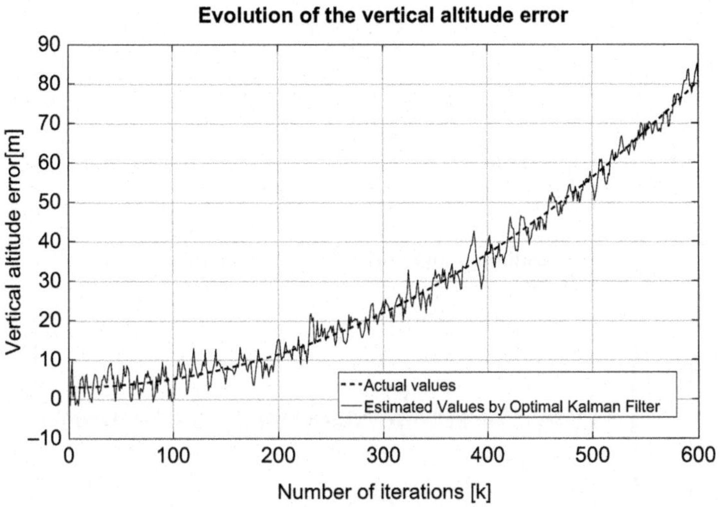

Fig. 18.10 Evolution of the altitude error given by IIG altimeter

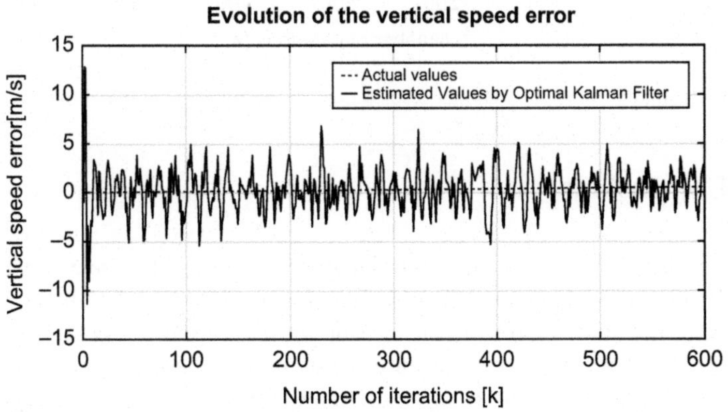

Fig. 18.11 Vertical speed error calculated via IIG altimeter

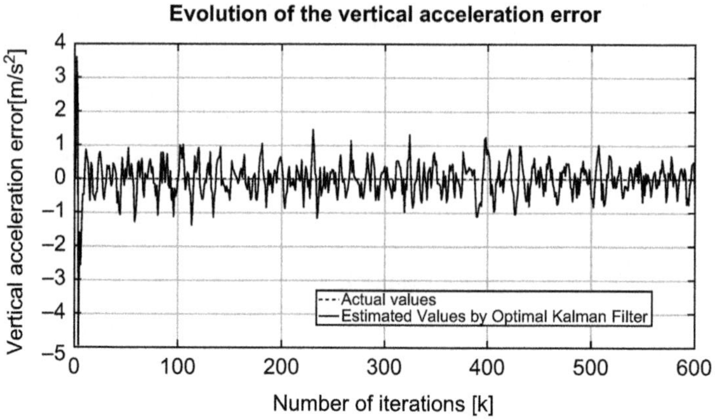

Fig. 18.12 Vertical acceleration error given by IIG altimeter

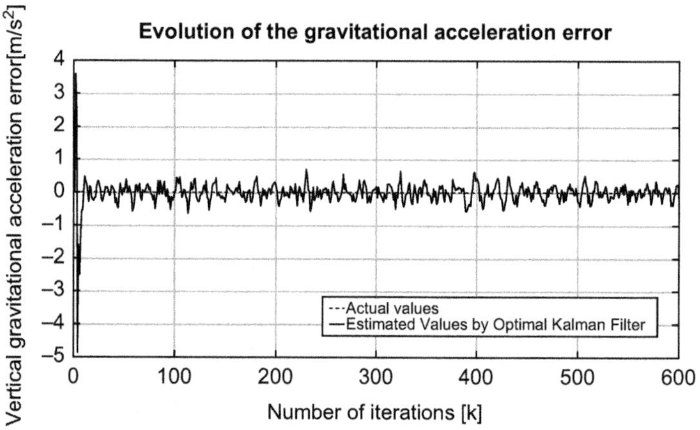

Fig. 18.13 Evolution of the gravitational acceleration error estimated via IIG altimeter

18.5 Conclusions

Throughout this research three different integrated systems have been studied: IBI, IIG and IBIG. All of them have been designed using an optimal Kalman filter. As stated during the study, barometrical measurements are more accurate than those ones tracked by GPS. However, the fused system gives better measurements than both sources separately. This is the most relevant and essential outcome of this research. As shown in the bibliography, there are similar studies which conclude with the same results. Hence, the present findings can be validated in agreement

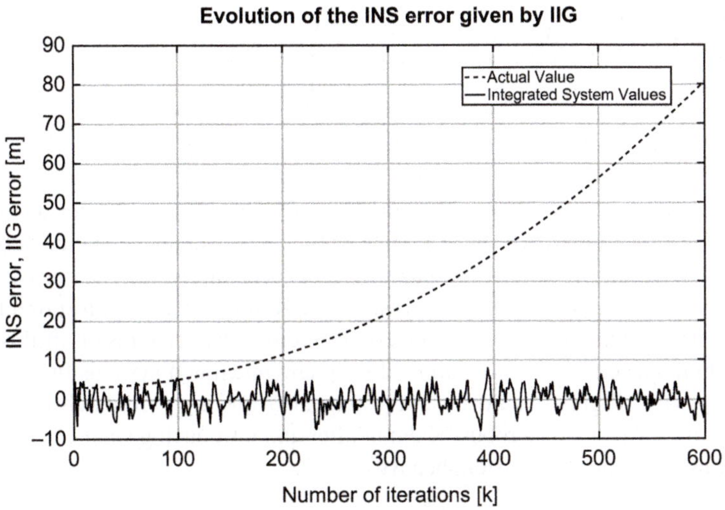

Fig. 18.14 Evolution of the INS error versus IIG altimeter errors

Table 18.3 Mean error values given by IIG (integrated inertial-GPS altimeter)

Variable	Value	Units	Definition
ΔH_I	2.13	m	Altitude error
ΔW_z	1.74	m/s	Vertical speed error
Δa_z	0.39	m/s^2	Vertical acceleration error
Δg	0.21	m/s^2	Gravitational error

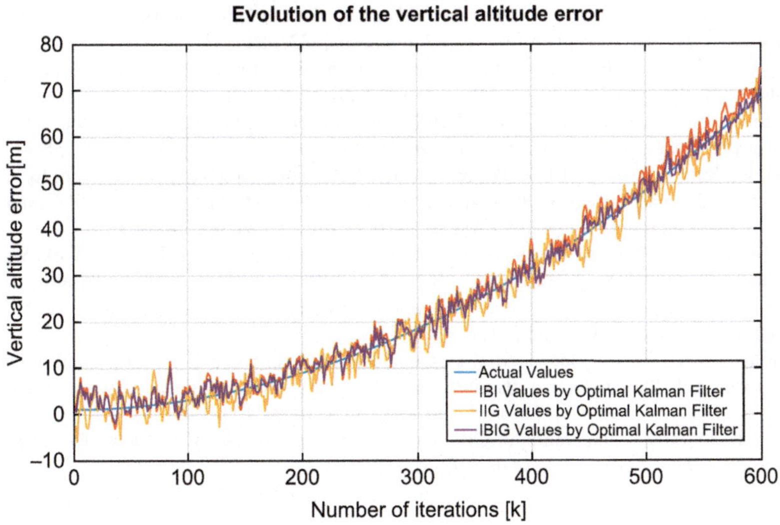

Fig. 18.15 Altitude measurements comparison between altimeters

Table 18.4 Comparison between altimeters in altitude measurements

Altimeter	Mean error value [m]
Baro-inertial (IBI)	1.99
Inertial-GPS (IIG)	2.26
Baro-inertial-GPS (IBIG)	1.52

with those studies. The obtained results show that the baro-altimeter augmented IIG navigation system can provide more reliable and accurate navigation solutions. The data acquired by the sensors pass through local filters and then arrive to the integration module, where they are fused in a Kalman filter before providing to the user the estimation of the parameters of interest. The conclusion of this study holds significance, for it can be applied into new integrated systems in the future, providing safer flights. Could be interesting also making a research about fault tolerant altimeters, using a robust Kalman filter and introducing some measurement errors, observing how the system behaves.

References

1. Biezad, D. J. (1999). *Integrated navigation and guidance systems*. Reston VA: American Institute of Aeronautics and Astronautics, Inc.
2. Dinc, M., & Hajiyev, C. (2015). Integration of navigation systems for autonomous underwater vehicles. *Journal of Marine Engineering and Technology, 14*(1), 32–43.
3. Gray, R. A., & Maybeck, P. S. (1995). Integrated GPS/INS/BARO and radar altimeter system for aircraft precision approach landings. In *Proceedings of the IEEE National Aerospace and Electronics Conference, Part 1, Dayton, KY, USA*.
4. Hajiyev, C. (2007). Adaptive filtration algorithm with the filter gain correction applied to integrated INS/Radar altimeter. *Proceedings of the Institution of Mechanical Engineers (IMechE), Part G, Journal of Aerospace Engineering, 221*, 847–855.
5. Hajiyev, C. (2012). Fault tolerant integrated radar/inertial altimeter based on nonlinear robust adaptive Kalman filter. *Aerospace Science and Technology, 17*(1), 40–49.
6. Hajiyev, C., & Saltoglu, R. (2004). RKF-based fault tolerant integrated INS/Radar altimeter. *Aircraft Engineering and Aerospace Technology: An International Journal, 76*(1), 38–46.
7. Jin, R., Sun, H., Sun, J., Chen, W., & Chu, J. (2016). Integrated navigation system for UAVs based on the sensor of polarization. In *Proceedings of the 2016 IEEE International Conference on Mechatronics and Automation*, August 7–10, Harbin, China, pp. 2466–2471.
8. Kayton, M., & Fried, W. R. (1997). *Avionics navigation systems* (2nd ed.). New York: Wiley.
9. Mañero Contreras, A., & Hajiyev, C. (2016). Integrated barometric inertial GPS altimeter. In *Proceedings of the International Symposium on Sustainable Aviation*, 29 May–1 June, 2016, Istanbul, Turkey, ISSA-2016-0221, pp. 386–389.
10. Molnar, D. O. (1994). Baro-inertial altitude accuracy improvement for GPS monitoring and integration. In *Proceedings of the 50th Annual Meeting-Institute of Navigation, Partnerships for Technology Conversion, Colorado Springs, CO, USA*, pp.129–138.
11. Rao, K. D. (1997). Integration of GPS and baro-inertial loop aided strapdown INS and radar altimeter. *IETE Journal of Research, 43*(5), 383–390.

12. Rogers, R. M. (2007). *Applied mathematics in integrated navigation systems* (3rd ed.). Reston, VA, USA: American Institute of Aeronautics and Astronautics, Inc. (AIAA).
13. Sokolovic, V., Dikic, G., & Stancic, R. (2014). Adaptive error damping in the vertical channel of the INS/GPS/Baro-Altimeter integrated navigation system. *Scientific Technical Review, 64*(2), 14–20.
14. Zhukovskiy, A. P., & Rastorguev, V. V. (1998). *Complex radio navigation and control systems of aircraft*. Moscow: MAI (in Russian).

Chapter 19
Testing the Determinant of the Innovation Covariance Matrix Applied to Aircraft Sensor and Actuator/Surface Fault Detection

Chingiz Hajiyev and Ulviye Hacizade

19.1 Introduction

The fault detection filter to detect and identify the sensor and actuator/surface faults can be developed by using analytical redundancy-based techniques. An analytical redundancy-based technique for detecting and isolating sensor and actuator/surface faults of the aircraft and spacecraft is presented in [1]. The standard generalized likelihood ratio (GLR) statistic is used in this study to detect the sensor and actuator/surface faults operatively. However, this method requires the statistical characteristics of the investigated system to be known before and after the fault occurs. As these characteristics are usually not available after the fault, the GLR approach has not been used in aerospace widely.

The fault detection and diagnosis algorithms based on the interacting multiple-model (IMM) estimation approach are given in Refs. [2, 3]. These algorithms, which are one of the most cost-effective adaptive estimation techniques, can detect and isolate the multiple sensor and actuator/surface failures in real time. The main advantage of IMM approach is to provide an integrated framework for fault detection, diagnosis, and state estimation. In Ref. [4], the multiple-model adaptive estimation (MMAE) methods were used in the flight control system design for the aircraft (VISTA) F-16. These methods allow to detect and compensate the sensor and control surface/actuator failures. In methods described in [2–4], the type of faults is assumed to be known previously. In this case, the Kalman filters are

C. Hajiyev (✉)
Faculty of Aeronautics and Astronautics, Istanbul Technical University, Maslak, 34469 Istanbul, Turkey
e-mail: cingiz@itu.edu.tr

U. Hacizade
Faculty of Engineering, Halic University, Istanbul, Turkey
e-mail: ulviyehaciyeva@halic.edu.tr

T.H. Karakoç et al. (eds.), *Advances in Sustainable Aviation*,
DOI 10.1007/978-3-319-67134-5_19

designed for the known sensor and control surface/actuator faults. As the approach requires several parallel Kalman filters for realization, and the type of faults should be known a priori, it has limited applications in practice.

In Refs. [5–8], the neural network-based approach to detect sensor and control surface/actuator faults is developed and investigated. These methods strongly depend on the data, which is used for the design of the neural network model. The different data corresponding to the same event in the neural network model obtains different results. Furthermore, the methods based on the artificial neural networks do not have physical bases. Consequently, the neural network model should be continuously trained by using the new data.

For the sensor and control surface/actuator fault detection and diagnosis purpose, the analysis of the innovation sequence of the Kalman filter can be used [9, 10]. These methods do not require that the statistical characteristics of the faults must be known a priori. It is the main advantage of this approach. If there is no any fault in the system, the normalized innovation sequence of the Kalman filter is a zero-mean Gaussian white noise with a unit covariance matrix. The sensor and control surface/actuator faults in the dynamic system such as abnormal measurements, biases in the measurement apparatus, sensor noise increment, decrease of measurement instrument accuracy, reduction in control surface/actuator effectiveness, etc. affect the statistical characteristics of the normalized innovation sequence by changing its white noise nature, displacing its zero mean, and varying unit covariance matrix. Thus, the problem is to detect operatively any change of these characteristics from their nominal values. Methods of testing the whiteness property of the innovation sequence and the detection of changing its mean value were discussed in details in the literature [10, 11]; therefore, they are not considered in this study.

Testing, in real time, the covariance matrix of the Kalman filter innovation sequence becomes very complicated, since there are difficulties in the determination of the confidence domain for a random matrix. The existing methods of high-dimensional statistical analysis [12, 13] cannot be used widely because they usually lead to asymptotic distributions; this sharply diminishes the operativeness of these methods. For this purpose, one of a scalar measure of the innovation covariance matrix such as the trace, the sum of the matrix elements, the determinant, and the maximal eigenvalue, each characterizing different geometrical parameters of the correlation ellipsoid, can be used.

In Ref. [10], the trace of the matrix is used for the scalar measure of the tested innovation covariance matrix. Testing of the trace of the sample covariance matrix might lead to incorrect decisions at detection of faults because in this case the off-diagonal matrix elements are not taken into consideration. In Ref. [9], a method is described for checking the sum of all elements of the inverted innovation covariance matrix. However, the trace and sum of all elements of the covariance matrix is strongly dependent on the units and is not scale invariant [14]. This aspect needs to be carefully considered when designing the fault detection algorithm.

A quadratic form of the random Wishart matrix, which has χ^2-distribution under the normal operation mode of the system, is used in [15] for testing the innovation covariance matrix. The optimization process of testing quality is reduced to the classical problem of maximization of a quadratic form on the unit sphere. As a result, two algorithms are proposed for testing the innovation covariance matrix of the Kalman filter. The sum of all the elements in the first algorithm and the largest eigenvalue in the second algorithm are used as the scalar measure of the tested Wishart matrix. The proposed approach is used for testing the innovation covariance matrix in the sensor/actuator fault detection problem.

The Tracy–Widom distribution-based innovation covariance matrix testing method is proposed in [16]. In this method, the maximal eigenvalue of the innovation covariance matrix is used as the fault detection statistic. As a result, the covariance matrix testing problem is reduced to determine the asymptotes for the largest eigenvalue of the Wishart matrix. The Tracy–Widom distribution-based largest eigenvalue testing algorithm is developed and applied to the F-16 fighter model of dynamics for detection of sensor and control surface/actuator faults.

A new approach for testing the innovation covariance matrix using the generalized Rayleigh quotient is proposed in [17]. The generalized Rayleigh quotient of the vector under the sample and theoretical innovation covariance matrices of Kalman filter is used in this study as the fault detection statistic. The testing quality optimization process in this work is reduced to the classical problem of maximization of the generalized Rayleigh quotient. The proposed innovation covariance matrix testing method is used in this work for the sensor and control surface/actuator fault detection problem in the AFTI/F-16 aircraft flight control system.

To obtain more accurate states, in [18], the conventional Kalman filter is enhanced to the multi-innovation Kalman filter. This approach is applied to the pitch sensor offset fault detection problem of the wind turbine. With regard to the large noise of sensors, the multi-innovation Kalman filter is adopted to increase the estimation accuracy and convergence speed. The fault diagnosis of the performance of the multi-innovation Kalman filter is better than the conventional Kalman filter; however, it requires more computational burden.

In this study, an algorithm for checking the generalized variance (determinant) of the innovation sequence, characterizing the volume of the correlation ellipsoid and application to aircraft dynamics, is investigated. This is the overall measure in terms of the volume of the error ellipsoid, which is proportional to the determinant of the innovation covariance matrix of the Kalman filter. In this case, the fault detection algorithm is invariant to scaling of the system states.

The structure of this chapter is as follows: In Sect. 2, the fault detection problem in multidimensional dynamic systems based on the Kalman filter innovation sequence is formulated. New results for testing the generalized variance of the innovation covariance are proposed in Sect. 3. The model of dynamics of the AFTI/F-16 fighter description and the extended Kalman filter (EKF) algorithm for estimation of the F-16 nonlinear mathematical model states are presented in Sect. 4. In Sect. 5 several simulations are carried out for the sensor and control surface/actuator fault detection problem in the AFTI/F-16 aircraft flight control system. The changes that affect the innovation covariance are considered. The fault detection algorithm based on the construction of confidence intervals for the Wishart determinants via its asymptotes is examined in this section. A brief summary of the obtained results and the conclusion are given in Sect. 6.

19.2 Statement of the Problem

The mathematical model of the dynamic system is described in discrete form as:

$$\begin{aligned} x(k+1) &= \Phi(k+1,k)\,x(k) + G(k+1,k)\,w(k) \\ z(k) &= H(k)x(k) + v(k), \end{aligned} \quad (19.1)$$

where $x(k)$ is the n-dimensional state vector of the system, $\Phi(k+1,k)$ is the $n \times n$ transition matrix of the system, $w(k)$ is the random n-dimensional vector of system noises, $G(k+1,k)$ is the $n \times n$ transition matrix of system noises, $z(k)$ is the s-dimensional measurement vector, $H(k)$ is the $s \times n$ measurement matrix of the system, and $v(k)$ is the random s-dimensional vector of measurement noises. It is assumed that the random vectors $w(k)$, $v(k)$, and $x(0)$ are mutually independent white Gaussian processes with zero expectations and covariance matrices defined by the relations:

$$\begin{aligned} E\left[w(k)w^T(j)\right] &= Q(k)\delta(kj), \\ E\left[v(k)v^T(j)\right] &= R(k)\delta(kj), \\ E\left[x(0)x^T(0)\right] &= P(0), \end{aligned}$$

where $\delta(kj)$ is the Kronecker symbol.

The estimation of the state vector $\hat{x}(k/k)$and the covariance matrix of the estimation errors $P(k/k)$ can be found via the optimal Kalman filter. Furthermore, if the estimation system is normally operating, then the normalized innovation of Kalman filter

$$\tilde{v}(k) = \left[H(k)P(k/k-1)H^T(k) + R(k)\right]^{-1/2} \times \left[z(k) - H(k)x(k/k-1)\right] \quad (19.2)$$

is a white Gaussian noise with zero mean and unit covariance matrix [4]:

$$E\left[\tilde{v}(k)\right] = 0, E\left[\tilde{v}(k)\tilde{v}^T(j)\right] = P_{\tilde{v}} = I\delta\,(\mathrm{kj})\,,$$

where $\hat{x}(k/k-1)$ is the extrapolation value by one step,

$$\begin{aligned} P(k/k-1) = {} & \Phi(k,k-1)P(k-1/k-1)\Phi^T(k,k-1) \\ & + G(k,k-1)Q(k-1)G^T(k,k-1) \end{aligned}$$

is the covariance matrix of extrapolation errors, $P(k-1/k-1)$ is the covariance matrix of the estimation errors in the preceding step, and I is the unit matrix.

Since the normalized innovations have the distribution $N(0,1)$, they will be in the confidence interval $[-3, +3]$ with probability 0.9986 in case of optimal operation of Kalman filter.

The changes in the statistical characteristics of the system or disturbances (faults of measuring instruments, abnormal measurements, changes in statistical characteristics of system noises, changes in statistical characteristics of measurements noises, the reduction in the actuator effectiveness, the friction between the moving parts of the control surfaces, etc. leading to a change in the covariance matrix of the normalized innovation sequence (19.2) are considered.

It is of interest to develop an operative method of testing the determinant of the covariance matrix of sequence (19.2).

19.3 Algorithm of Solution

Let us take a look at the physical significance of the generalized variance. If the columns of the innovation covariance matrix are independent, the determinant of the matrix will have a maximum value subject to the given constraints in terms of the postulated model. On the other hand, if parameters are correlated, the determinant will be smaller; in the worst case, linear dependency of two parameters results in a zero determinant, indicating that the two parameters cannot be estimated independently. Thus, the generalized variance minimizes the redundancy and leads a better fault detection performance. Furthermore, differently from the trace and sum of all elements of the tested covariance matrix, the generalized variance is invariant to scaling of the system states. For these reasons the generalized variance is used in this study as a scalar measure of the tested innovation covariance matrix.

Two hypotheses are introduced:

γ_0: no sensor/actuator fault occurs.
γ_1: a sensor/actuator fault occurs.

The expression for the sample covariance matrix of the sequence $\tilde{\upsilon}(k)$is written as:

$$\widehat{S}(k) = \frac{1}{M-1} \sum_{j=k-M+1}^{k} \left[\tilde{\upsilon}(j) - \overline{\tilde{\upsilon}}(k)\right]\left[\tilde{\upsilon}(j) - \overline{\tilde{\upsilon}}(k)\right]^T \tag{19.3}$$

where

$$\overline{\tilde{\upsilon}}(k) = \frac{1}{M} \sum_{j=k-M+1}^{k} \tilde{\upsilon}(j) \tag{19.4}$$

is the sample mean; M is the number of realizations used (the width of the sliding window). As known [12], under the validity of the hypotheses γ_o, the random matrix

$$A(k) = (M-1)\widehat{S}(k) \tag{19.5}$$

has the Wishart distribution with M degrees of freedom and is denoted by $W_s(M, P_{\tilde{\nu}})$:

$$A \sim W_s(M, P_{\tilde{\nu}}), \tag{19.6}$$

where s and $P_{\tilde{\nu}}$ are the dimension and covariance matrix of the normalized innovation sequence $\tilde{\nu}$, respectively. In testing statistical hypotheses, the testing of the Wishart statistic (19.6) is complicated and not well developed in view of the difficulty of constructing the confidence domain for a random matrix. For this purpose, one of the scalar measures of the Wishart matrix can be used for testing the covariance of the normalized innovation (19.2). In this study, construction of confidence intervals for the Wishart determinants via its asymptotes is considered. In [19] the following asymptotic result for determinants of Wishart matrices is obtained.

Theorem Let $A \sim W_s(M, P_{\tilde{\nu}}), M = 1, 2, \ldots,$ and then

$$\left(\frac{\det(A)}{\det(P_{\tilde{\nu}})(M-1)!}\right)^{\frac{1}{\sqrt{2\log(M)}}} \xrightarrow{d} e^N, \tag{19.7}$$

where $\xrightarrow{d}$ stands for convergence in distribution and N is a standard normal random variable.

Taking the logarithm of the left- and right-hand sides of the formula (19.7), the following expression can be written as:

$$\frac{1}{\sqrt{2\log(M)}}\log\left(\frac{\det(A)}{\det(P_{\tilde{\nu}})(M-1)!}\right) \xrightarrow{d} N \tag{19.8}$$

The left-hand side of the expression (19.8) is denoted as:

$$\mathrm{WD} = \frac{1}{\sqrt{2\log(M)}}\log\left(\frac{\det(A)}{\det(P_{\tilde{\nu}})(M-1)!}\right) \tag{19.9}$$

and called as the statistic of the Wishart determinants. Taking into account that the covariance matrix of the normalized innovation sequence in the case of normally operating the system is the identity matrix, e.g., $P_{\tilde{\nu}} = I$, and combining the formulas (19.8) and (19.9), the following expression can be written [20]:

$$\mathrm{WD} = \frac{1}{\sqrt{2\log(M)}}\log\left(\frac{\det(A)}{(M-1)!}\right) \xrightarrow{d} N \tag{19.10}$$

If the confidence probability β is selected as:

$$P\left\{|\mathrm{WD}| \le u_{\frac{1+\beta}{2}}\right\} = \beta;\ \ 0 < \beta < 1 \tag{19.11}$$

the threshold value $u_{\frac{1+\beta}{2}}$ can be determined ($u_{\frac{1+\beta}{2}}$ is the quantile of the standard normal distribution). Hence, when the hypothesis γ_1 is true, the absolute value of the statistic *WD* will be greater than the threshold value $u_{\frac{1+\beta}{2}}$. Then the decision rule on the current state of the system of estimation with respect to the introduced hypotheses will be written in the following form:

$$\begin{aligned} \gamma_0 &: |WD(k)| \le u_{\frac{1+\beta}{2}}, \forall k \\ \gamma_1 &: |WD(k)| > u_{\frac{1+\beta}{2}}, \exists k \end{aligned} \tag{19.12}$$

Consequently, by comparing the above-defined statistic of the Wishart determinants *WD(k)* with the confidence limits for the corresponding standard normal distribution, it is possible to detect sensor and control surface/actuator faults using decision rule (19.12).

19.4 EKF for the Estimation of the F-16 Aircraft Model of Dynamics

The proposed innovation of covariance matrix testing algorithm in this study is applied to multi-input multi-output model of an AFTI/F-16 fighter. The mathematical model of the nonlinear dynamics of F-16 fighter is written in the following form:

$$x(k+1) = \mathrm{Ax}(k) + \mathrm{Bu}(k) + F(x(k)) + \mathrm{Gw}(k) \tag{19.13}$$

where $x(k)$ is the nine-dimensional aircraft state vector, A is the 9×9 transition matrix of the aircraft, B is the 9×6 control distribution matrix of the aircraft, $u(k)$ is the six-dimensional control input vector, $F(x(k))$ is the nine-dimensional vector of nonlinear elements of the aircraft, G is the 9×9 transition matrix of the system noises, and $w(k)$ is the nine-dimensional random system noise vector with zero mean and the covariance matrix of $E[w(k)w^T(j)] = Q(k)\delta(\mathrm{jk})$. The state variables of fighter are

$$x = [v, \alpha, q, \theta, \beta, p, r, \varphi, \psi]^T,$$

where v is the forward velocity, α is the angle of attack, q is the pitch rate, θ is the pitch angle, β is the sideslip angle, p is the roll rate, r is the yaw rate, ϕ is the roll angle, and ψ is the yaw angle. The control inputs of fighter are

$$u = [\delta_{\mathrm{HR}}, \delta_{\mathrm{HL}}, \delta_{\mathrm{FR}}, \delta_{\mathrm{FL}}, \delta_C, \delta_R],$$

where δ_{HR} and δ_{HL} are the deflections of the right and left horizontal stabilizers, δ_{FR} and δ_{FL} are the deflections of the right and left flaps, and δ_C and δ_R are the canard and rudder deflections. The measurement equations can be expressed as:

$$z(k) = \text{Hx}(k) + V(k), \tag{19.14}$$

Here H is the measurement matrix, which is a 9×9 unit matrix in case, and $V(k)$ is the nine-dimensional vector of measurement noises with the following statistical characteristics:

$$E\left[V(k)\right] = 0; E\left[V(k)V^T(j)\right] = R(k)\delta\left(\text{jk}\right).$$

After linearization of the Eq. 19.13 via quasi-linearization method, the following equation can be obtained:

$$\begin{gathered} x(k) = A\widehat{x}\left(k-1\right) + B\widehat{u}\left(k-1\right) + F\left(\widehat{x}\left(k-1\right)\right) \\ + A\left[x\left(k-1\right) - \widehat{x}\left(k-1\right)\right] + F_x\left(k-1\right)\left[x\left(k-1\right) - \widehat{x}\left(k-1\right)\right] \\ + B\left[u\left(k-1\right) - \widehat{u}\left(k-1\right)\right] + \text{Gw}\left(k-1\right) \end{gathered} \tag{19.15}$$

where $F_x = \left[\frac{\partial F}{\partial x}\right]_{\widehat{x}(k-1)}$.

In [21] the recurrent EKF algorithm for the state vector estimation of the F-16 fighter motion is obtained as:

$$\begin{aligned} &\widehat{x}(k) = A\widehat{x}\left(k-1\right) + B\widehat{u}\left(k-1\right) + F\left(\widehat{x}\left(k-1\right)\right) + P(k)H^T R^{-1}(k) \times \nu(k) \\ &\nu(k) = z(k) - H\left[A\widehat{x}\left(k-1\right) + B\widehat{u}\left(k-1\right) + F\left(\widehat{x}\left(k-1\right)\right)\right] \\ &P(k) = M(k) - M(k)H^T\left[R(k) + \text{HM}(k)H^T\right]^{-1}\text{HM}(k) \\ &M(k) = \text{AP}\left(k-1\right)A^T + \text{BD}_u\left(k-1\right)B^T + F_x\left(k-1\right)P\left(k-1\right)F_x^T\left(k-1\right) \\ &\qquad + \text{GQ}\left(k-1\right)G^T \end{aligned} \tag{19.16}$$

where $P(k)$ is the covariance matrix of the estimation error, $M(k)$ is the covariance matrix of the extrapolation error, and D_u is the covariance matrix of the control input error.

19.5 Sensor and Control Surface/Actuator Fault Detection Simulation Results

It can be shown that the algorithm for testing the innovation covariance matrix proposed in this study can detect the faults appearing in the measuring channel operatively. Measurements were processed using the EKF (19.16). The expressions for the innovation sequence and the normalized innovation sequence of EKF, respectively, are:

$$\nu(k) = z(k) - H\left[A\widehat{x}(k-1) + B\widehat{u}(k-1) + F(\widehat{x}(k-1))\right] \tag{19.17}$$

$$\tilde{\nu}(k) = \left[R(k) + \mathrm{HM}(k)H^T\right]^{-1/2} \nu(k) \tag{19.18}$$

To detect failures which change the innovation covariance, the generalized variance of the random Wishart matrix $A(k)$ is proposed to be used. In the simulations, $M = 20$, $s = 9$, and $\beta_2 = 0.9986$ are taken, and the threshold value $u_{\frac{1+\beta}{2}}$ is found as 3. It is assumed that fault is detected if the statistic $WD(k)$ exceeds the admissible limits for three successive iterations. This rule allows to reveal the random instantaneous abnormal situations from the faulty cases.

The proposed fault detection algorithm is based on the computation of the generalized variance of the matrix $A(k)$. The sequence of computations performed at each step of measurements has the following form:

- The estimates of the EKF and the value of the normalized innovation sequences are computed according to the (19.16–19.18).
- The random Wishart matrix $A(k)$ is determined via expressions (19.3–19.5).
- The generalized variance of the matrix $A(k)$ is found, and the statistic of the Wishart determinants $WD(k)$ is calculated via expression (19.9).
- The decision on the current state of the estimation system is made according to the decision rule (19.12).
- The sequence of computations is repeated starting from item 1 for the next moment of instant $k + 1$.

19.5.1 Sensor Fault Detection Results

The sensor fault detection results are given in Figs. 19.1, 19.2, 19.3, and 19.4. Figure 19.1 shows the admissible bounds of the statistic $WD(k)$ and the plots of their behaviors in the case of normal operating mode of all the measurement channels of system. As expected, the values of statistic $WD(k)$ lay between the admissible limits (except the rare and instantaneous abnormal situations). The corresponding normalized innovation sequence in the third measurement channel (pitch rate gyroscope channel) $\tilde{\nu}_q(k)$ is shown in Fig. 19.2. The graphs of the normalized innovation sequences in the other measurement channels are very similar to the ones in Fig.19.2.

To verify the proposed algorithm in case of sensor fault, beginning from the step $k = 30$, a fault in the third measurement channel (pitch rate gyroscope fault) is simulated (the noise variance in the pitch rate gyroscope is increased by multiplying with 1,5). The simulation results corresponding to this case are presented in Figs. 19.3 and 19.4. Figure 19.3 shows that the amplitude range of statistic $WD(k)$ increases after the 30th step with increasing iterations. According to the rule (19.12),

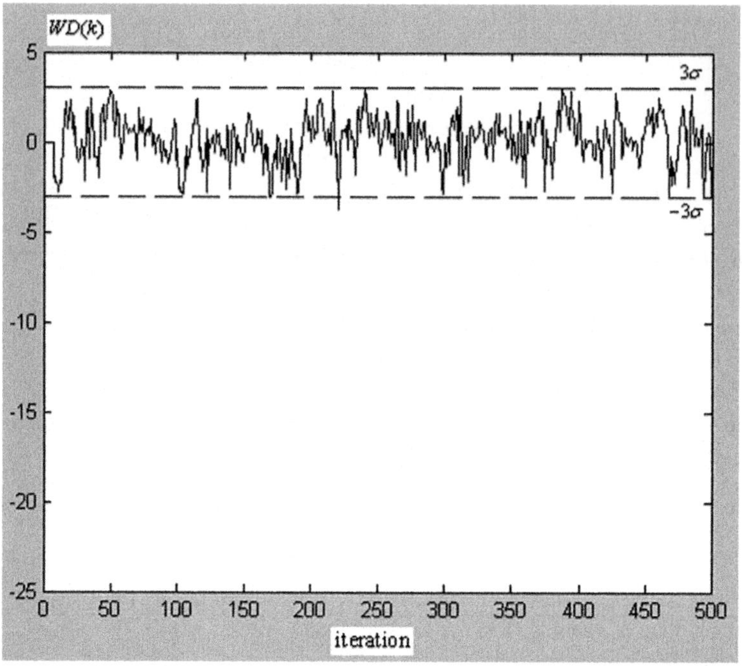

Fig. 19.1 Graph of the statistic *WD*(*k*) for normal operating mode of the measurement channels

the fault is detected at the iteration $k = 84$ (54 iterations or 1,62 s after the fault occurs). Due to randomness of measurement noises, the simulated measurements of the related sensor may be from the "good" sample range (e.g., corresponding to normal standard deviation) or "bad" one (corresponding to increased standard deviation). Consequently, the realizations of the innovation sequences entering to the sample innovation covariance matrix may be from the "good" or "bad" samples. As a result, the values of the fault detection statistic of Wishart determinants increase when the "bad" sample is used and decrease when the "good" sample is used. In innovation covariance matrix testing problem, this effect is natural. Therefore, the fault detection statistic may exceed their admissible bound and after certain iterations may drop significantly below the admissible bound. This movement can be observed from Fig. 19.3.

The behavior of the appropriate normalized innovation sequences $\tilde{\nu}_q(k)$ is presented in Fig. 19.4.

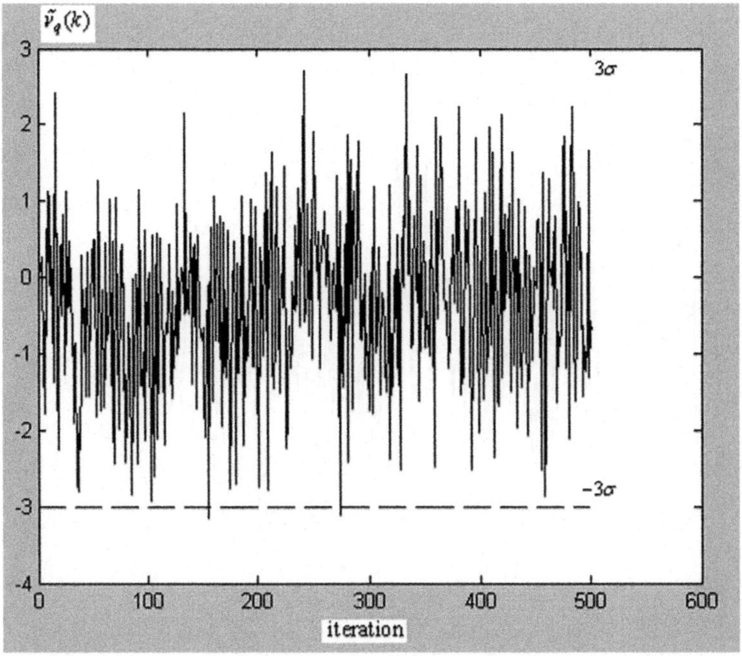

Fig. 19.2 Behavior of the $\tilde{\nu}_q(k)$ in the case of normal operating mode of the measurement channels

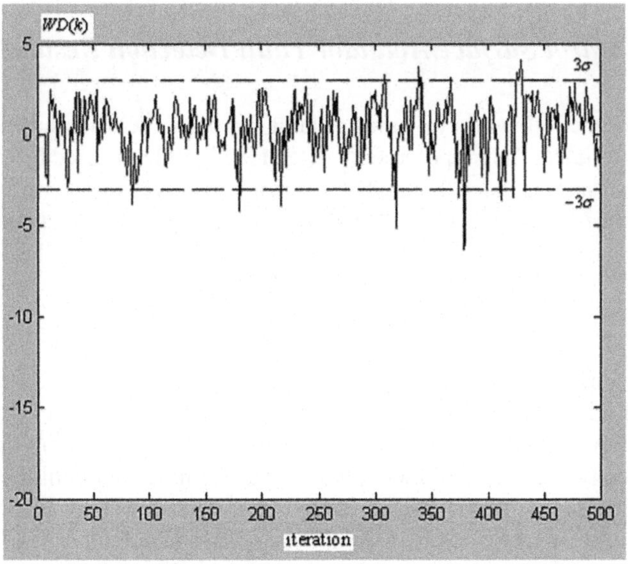

Fig. 19.3 Behavior of the statistic *WD*(*k*) in case of changes in noise variance in the pitch rate gyroscope

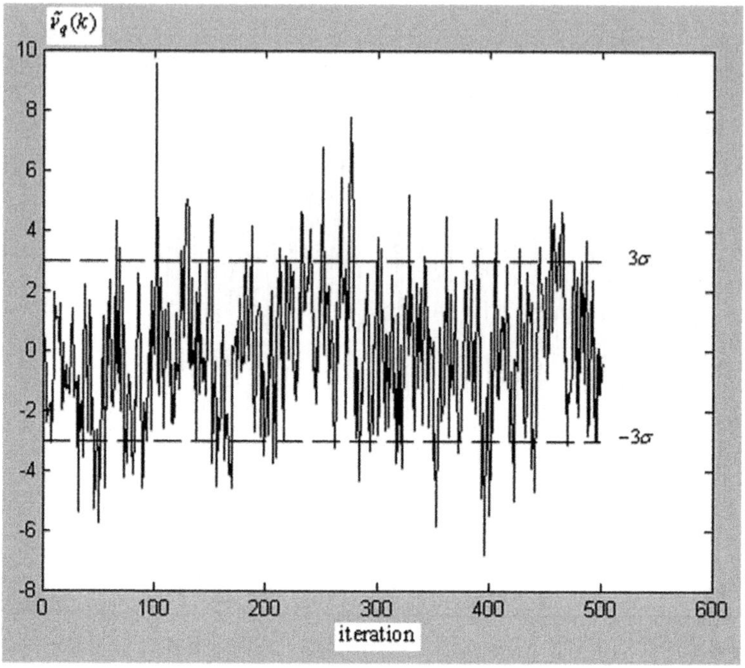

Fig. 19.4 Behavior of the normalized innovation sequence $\tilde{\nu}_q(k)$ in case of changes in noise variance in the pitch rate gyroscope

19.5.2 Control Surface/Actuator Fault Detection Results

In an actuator the following two kinds of fault can occur: actuator surface fault and actuator motor fault. The proposed fault detection algorithm is used below to detect the control surface/actuator faults. The control derivatives that correspond to the first control surface (right horizontal stabilizer) have been changed at iteration 30 as follows:

$$B^f(j,1) = B(j,1) \times 0.5, \quad j = \overline{1,9} \tag{19.19}$$

As seen from expression (19.19), in this case the reduction in effectiveness of the faulty right horizontal stabilizer is 50 %. The behavior of the statistic of the Wishart determinants $WD(k)$ when the control derivatives change is given in Fig. 19.5. As it is presented in Fig. 19.5, $WD(k)$ is lower than the threshold until the change of the control derivatives. In contrast, when the control derivatives are changed, $WD(k)$ increases, and at the step $k = 48$ (18 iterations or 0.54 s after fault occurs), it exceeds its admissible bound. As a result, the fault in the system is detected according to the decision rule (19.12). This fault causes a change in the innovation mean and the innovation covariance. The change of the control derivatives (elements of the

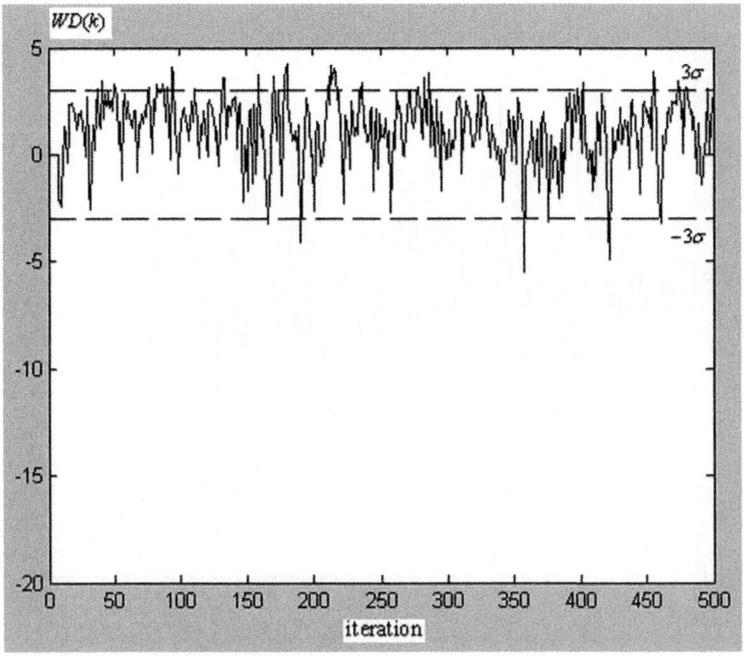

Fig. 19.5 Behavior of the statistic $WD(k)$ in case of control surface fault

control distribution matrix B) may be caused by the faults such as the reduction in the control surface/actuator effectiveness, the friction between the moving parts of the control surfaces, the partial loss of a control surface (break off of a part of control surface), etc. The behavior of the appropriate normalized innovation sequence $\tilde{\upsilon}_q(k)$ is presented in Fig. 19.5.

The application of developed sensor and actuator/surface fault detection algorithm does not distort the EKF estimation results and has no influence on their accuracy. On the whole, the simulation results justify the theoretical calculation obtained and show the practical applicability of the proposed fault detection algorithm.

19.6 Conclusion

In this study operative method for sensor and control surface/actuator fault detection based on testing the Kalman filter innovation covariance is proposed. The generalized variance (determinant) of the random Wishart matrix is used in this process as a fault detection statistic, and the testing problem is reduced to determination of the asymptotes for Wishart determinants. An algorithm for testing the innovation

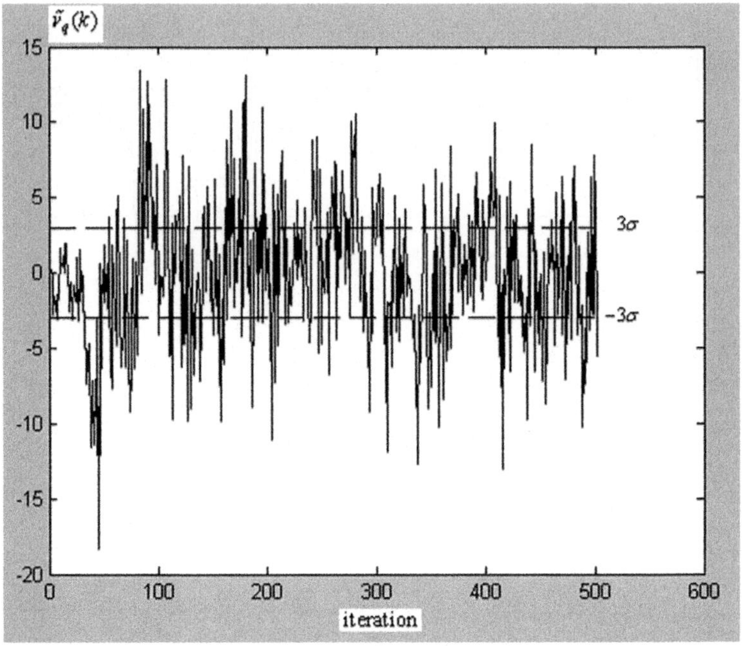

Fig. 19.6 Behavior of the normalized innovation sequence $\tilde{\nu}_q(k)$ in case of control surface fault

covariance matrix based on construction of confidence intervals for the Wishart determinants is proposed.

The generalized variance of the random Wishart matrix is the overall measure in terms of the volume of the error ellipsoid, which is proportional to the determinant of innovation covariance matrix. The fault detection algorithm is based on the evaluation of the generalized variance of the Wishart matrix and its comparison with the confidence bounds of the standard normal distribution. In this case the fault detection algorithm is invariant to scaling of the system states.

The suggested approach to the innovation covariance testing is used for the sensor and control surface/actuator fault detection problem in the AFTI/F-16 aircraft flight control system. An EKF has been applied for estimation of the nonlinear aircraft flight dynamics. The sensor faults that affect the innovation covariance have been considered. The simulated control surface/actuator faults cause a change in the mean and the covariance of the innovation sequence. The theoretical results are confirmed by the simulations carried out on a nonlinear dynamic model of the F-16 fighter.

The proposed approach permits real-time checking of the innovation mean and covariance and does not require a priori information about the faults and statistical characteristics of the system in faulty cases. This approach can be used in the problems of sensor and control surface/actuator fault detection for the multidimensional dynamic systems.

References

1. Larson, E.C., et al. (2002). Model-based sensor and actuator fault detection and isolation. In *Proceeding American Control Conference, 5*, 4215–4219.
2. Rago, C., et al. (1998). Failure detection and identification and fault tolerant control using the IMM-KF with applications to the Eagle Eye UAV. *Proceedings of the IEEE Conference on Decision and Control, 4*, 4208–4213.
3. Zhang, Y., & Li, X. R. (1997). Detection and diagnosis of sensor and actuator failures using Interacting Multiple-Model estimator. *Proceedings of the IEEE Conference on Decision and Control, 5*, 4475–4480.
4. Maybeck, P. S. (1999). Multiple model adaptive algorithms for detecting and compensating sensor and actuator/surface failures in aircraft flight control systems. *International Journal of Robust and Nonlinear Control, 9*(14), 1051–1070.
5. Alessandri, A. (2003). Fault diagnosis for nonlinear systems using a bank of neural estimators. *Computers in Industry, 52*(3), 271–289.
6. Borairi, M., & Wang, H. (1998). Actuator and sensor fault diagnosis of non-linear dynamic systems via genetic neural networks and adaptive parameter estimation technique. *Proceedings of the IEEE Conference on Control Applications, 1*, 278–282.
7. Napolitano, M. R., et al. (1996). Online learning neural architectures and cross-correlation analysis for actuator failure detection and identification. *International Journal of Control, 63*(3), 433–455.
8. Michail, K., Deliparaschos, K. M., Tzafestas, S. G., & Zolotas, A. C. (2016). AI-based actuator/sensor fault detection with low computational cost for industrial applications. *IEEE Transactions on Control Systems Technology, 24*(1), 293–301.
9. Gadzhiev, C. M. (1992). Dynamic systems diagnosis based on Kalman filter updating sequences. *Automation and Remote Control, 53*(1), 147–150. Conference, vol. 5, pp 4215–4219.
10. Mehra, R. K., & Peschon, J. (1971). An innovations approach to fault detection and diagnosis in dynamic systems. *Automatica, 7*, 637–640.
11. Hajiyev, C., & Caliskan, F. (2003). *Fault diagnosis and reconfiguration in flight control systems*. Boston: Kluwer Academic Publishers.
12. Anderson, T. W. (1984). *An introduction to multivariate statistical analysis* (2nd ed.). New York: Wiley.
13. Kendall, M. G., & Stuart, A. (1969). *The advanced theory of statistics*. London: Griffin.
14. Jategaonkar, R. V. (2006). *Flight vehicle system identification. A time domain methodology*. Reston: American Institute of Aeronautics and Astronautics.
15. Hajiyev, C. (2010). Testing the covariance matrix of the innovation sequence with sensor/actuator fault detection applications. *International Journal of Adaptive Control and Signal Processing, 24*(9), 717–730.
16. Hajiyev, C. (2012). Tracy–Widom distribution based fault detection approach: Application to aircraft sensor/actuator fault detection. *ISA Transactions, The Journal of Automation, 51*(1), 189–197.
17. Hajiyev, C. (2014). Generalized Rayleigh quotient based innovation covariance testing applied to sensor/actuator fault detection. *Measurement, 47*, 804–812.
18. Wu, D., Gao, C., Zhai, Y., Shen, Y., Ji, Zh. (2016). Fault diagnosis of pitch sensor bias for wind turbine based on the multi-innovation Kalman filter. In *Proc. of the 35th Chinese Control Conference*. Chengdu.
19. Rempala, G., & Wesolowski, J. (2005). Asymptotics for products of independent sums with an application to Wishart determinants. *Statistics & Probability Letters, 74*, 129–138.
20. Hajiyev, C., Hacizade, U. (2015). Sensor fault detection by testing the generalized variance of the innovation covariance. In *Proceeding XXI IMEKO World Congress "Measurement in Research and Industry"*, 1181–1186.
21. Caliskan, F., Hajiyev, C. (2003). Actuator failure detection and reconfigurable control for F-16 aircraft model. In *Proc. 3rd IFAC Workshop "Automatic Systems for Building the Infrastructure in Developing Countries" (DECOM-TT 2003), Istanbul, Turkey*, pp. 231–236.

Chapter 20
New Approach to the ATC-Pilot Radio Communication Phraseology Phonetic Confusion

Orhan Ertugrul Guclu

Nomenclature

ANSP	Air navigation service provider
ATC	Air traffic control
ATCOs	Air traffic control officers
NATO	North Atlantic Treaty Organization

20.1 Introduction

The essential way to train ATC is to use voice communication over radio to guide the aircraft pilots [1, 2]. The voice communication system is used to connect the ATCs to voice communication technologies such as a radio and the public switched telephone network (PSTN). This system structure is usually a proprietary circuit-switched network where the work stations of the air traffic controllers are connected via an electronic line exchange to the radio and the telephone network.

Comprehensibility of communication between the ATCO and the pilot is one of the essential principles of air traffic control [3]. ATCO-pilot interaction is performed by radio communication conventionally and accessed through hardware such as headset, microphone, etc. The communication loop is composed of a confirmation and correction process between the pilot and the ATCO in order to ensure safety, orderliness, and quickness [4]. This commonly used communication method has numerous incapabilities. The main disadvantage is radio communication failures,

O.E. Guclu (✉)
Department of Air Traffic Control, Anadolu University, Iki Eylul Campus, 26470, Eskisehir, Turkey
e-mail: oeguclu@anadolu.edu.tr

T.H. Karakoç et al. (eds.), *Advances in Sustainable Aviation*,
DOI 10.1007/978-3-319-67134-5_20

due to background noise (sometimes called static), atmospheric disturbances such as electrical storms and rains, and high-power RF sources (such as radar equipment and broadcast equipment). Communication errors can be divided into two faulty categories: between a pilot and an ATCO and between the ATCOs. These errors occur in the form of incorrect readback, using wrong call signs and nonstandard phraseology and clipping the call sign [5].

There are two main issues regarding classic radio communications. Channel congestion and the language-related problems can be accepted as classic radio communications. Geacăr has shown that according to their statistics, almost 80% of all pilot radio communications contain one or more errors. However, the same statistics show that 30% of all incidents are caused, in the case of the air traffic controllers, among others, by communication errors and 23% of flight level intrusions are caused by communications errors (40% in the case of runway incursions).

According to the Boeing Company, it turned out that of the commercial aircraft accidents for the past 10 years, 55% were caused by pilot error, 17% by aircraft defect, 13% by weather condition, 5% by airport and ATC, 3% by maintenance, and 7% by miscellaneous matters [6]. ATC accounted ones are only 5% of commercial aircraft accidents, which is comparatively lower than other factors. But it should not be overlooked that the 55% of pilot error caused accidents are related with ATC interaction, since the cooperation between a pilot and ATCO is the essential part of aircraft operations.

While the problem attracted the interest of Prinz et al. and was identified as threat in a study by Eurocontrol, it has not really attracted the air navigation service provider (ANSP) attention so far [7, 8]. However, this ensues from not causing crucial damages up until now; on the contrary, this is caused by the cautious policy of ANSPs. However, Emilio showed that there are a significant number of cyber attacks, and examples demonstrate that they induce a real danger for confusing conversations between air traffic controllers and pilots [9–11].

20.2 Methodology

The voice communications were obtained during a set of exercises. These exercises are created from the real-life scenarios of Istanbul Ataturk Airport Approach (LTBA-Approach). The selection of application airport depends on its traffic density.

The traffic density of LTBA is the busiest one in Turkey, and it is in the third place in European airports. Approximately, over 1400 operations are completed in a day. This means that in an hour, around 60 departure and arrival operations are held. This number can be reached by using two independent runways. Runway 05 is used for arrivals, and runway 35R/L is used for departures. With the help of this configuration, LTBA can reach to its highest capacity.

The experimental voice recordings are taken from real ATC conversations in LTBA approach. These recordings are identical to the ATCO and pilot heard in the radio communication. Therefore, these recordings have a lot of noise and

Table 20.1 Communication error classification [5]

Communication error	
Difficulties in communicative interaction or aeronautical operations	Incorrect readback not challenged
	Wrong call sign used
	Nonstandard phraseology
	Call sign omission/truncation
	Missed call
	Clipped call

disturbances. The voice disturbances depend on the aircraft's voice communication unit, whereas the noises depend on bad weather conditions. The recordings are taped on 14th of November 2015. The weather over LTBA at that day was cloudy. So, the recordings have noise and disturbances.

The recordings are classified with their understandability. Besides, their instruction from controller and readback from pilot pairs aren't separated. Also, the same pairs are repeated during the conversation with different instructions and readbacks.

However, they were analyzed according to their grammatical and semantical features and later correlated with the information extracted from the structure, observing the controller works [12, 13]. Interaction ability with the structure in terms of comprehension of control commands and convenience of pseudo-pilot's text-to-speech conversion were also provided by Cordero et al. [14]. These controller works obtained from any of the two sources may feed different workload calculation models based on controller events.

Six types of communication errors are defined by Moon et al. (Table 20.1). Radio communication disturbance can be the reason of these errors such as call sign truncation, incorrect readback, clipped call, etc. These reasons are also used to evaluate the test results applied to the applicants.

The main aim of the study is to find out the misunderstood words in the communication loop between ATC and pilotage students. The survey is taken by listening to what is said in the recorded voice. The survey indicates that the misunderstanding is the result of false assumptions caused by the words or numbers that were half-heard or guessed wrongfully.

20.3 Results and Discussions

The call signs in the conversations have mostly misunderstood letter pairs (Table 20.2). The 20 applicants are listened with 15 conversations. The applicants are selected from the Department of Air Traffic Control and Department of Pilot Training. We haven't classified the applicants with their gender and other features. Through these conversation recordings, they tried to catch and write down the call signs of related aircraft in the conversations. The valid and missing answer numbers are shown in Table 20.3.

Table 20.2 Call signs used in the conversations

Conversation number	Call Sign
1	THY6DT
2	THY7WU
3	THY7JT
4	KAC153
5	THY7JT
6	THY8DP
7	THY5220
8	KAC153
9	THY7WU
10	THY6DT
11	THY8MY
12	THY831
13	THY9PE
14	THY10Z
15	THY22R

Table 20.3 Applicant statistics

	Valid answer amount	Missing answer amount
Applicant # 1	8	7
Applicant # 2	4	11
Applicant # 3	6	9
Applicant # 4	11	4
Applicant # 5	6	9
Applicant # 6	7	8
Applicant # 7	8	7
Applicant # 8	8	7
Applicant # 9	9	6
Applicant # 10	6	9
Applicant # 11	12	3
Applicant # 12	4	11
Applicant # 13	15	0
Applicant # 14	13	2
Applicant # 15	9	6
Applicant # 16	8	7
Applicant # 17	9	6
Applicant # 18	12	3
Applicant # 19	10	5
Applicant # 20	8	7

When we look at Table 20.2, we see that some of the conversations have the same call signs. However, although they have the same call signs, they were misunderstood since the voice recordings have different disturbances and noises. Main misunderstood alphanumeric alphabets are as shown:

Letter "J" is understood as number "8."
Number "2" is understood as letter "Z."
Letter "Z" is understood as number "2."
Letter "Z" is understood as number "0."
Number "0" is understood as letter "Z."
Letter "E" is understood as letter "K."
Letter "M" is understood as number "9."

The study shows that the letters and numbers that were affected by the noise and disturbance mostly are "J, Z, E, K, and M" and "8, 2, and 9." Hence, the usage of these letters and numbers should be restricted.

20.4 Conclusions

Misunderstandings between ATCOs and pilots are a significant problem and situation, which needs to be researched thoroughly. The main aim of the study is to find out the misunderstood words in the communication loop between ATC and pilotage students.

The survey indicates that the misunderstanding is the result of false assumptions caused by the words or numbers that were half-heard or guessed wrongfully.

The survey puts forward that the misunderstanding arises mostly with the letters of J, Z, E, K, and M and with the numbers of 8, 2, 9, etc. in the alphanumeric alphabet. As a result of these similarities, it is shown that the NATO phonetic alphabet and numbers – also known as ICAO radiotelephonic, phonetic, and spelling alphabet [15] – cause some misunderstanding in some letters and numbers in pairs such as "Zulu-Zero," "Mike-Nine," etc.

According to the necessity for the enhancement of voice communication between the ATCO and the pilot, some researches are kept on up until now. The GAMMA (Global ATM Security Management) project the German Aerospace Center (DLR) is developing a prototype to support air traffic control officers (ATCOs) in detecting intrusions into the air ground voice system and therefore allow subsequent mitigating actions to be conducted [16]. Besides this, Cordero et al. [14] described a prototype developed to perform this automated speech recognition and controller event detection, as well as the methodology used to reach it.

References

1. Bardach, H., Kampichler, W., Prinz, J., & Göschka, K. M. (2003). IP communications in air traffic control (ATC). *e & i: Elektrotechnik und Informationstechnik, 120*(11), 405–407.
2. Miller, C. A. & Greenspan, R. L. (2008). Air-traffic control. In *AccessScience*. McGraw-Hill Companies.
3. Hopkin, V. D. (1995). *Human factor in air traffic control*. Bristol: Taylor & Francis.

4. Airbus Company. (2004). *Flight operations briefing notes/Human Performance "CRM Aspects in Incidents"*. Section 02, Revision 03, 1-10. FLT_OPS-HUM_PER-SEQ02-REV0
5. Moon, W. C., Yoo, K. E., & Choi, Y. C. (2011). Air traffic volume and air traffic control human errors. *Journal of Transportation Technologies, 1*(03), 47.
6. Boeing Company. (2006). *Statistical summary of commercial jet airplane accidents worldwide operation*. Aviation Safety Boeing Commercial Airplanes, Chicago.
7. Eurocontrol, VHF Security Study, Final Report, available: www.icao.int/safety/acp/Inactive%20working%20gro ups%20library/ACP-WG-N-SWG4-1/sgn04-01- misc01.doc.
8. Prinz, J., M. Sajatovic, B. Haindl. (2005) S/sup 2/EV-safety and security enhanced ATC voice system. Aerospace Conference, IEEE. IEEE, 2005.
9. Iasiello, E. (2013). Getting ahead of the threat: Aviation and cyber security. *Aerospace America, 51*(7), 22–25.
10. LiveATC. (2011). 25-MAY-2011 Fake ATC in Action (LTBA-ISTANBUL). Available: http://www.liveatc.net/forums/atcaviation-audio- clips/25-may-2011-fake-atc-in-action-%28ltba- istanbul%29.
11. Chivers, H., Hird, J. (2013). Security blind spots in the ATM safety culture. Presented at 2013 International Conference on Availability, Reliability and Security, Regensburg, Germany.
12. Manning, C., Mills, S. H., Fox, C. M., et al. (2002). *Using air traffic control taskload measures and communication events to predict subjective workload, DOT/FAA/AM- 02/4*. Oklahoma City: Federal Aviation Administration.
13. Manning, C., & Pfleiderer, E. (2003). Relationships between measures of air traffic controller voice communications, taskload, and traffic complexity. 5th USA/EUROPE ATM R&D Seminar.
14. Cordero, J.M., Dorado, M., Pablo J. M. (2012). Automated speech recognition in ATC environment, ATACCS'2012, Research Paper, London.
15. ICAO. http://www.icao.int/Pages/AlphabetRadiotelephony.aspx. Accessed on 15 May 2016.
16. Stelkens-Kobsch T. H., Hasselberg A., Mühlhausen T., Carstengerdes, N., Finke, M. and Neeteson, C., German Aerospace Center (DLR). (2015). Braunschweig, Germany, 34th Digital Avionics Systems Conference.

Index

T.H. Karakoç et al. (eds.), *Advances in Sustainable Aviation*,
DOI 10.1007/978-3-319-67134-5

Printed by Printforce, the Netherlands